NUCLEAR ENERGY AND NUCLEAR WEAPON PROLIFERATION

STOCKHOLM INTERNATIONAL PEACE
RESEARCH INSTITUTE

Routledge
Taylor & Francis Group

LONDON AND NEW YORK

First published in 1979 by Taylor & Francis Ltd

This edition first published in 2021
by Routledge
2 Park Square, Milton Park, Abingdon, Oxon OX14 4RN

and by Routledge
52 Vanderbilt Avenue, New York, NY 10017

Routledge is an imprint of the Taylor & Francis Group, an informa business

British Library Cataloguing in Publication Data
A catalogue record for this book is available from the British Library

ISBN: 978-0-367-50682-7 (Set)
ISBN: 978-1-00-309763-1 (Set) (ebk)
ISBN: 978-0-367-51339-9 (Volume 4) (hbk)
ISBN: 978-1-00-305346-0 (Volume 4) (ebk)

Publisher's Note
The publisher has gone to great lengths to ensure the quality of this reprint but
points out that some imperfections in the original copies may be apparent.

Disclaimer
The publisher has made every effort to trace copyright holders and would welcome
correspondence from those they have been unable to trace.

ROUTLEDGE LIBRARY EDITIONS:
NUCLEAR SECURITY

Volume 4

NUCLEAR ENERGY AND NUCLEAR WEAPON PROLIFERATION

Nuclear Energy
and Nuclear Weapon Proliferation

sipri

Stockholm International Peace Research Institute

SIPRI is an independent institute for research into problems of peace and conflict, especially those of disarmament and arms regulation. It was established in 1966 to commemorate Sweden's 150 years of unbroken peace.

The Institute is financed by the Swedish Parliament. The staff, the Governing Board and the Scientific Council are international. As a consultative body, the Scientific Council is not responsible for the views expressed in the publications of the Institute.

sipri

Stockholm International Peace Research Institute
Sveavägen 166, S-113 46 Stockholm, Sweden
Cable: Peaceresearch, Stockholm
Telephone: 08-15 09 40

Nuclear Energy and Nuclear Weapon Proliferation

sipri

Stockholm International Peace Research Institute

Taylor & Francis Ltd
London
1979

First published 1979 by Taylor & Francis Ltd
10-14 Macklin St, London WC2B 5NF

Copyright © 1979 by SIPRI
Sveavägen 166, S-113 46 Stockholm

Distributed in the United States of America by
Crane, Russak & Company, Inc.,
3 East 44th Street, New York, N.Y. 10017
and in Scandinavia by
Almqvist & Wiksell International,
26 Gamla Brogatan,
S-101 20 Stockholm, Sweden

British Library Cataloguing in Publication Data

Nuclear energy and nuclear weapon proliferation.
   1. Atomic power
   2. Atomic weapons and disarmament
   I. Stockholm International Peace Research
   Institute
   621.48   TK9145

   ISBN 0-85066-184-6

Typeset by Red Lion Setters, 22 Brownlow Mews, London WC1N 2LA
Printed and bound in the United Kingdom by
Taylor & Francis (Printers) Ltd., Rankine Road,
Basingstoke, Hampshire RG24 0PR

Preface

In mid-1980 a second conference for the review of the Non-Proliferation Treaty (NPT) will take place in Geneva. Given the importance cf preventing, or at least slowing down, nuclear weapon proliferation, this conference will be a crucial event in the field of arms control and disarmament. For many countries the technical and economic barriers to proliferation have disappeared, and the only remaining barriers are political.

In an attempt to contribute to the discussions at the NPT Review Conference, SIPRI assembled a group of experts from a number of countries to discuss the technical aspects of the control of fissionable materials in non-military applications. The meeting took place in Stockholm, 12–16 October 1978. This book on nuclear energy and nuclear weapon proliferation contains the papers presented at the symposium and reflects the discussions at the meeting.

In SIPRI's opinion the following steps could and should be taken to implement the aims and purposes of the NPT:

(a) The nuclear weapon powers should clearly commit themselves to reversing the arms race; they could start by halting all nuclear weapon tests and undertaking to reduce significantly their strategic and tactical nuclear armaments;

(b) Participation in the treaty should be made more attractive by the provision of internationally agreed, legally binding security assurances to non-nuclear weapon parties;

(c) Pressure should be brought to bear upon non-parties by denial of supplies of nuclear materials and equipment, while outright defiance of the treaty should be met with more stringent measures;

(d) The obligation not to assist others to manufacture nuclear weapons should apply to all states without exception and, consequently, all exports of nuclear material and equipment to nuclear weapon powers should be subject to IAEA safeguards so as to avoid their use for weapon purposes; and

(e) Safeguards procedures should be improved, and IAEA authority strengthened, to enable rapid detection of any diversion of fissionable material for weapon purposes, and quick subsequent action.

Insofar as the peaceful use of nuclear energy is concerned, the cause of non-proliferation would best be served if:

(a) the sensitive parts of the nuclear fuel cycle, that is, uranium enrichment, fuel fabrication and reprocessing, were managed on an international scale and operated only under the authority of an international agency with full responsibility for the security of the plants and their sites;

(b) an international repository of spent fuels and a bank of fresh fuels were to be established; and

(c) encouragement, including financial support, were given to countries wishing to rely on non-nuclear sources of energy. This might best be achieved by the setting up of a specialized international body to deal with energy matters.

The editors of this book are Dr Frank Barnaby, Director of SIPRI, Jozef Goldblat and Dr Bhupendra Jasani, members of the SIPRI research staff, and Professor Joseph Rotblat, a visiting scholar at SIPRI.

Acknowledgements

Dr Lars Kristoferson, Royal Swedish Academy of Sciences, acted as rapporteur during the symposium and contributed to the preparation of this book. The editorial assistance of Billie Bielckus is gratefully acknowledged.

SIPRI
July 1979

Contents

Tables and figures

Paper 9.   The role of the breeder reactor

Chapter 7.   Hybrid reactors

Paper 10.   Fusion–fission hybrid reactors

**Paper 11.   Laser fusion and fusion hybrid breeders:
proliferation implications**

Part IV   Introduction

Chapter 11.   Peaceful nuclear explosions

Paper 19.   Peaceful applications of nuclear explosions

List of symposium participants

Professor Dean E. Abrahamson
Humphrey Institute of Public
   Affairs
University of Minnesota
Minneapolis, Minn. 55455
USA

Dr Adolf von Baeckmann
Director
Division of Development and
   Technical Support
IAEA
P.O. Box 590
A-1011 Vienna
Austria

Dr Frank Barnaby
Director
SIPRI
Sveavägen 166
S-113 46 Stockholm
Sweden

Mr Bertrand Barré
Commissariat à l'Energie Atomique
BP 510
Paris 75015
France

Dr Peter Boskma
Technische Hogeschool Twente
P.O. Box 217
Enschede
The Netherlands

Dr Jean H. Coates
Commissariat à l'Energie Atomique
BP 510
Paris 75015
France

Dr David Davies
Nature
4 Little Essex Street
London WC2R 3LF
England

Dr Warren H. Donnelly
Congressional Research Service
The Library of Congress
Washington, D.C. 20540
USA

Acad. Vassili S. Emelyanov
Academy of Sciences of the USSR
Leninsky Prospekt 14
Moscow
USSR

Dr Hubert Eschrich
Euro-Chemic
B-2400 Mol
Belgium

Professor Ugo Farinelli
Laboratorio CNEN-CSN Casaccia
00060 Santa Maria di Galeria
Rome
Italy

Professor Bernard T. Feld
6-308
Department of Physics
MIT
Cambridge, Mass. 02139
USA

Dr Richard L. Garwin
IBM Thomas J. Watson Research
    Center
P.O. Box 218
Yorktown Heights, N.Y. 10598
USA

Mr Jozef Goldblat
SIPRI
Sveavägen 166
S-133 46 Stockholm
Sweden

Dr Kåre Hannerz
Asea-Atom
Box 53
S-721 04 Västeras 1
Sweden

Dr William A. Higinbotham
Brookhaven National Laboratory
Associated Universities, Inc.
Upton, N.Y. 11973
USA

Dr Frank von Hippel
School of Engineering/
    Applied Sciences
Center for Environmental Studies
Engineering Quadrangle
Princeton University,
Princeton, N.J. 08540
USA

Dr Bhupendra M. Jasani
SIPRI
Sveavägen 166
S-113 46 Stockholm
Sweden

Professor Karl L. Kompa
Projektgruppe für Laserforschung
    der Max-Planck-Gesellschaft
    zur Förderung der
    Wissenschaften e.V.
D-8046 Garching bei München
Federal Republic of Germany

Professor Vitali Kuleshov
Academy of Sciences of the USSR
Leninsky Prospekt 14
Moscow
USSR

Professor Derek Paul
Department of Physics
University of Toronto
Toronto, Ont. M5S 1A7
Canada

Professor Joseph Rotblat
8 Asmara Road
London NW2 3ST
England

Mr Ben Sanders
Centre for Disarmament
United Nations, Room 2755
New York, N.Y. 10017
USA

Mr Fredrik Segerberg
Asea-Atom
Box 53
S-721 04 Västeras 1
Sweden

Dr Donald R. Westervelt
Los Alamos Scientific Laboratory
    MS-668
P.O. Box 1663
Los Alamos, N.M. 87545
USA

Dr A. R. W. Wilson
Australian Atomic Energy
    Commission
P.O. Box 41
Coogee, N.S.W. 2034
Australia

Abbreviations, acronyms, units and conversions

Abbreviations and acronyms

ACDA	(US) Arms Control and Development Agency
ACR	Advanced converter reactor
AEC	(US) Atomic Energy Commission
AGR	Advanced gas-cooled reactor
BWR	Boiling water reactor
CANDU	Canadian heavy water reactor
CCTV	Closed circuit television
CEA	Commissariat à l'Energie Atomique
CTR	Controlled thermonuclear reactor
CECo	Commonwealth Edison Company
DA	Chemical analysis
DOE	(US) Department of Energy
EPRI	Edison Power Research Institute
EURATOM	European Atomic Energy Community
FBR	Fast breeder reactor
HEU	Highly enriched uranium
HTGR	High temperature gas-cooled reactor
HWR	Heavy water reactor
HTR	High temperature reactor
IAEA	International Atomic Energy Agency
ICF	Inertial confinement fusion
INFA	International Nuclear Fuel Authority
INFCE	International Nuclear Fuel Cycle Evaluation
LEU	Low-enriched uranium
LIS	Laser isotope separation
LMFBR	Liquid metal fast breeder reactor
LWR	Light water reactor
LWGR	Light water gas-cooled reactor
MBA	Material balance area
MEU	Medium-enriched uranium
MHF	Massive hydraulic fracturing
MSBR	Molten-salt breeder reactor

MTR	Materials testing reactor
MUF	Material unaccounted for
NDA	Non-destructive assay techniques
NPT	Treaty on the Non-Proliferation of Nuclear Weapons
NRC	Nuclear Regulatory Commission
PLBR	Prototype large breeder reactor
PNE	Peaceful nuclear explosion
PNET	Peaceful Nuclear Explosions Treaty
PTBT	Partial Test Ban Treaty
PWR	Pressurized water reactor
RTG	Radioisotope thermoelectric generator
SAM-II	Stabilized assay meter
SWU	Separative work unit
TCC	Technical coordinating committee
TCR	Thermal converter reactor
THTR	Thorium high temperature reactor
UKAEA	United Kingdom Atomic Energy Authority
UNCPUOS	UN Committee on Peaceful Uses of Outer Space
UNGG	Natural uranium gas/graphite reactor

Units

eV	electronvolt
GW	gigawatt ($1GW = 10^9$ watts)
GW(e)	gigawatts of electric power
GeV	giga-electronvolt
J	joule
kt	kiloton
kW(th)	kilowatts of thermal power
kW-h	kilowatt-hour
keV	kilo-electronvolt
mill	one-thousandth of a dollar (US)
MeV	mega-electronvolt
Mg	megagram
mSv	millisievert
MW(e)	megawatts of electricity
MW(th)	megawatts of thermal power
MWd/kg	megawatt-day per kilogram
SWU	separative work unit (see glossary)
t	tonne (1 000 kg)
TW-year	terawatt-year ($1TW = 10^{12}$ watts)
TW-h	terawatt-hour ($1TWh = 3.6 \times 10^{15}$ J)

Conversions

Units of mass

1 ton = 1 000 kilograms (tonne) = 2 205 pounds, avoirdupois
    = 0.98 long ton = 1.1 short tons
1 short ton = 2 000 pounds = 0.91 ton = 0.89 long ton
1 long ton = 2 240 pounds = 1.1 tons = 1.12 short tons
1 kiloton = 1 000 tons
1 megaton = 1 000 000 tons
1 kilogram = 2.2 pounds
1 pound = 0.45 kilogram

Part I

Introduction

B. JASANI

Square-bracketed numbers, thus [1], refer to the list of references on page 9.

Energy, whether it is derived from oil or from nuclear power reactors, has become an issue of national security and economic progress of most countries. However, international concern about nuclear energy used for generating power is caused by the fact that the technology not only provides an insight into an understanding of nuclear weapon technology, but it also makes available the required fissile material for nuclear explosives. It is generally argued that the availability of fissile material increases the risk of proliferation of nuclear weapons. Recently this concern has focused international attention on the problem of making the nuclear fuel cycle proliferation-resistant. In this Part, the fuel cycle in general and enrichment and reprocessing of nuclear fuel in particular are discussed. The proliferation implications of fast reactors are considered in Part II.

Of the three useful fissile materials—uranium-233 (U-233), uranium-235 (U-235) and plutonium-239 (Pu-239)—generally only U-235 occurs in nature; about 0.711 per cent natural uranium is U-235, the remainder being U-238. The other fissile materials can be artificially produced by neutron bombardment of thorium-232 (Th-232) to produce U-233, or U-238 to produce Pu-239. It is this last fact which gives rise to international concern over the use of nuclear energy for power production.

Only two, U-235 and Pu-239, of the three fissile materials have been used so far as nuclear explosives. China is the only country which carried out its first nuclear explosion using U-235; the other four nuclear weapon powers used Pu-239 (table I.1). India too used Pu-239 in its 1974 underground nuclear explosion. The first nuclear weapon used at Hiroshima in 1945, however, employed highly enriched U-235 as the fissile material. Most of the fission weapons in possession of the nuclear weapon powers use Pu-239. Uranium-235 appears to have been used as a fission trigger in the first thermonuclear explosions of all the five nuclear weapon powers (table I.1). However, when a B-52 aircraft and a KC-135 refuelling tanker aircraft collided in mid-air in January 1966 near Palomares, Spain, four hydrogen bombs separated from the aircraft and two of these released radioactive material [1]. It has been reported that the radioactive material contained Pu-239, suggesting that this fissile material is used as a trigger in a thermonuclear

1

Table I.1. First nuclear explosions

Fission devices				Thermonuclear devices		
Country	Year of first explosion	Fissile material	Source of fissile material	Year of first explosion	Fissile material	Source of fissile material
USA	1945	Pu-239	Reactor	1952	U-235	Gaseous diffusion
USSR	1949	Pu-239	Reactor	1952	U-235	Gaseous diffusion
UK	1952	Pu-239	Reactor	1957	U-235	Gaseous diffusion
France	1960	Pu-239	Reactor	1968	U-235?	Gaseous diffusion
China	1964	U-235	Gaseous diffusion	1967	U-235	Gaseous diffusion
India	1974	Pu-239	Reactor	–	–	–

Table I.2. Plutonium production in various types of reactor

Type of reactor	Irradiation level of heavy metal (MWd/kg)	Average enrichment (% U-235)	Initial fuel inventory (kg/MW(e))		Pu-239 production (g/MW(e) yr)
			Core (natural U)	Blanket (depleted U)	
BWR	17	2	434	–	250
PWR	22.6	2.3	365	–	255
AGR	–	1.6	620	–	100
HWR (CANDU)	6	0.711	143	–	490
HTGR (USA)	54.5	~93	326	–	–
SGHWR	15.5	1.8	520	–	150
FBR (UK)	~70	–	9.5 (depleted U) 2.8 (Pu)[a]	16.9 (depleted U)	2850 (core Pu)[b] 409 (blanket Pu)[b]

[a] Plutonium composition is 57 per cent Pu-239, 24 per cent Pu-240, 14 per cent Pu-241 and 5 per cent Pu-242.
[b] Total of 2 980 kg/yr containing 58 per cent Pu-239, 28 per cent Pu-240, 9 per cent Pu-241 and 5 per cent Pu-242.

Sources: See References [2] and [3].

weapon. It is therefore obvious that the most used fissile material in nuclear weapons is Pu-239.

At present the trend is for increased use of thermal reactors using natural or slightly enriched uranium as fuel (see Chapter 14, table 1). Any reactor fuelled with uranium produces plutonium as a by-product. Plutonium

Table I.3. Critical masses of various fissile materials

Enrichment (% of U-235 or Pu-239)	Critical mass (kg) in metal form					
	Uranium-235		Plutonium		Uranium-233	
	Without reflector	With Be reflector	Without reflector	With Be reflector	Without reflector	With Be reflector
0	–	–	–	–	–	–
10	–	~ 1300	–	–	–	–
20	–	~ 250	–	–	–	–
30	–	–	–	–	–	–
40	–	75	–	–	–	–
50	145	50	–	9.6	–	–
60	105	37	–	7.8	–	–
70	82	–	23	6.7	–	–
80	66	21	–	5.6	–	–
90	54	–	–	5.0	–	–
100	50	15	15	4.4	17	4 to 5

Sources: See references [3] and [4].

production in various types of reactors is shown in table I.2. The quantity and quality of this plutonium may, however, vary. The significance of the quantities of Pu-239 produced from these reactors can be seen from table I.3, which shows the critical masses of various fissile materials, in metallic forms. The critical mass also depends on the physical form. For example, the critical mass of a sphere of plutonium in oxide form can be greater than that for metal by a factor of about five, depending on the density of plutonium oxide. As far as U-233 is concerned, it is more like Pu-239 than U-235 from the point of view of its use as an explosive. However, when, for example, one kilogram of U-233 is extracted from reprocessed reactor fuel, the rate at which the neutrons are released from it is several orders of magnitude smaller than that extracted from the same quantity of plutonium [4]. The critical mass of U-233 metal or compounds is about 30 per cent greater than that of Pu-239 in the same form.

This brief discussion illustrates the concern about the use of nuclear energy for generating power. Are there any solutions, for example, technical ones, which can be introduced to make the nuclear fuel cycle proliferation-resistant, or will it be, in the end, a political question? The issue of proliferation is a complex one since a number of questions—technological, economic, environmental and political—must be considered. Here, only some of the technical solutions recently put forward are considered. Such solutions are directed at making the nuclear fuel cycle as a whole proliferation-resistant or making some of the sensitive processes within the cycle proliferation-resistant. The fuel cycle as a whole will be considered first.

I. A proliferation-resistant nuclear fuel cycle

Several technical solutions have been proposed to make the present nuclear fuel cycle proliferation-resistant. These are discussed in Paper 1. The idea here is to operate nuclear power plants on once-through fuel cycles; that is, the spent fuel rods containing an appreciable amount of the unused fissile materials, U-235 and Pu-239, are stored without reprocessing. Since the fuel for burner reactors is low-enriched uranium (2−3 per cent U-235), the fuel cannot be used for weapons without separating the unburned U-235 to high concentrations. The Pu-239 produced in the fuel element as a result of neutron bombardment of U-238 in the fuel remains relatively inaccessible, but only so long as the fuel elements contain very high levels of radio-activity, at which stage Pu-239 can be removed only in a sophisticated, expertly operated, well-designed reprocessing plant. As the level of radio-activity decreases, however, the degree of sophistication needed in handling the fuel elements also declines. Thus, the once-through cycle is not sufficient to prevent proliferation, except for countries which do not have a well-developed nuclear programme.

The authors of Paper 1 have suggested that in a decade or so the current burner reactors could be replaced by more uranium-efficient burner (or so-called advanced converter) reactors. Technology for such advanced converter reactors already exists in the Canadian HWR and the high-temperature gas-cooled reactors developed both in the United States and in the Federal Republic of Germany.

In a second context, the same authors have suggested that the advanced converter reactors would use a denatured uranium−thorium fuel cycle. The fuel, in this case, is a mixture mainly of Th-232 and U-238. The thorium resources are about twice to three times as large as uranium [2]. Most of the new fissile material in such a mixture is U-233, with some Pu-239 produced from U-238. It is proposed that the spent fuel is reprocessed to retrieve uranium and thorium for recycle and the Pu-239 is stored. The U-233 and U-235 thus separated are denatured with the recycled uranium. Although such an advanced converter reactor concept may have the advantage of postponing for some time decisions regarding fuel reprocessing or deploying fast breeder reactors, the problems of proliferation of nuclear weapons are not solved.

What other technical solutions are there to eliminate or even to reduce the proliferation risks? The two most sensitive elements of the nuclear fuel cycle from the point of view of proliferation are the enrichment process and the reprocessing of the spent fuel. These are considered briefly below.

Enrichment

The uranium enrichment process has been the only part of the nuclear fuel cycle which has been kept under strict secrecy. The reasons are obvious. It can be seen from table I.1 that U-235 is not a preferred fissile material for fission weapons and, as far as thermonuclear weapons are concerned, it has been used to start the fission reaction in such weapons. However, Pu-239 would also trigger such a reaction in fusion weapons. Should a country decide to start its nuclear weapon (fission weapon) programme with U-235 as the fissile material, then the acquisition of an enrichment facility becomes essential. The uranium for a nuclear weapon is enriched to at least 50 per cent and it is generally assumed that the fissile material used by nuclear weapon powers in their uranium weapons contains uranium enriched to at least 90 per cent. Even if a country does not embark upon a nuclear weapon programme, the development and the possession of an enrichment facility would make the country independent of other sources of enriched uranium fuel for its nuclear power programme, particularly if the country has uranium resources of its own.

Moreover, uranium enrichment is an essential part of the fuel cycle of light water reactors (LWRs), advanced gas-cooled reactors (AGRs) and high-temperature reactors (HTRs). Of these, the LWRs are most widely used. In 1976, the capacity of enriched uranium reactors was 87 per cent of all operating reactors and it is predicted that this will rise to about 90 per cent by the year 2000 [5]. It is, therefore, important to examine the status of enrichment technology.

Several methods for isotope separation were known even before nuclear fission was discovered. The most common ones are the gaseous diffusion and centrifuge processes. Two other technologies which are proliferation-prone are the jet nozzle process developed in the Federal Republic of Germany and the advanced vortex tube process which was patented in the United States in 1968 and developed further by South Africa. These technologies are described in Paper 3. However, they are proliferation-prone since, among other things, firstly, it is suggested that they require a lower level of technology and are therefore attractive to countries less technologically developed; and secondly, they have been developed by two states—one (South Africa) which is not a party to the NPT and the second (FR Germany) which plans to sell a jet-nozzle enrichment plant to Brazil, a non-NPT state.

The availability of a variety of lasers has now made laser isotope separation possible. Theoretically it is possible to achieve a very high degree of separation of uranium isotopes in a single step with low energy consumption. The process has been demonstrated on a laboratory scale. Paper 4 discusses enrichment through laser. It has been estimated that the running costs of a laser enrichment plant would be lower than other techniques and the capital costs are competitive. The technique will probably come to fruition in about a decade.

Another possible enrichment process is the recently proposed French method. This is a chemical process, the precise details of which have not yet been made available. However, a feature of the process which makes it proliferation-proof is that high U-235 concentrations are unattainable. The process is discussed in Paper 2.

What are the non-proliferation measures available as far as enrichment technology is concerned? Internationalization of the enrichment facilities, low price for enriched uranium for reactors so as to discourage new national enrichment facilities and a moratorium on individual processes have been suggested. It has been pointed out in discussion on the subject that a moratorium on the jet nozzle and the vortex tube methods would not affect the availability and particularly the cost of enriched uranium since the two technologies are no more efficient than the gas centrifuge process.

It must be realized, however, that enrichment is not the easiest road to proliferation. The technology is probably of interest from the point of view of clandestine effort or the desire to keep one's nuclear options open. In this respect, laser separation is attractive since the general development of laser technology needs only minor adjustments to create the option for highly enriched uranium.

The enriched fuel is made into reactor fuel rods and placed into the reactor. The fissile material that may be produced in the reactor is safe from theft or governmental diversion as long as the fuel elements are in the reactor. Outside the reactor, the fissile material can be diverted if the spent fuel is reprocessed. This second sensitive part of the nuclear fuel cycle is considered below.

Reprocessing

Fuel reprocessing is part of the fuel cycle in which the fission products are separated from unburned uranium and plutonium that is produced in the reactor and plutonium from uranium. However, it can be seen from table I.2 that all reactors in operation today produce plutonium and if the fuel from such reactors is reprocessed, then a considerable amount of plutonium will become available, thus increasing the risk of nuclear weapon proliferation. A number of technical solutions have been put forward and one of these, the once-through fuel cycle, has already been discussed. This subject and the subject of making the reprocessed fuel proliferation- and theft-proof by the so-called spiking technique are discussed in Paper 5. The question of making plutonium proliferation-resistant by denaturing it is discussed in Paper 7.

The Purex process, developed during World War II to produce plutonium for weapons, has continued to be the basis of present reprocessing technology. As an alternative to this, a Civex (for civilian fuel cycle) process which would make weapon-grade plutonium much less easily available has recently been proposed. This is also discussed later in connection with fast breeder reactors (see page 15). In this method, the fuel is reprocessed while its radioactivity is high, containing mainly the short-lived isotopes. So the plutonium going into the reprocessing plant and that coming out of it is highly radioactive because of the presence of some fission products, thus making it inaccessible. It will be seen later that this does not reduce the proliferation risks. Moreover, fuel fabrication would be more difficult and expensive. Nevertheless, Civex will be seriously considered during the International Fuel Cycle Evaluation (INFCE) deliberations.

Other methods of denaturing plutonium consist of using plutonium isotopes (see Paper 7). In one method, it is proposed that higher than usual concentrations of Pu-238 could be used. A typical reactor spent fuel contains about 1.5 to 2.0 per cent Pu-238. This level of plutonium can be increased by adding to fresh fuel neptunium-237 and U-236 which, on capturing neutrons, produce Pu-238. In this way concentrations of 5 to 10 per cent Pu-238 can be achieved in the fuel. With such concentrations, the surface temperatures of a critical mass of such a mixture of plutonium would be 875°C and 1 660°C, respectively [6]. At these temperatures the chemical explosive required in a fission bomb becomes unstable. Moreover, the melting point of plutonium is about 500°C so that it would be difficult to make a nuclear explosive from such a mixture. However, if a method were available for removing objectionable isotopes from the Pu-239 for the purpose of weapon production, then denaturing would not reduce the risk of proliferation of nuclear weapons. Laser technology may well provide such a method. As mentioned above, the technology has shown promise for uranium enrichment. There is no reason why the laser enrichment technique could not be applied to purify Pu-239 [7]. Of course, if the technique is perfected, then it would mean that no power reactor based on the U−Pu cycle is proliferation-resistant.

Although schemes such as Civex may make it difficult for a small group of people or even a country with a less developed technological base to divert plutonium and use it for weapon purposes, it is always possible for a sophisticated organization to extract weapon-grade plutonium if it is really determined to do so. No technical measures can solve such proliferation dangers.

The situation is further complicated by the development of new, uncontrolled methods for producing fissile materials. One such method is to use accelerators already extensively employed in high-energy physics research. An accelerated beam of protons can be used to breed fissile materials [8]. It is at present argued that in a U−Th cycle, sufficient new fissile U-233 could not be bred. Therefore, the required extra fissile material could be produced by bombarding a suitable target (lead−bismuth) by protons produced in an accelerator. The target could be surrounded by fertile Th-232 or depleted U-238 in which U-233 or Pu-239, respectively, is then produced. It has been shown, for example, that a beam of 300 mA, 1 GeV protons could produce about 1 000 kg of fissile material per year, or a less intense beam of 25 mA could be used to top up the fissile U-233 production in a Th−U thermal reactor blanket [8]. It is interesting to note that while all nuclear facilities in a state party to the NPT and some facilities in states not party to the NPT are under strict international control, the accelerators used in high-energy physics research are not subject to any international control, nor is there any debate about this technology from the point of view of proliferation of nuclear weapons.

The last component of the nuclear fuel cycle is waste disposal. This is considered briefly below.

Waste disposal

Apart from the problem of proliferation of nuclear weapons, the nuclear fuel cycle presents another critical problem—the question of nuclear waste disposal. Paper 6 deals with this subject. In the course of the reactor's operation, a variety of fission fragments, and plutonium and other actinides formed by transmutation of uranium and plutonium, are produced and continue to build up. Because of the build-up of these materials, they begin to interfere with further nuclear reaction. The nuclear wastes, therefore, contain the fission products, the unburned uranium and plutonium and the actinides if the spent fuel is not reprocessed. If, however, the spent fuel is reprocessed and the plutonium and the unburned uranium are removed, then the residue contains essentially the fission products and some additional material from fuel rod cladding, traces of unseparated plutonium and uranium and most of the transuranic elements.

Once the irradiated fuel is removed from the reactor, the usual practice is to store the fuel elements close to the reactor, normally under water until the initial high radioactivity and heat output have died down. After about a year or so, the fuel rods may be transported to reprocessing plants where they are usually allowed to cool for a further period of up to 10 years.

Whether the spent fuel is reprocessed or not, a number of methods have been proposed for disposing or storing the nuclear waste. Methods such as orbiting the waste into the Sun or out of the solar system by rockets, burial under the Antarctic ice-cap, or transfer to the Earth's mantle via an oceanic subduction zone have been discarded on the grounds of safety, cost or inadequacy of present-day technology.

According to a scheme suggested in Sweden, the unreprocessed spent reactor fuel would first be stored in water pools for some 40 years. The fuel element would then be placed into lead-lined copper containers which would be buried into underground rocks. In the case of reprocessed fuel, it is planned to glassify (the vitrification process) the waste products and then bury them sufficiently deep in rock [9]. Yet another suggestion is to incorporate the nuclear waste into synthetic mineral (Synroc) which is more stable than the glass [10].

Disposal of low- and intermediate-level of radioactive waste to shallow depths (underground) has been carried out for a number of years. In addition, for more than a decade, deep geological formations have been used for the disposal of radioactive waste. However, there are problems with this latter technique. For example, between 1951 and 1965, intermediate-level radioactive liquid waste at Oak Ridge National Laboratory was disposed of in several different pits and trenches. Moreover, since 1944, some solid radioactive waste has continually been buried in shallow trenches. It has been found that some radioactive nuclides are migrating from their original site of burial [11].

References

1. 'Accidents of nuclear weapon systems', *World Armaments and Disarmament, SIPRI Yearbook 1977* (Almqvist & Wiksell, Stockholm, 1977, Stockholm International Peace Research Institute), pp. 52−85.
2. Marshall, W., 'Nuclear power and the proliferation issue', *Atom*, No. 258, April 1978, p. 101.
3. 'Nuclear proliferation factbook', *US Library of Congress*, 23 September 1977, p. 382.
4. Taylor, T. B., 'Nuclear safeguards', *Annual Review of Nuclear Science*, Vol. 25, 1975, pp. 407−26.
5. Pedersen, O., 'Developments in the uranium enrichment industry', *IAEA Bulletin*, Vol. 19, No. 1, February 1977, p. 40.
6. 'AGNS proposes proliferation-resistant fuel cycle', *Nuclear Industry*, Vol. 25, No. 9, September 1978, pp. 24−25.
7. Casper, B. M., 'Laser enrichment: a new path to proliferation', *Bulletin of the Atomic Scientists*, Vol. 33, No. 1, January 1977, p. 32.
8. 'Could acceleration replace fast breeder reactor?' *CERN Courier*, Vol. 18, No. 5, May 1978, pp. 152−54.
9. Hill, J., 'Nuclear waste disposal', *Atom*, No. 259, May 1978, p. 126.
10. Walgate, R., 'Nuclear waste may be stored in synthetic rock', *Nature*, Vol. 274, No. 5470, 3 August 1978, p. 413.
11. Means, J.L., Crerar, D.A. and Borcsik, M.P., 'Radionuclide adsorption by manganese oxides and implications for radioactive waste disposal', *Nature*, Vol. 274, No. 5666, 6 July 1978, pp. 44−47.

Chapter 1. Fuel cycles

Paper 1. An evolutionary strategy for nuclear power

F. VON HIPPEL, H. A. FEIVESON and R. H. WILLIAMS

Square-bracketed numbers, thus [1], refer to the list of references on pages 41-44.

I. Introduction

In the past few years there has been a reawakening to the fact that the spread of civilian nuclear power technology is being accompanied by the spread of knowledge and material required for the production of nuclear weapons. Concern has focused especially on the plutonium fuel cycle—in the short term on efforts to commercialize nuclear fuel reprocessing for the recovery of plutonium, and in the longer term on the related efforts to commercialize the plutonium breeder reactor.

In response to these concerns, the Carter Administration in 1977 called for a deferral of US programmes aimed at the commercialization of the plutonium fuel cycle and initiated national and international studies of alternative nuclear power systems which might be more nuclear weapon proliferation-resistant.[1] This paper will explore the role of such alternative systems in the development of nuclear power in the United States and abroad.

We do not pretend that the problem of nuclear proliferation can be solved simply by developing an alternative to the plutonium breeder. Many nations could develop a nuclear weapon capability outside their nuclear power programmes [2]. One cannot, therefore, expect a non-proliferation strategy to succeed if it does not grapple with the political and security incentives and disincentives for countries to acquire nuclear weapons. More fundamentally, it appears unlikely that in the long term, proliferation of nuclear weapons can be stopped while the nuclear weapon states act as if nuclear weapons are politically useful to have.

Nevertheless, proliferation resistance should be an important criterion guiding the choice of future nuclear power technologies, because some of these technologies, by providing nations access to weapons-usable material,

---

[1] The "Non-Proliferation Alternative Systems Assessment Program" was established within what is now the US Department of Energy, following President Carter's decision in April 1977 not to continue the process of commercializing the plutonium fuel cycle in the USA. Forty nations agreed to establish a two-year International Fuel Cycle Evaluation (INFCE) following a meeting in Washington, D.C. in October 1977. See, for example, reference [1].

Figure 1.1. Three nuclear fuel cycles

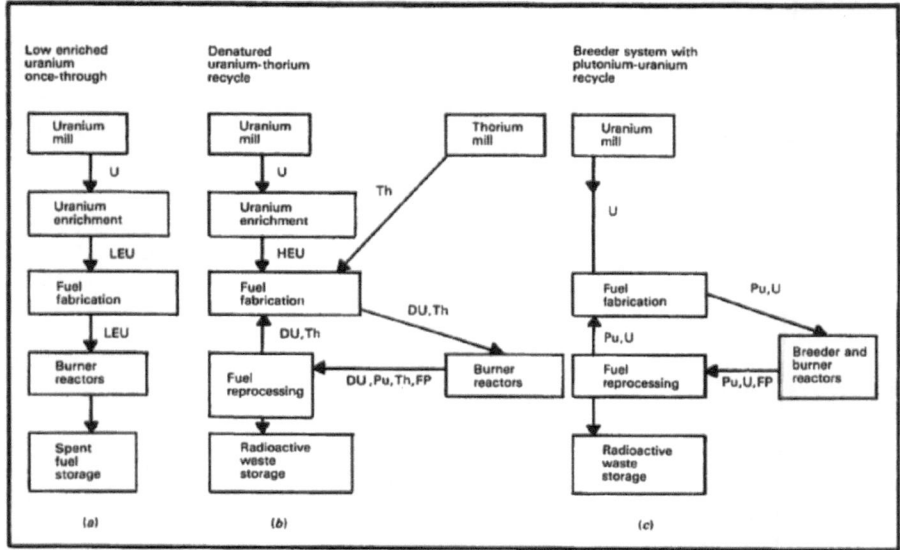

Note: The denatured uranium–thorium fuel cycle used with burner reactors lies between the once-through and plutonium breeder fuel cycles in both uranium utilization efficiency and in the technological barriers which it offers to the diversion of weapons-usable material.

In the once-through cycle (*a*), the uranium fuel contains low-enriched uranium (LEU), which cannot be used for weapon purposes without isotope separation. The spent fuel contains plutonium (Pu), which has been produced by neutron bombardment of the U-238 in the fuel. This spent fuel is not of commercial interest for the short term at least and may be stored or disposed of without separating the plutonium from the highly radioactive fission products (FP).

The denatured uranium–thorium recycle fuel cycle (*b*) uses isotopically denatured uranium (DU) as the reactor fuel. In this cycle, the spent fuel is chemically reprocessed to retrieve for recycle the uranium and thorium (Th). Most of the new fissile material produced is U-233, although some plutonium is still produced by neutron capture on the U-238 denaturant. It is assumed here that this plutonium is not recycled but is instead either stored or disposed of with the radioactive wastes. The 'make-up' U-235 required to augment the recycled U-233 and U-235 is contained in highly enriched uranium (HEU), a weapons-usable material which is denatured once it is mixed with the recycled uranium.

The plutonium fuel cycle (*c*) involves the recycle of both plutonium and uranium. Both fresh and spent reactor fuel contain large quantities of chemically separable nuclear weapons-usable material—of the order of 100 fast critical masses per year per reactor. With breeder reactors eventually no externally supplied fissile material would be required—hence the absence of a uranium enrichment plant.

contribute to a process of latent proliferation, whereby nations move closer to a weapon capability without having to declare or decide in advance their actual intentions. Reprocessing for the recycling of plutonium is one of these technologies. It involves a commitment to commerce in a weapons-usable material separated out of spent reactor fuel—an activity which does not occur with today's once-through fuel cycle. (See figure 1.1 for a description of the fuel cycles discussed most extensively in this paper.)

Despite the recent concern about the nuclear weapon proliferation implications of the deployment of a plutonium fuel cycle, the plutonium breeder reactor has been for about a decade the highest priority objective of the energy research and development programmes of the major industrialized nations. This technology has been seen to be essential to the preservation of the fission energy option for the long term. In contrast to today's nuclear technology, which exploits primarily the fission energy in the rare 'fissile' (chain-reacting) uranium isotope U-235, a breeder reactor would effectively tap in addition the fission energy in the far more abundant 'fertile' uranium isotope U-238, which is transmuted following neutron capture into fissile plutonium.[2]

However, the plutonium breeder strategy is not the only solution to the uranium resource problem. It is certainly not the most proliferation-resistant strategy, nor is it likely to be the most economical.

As an alternative to the breeder strategy, we set forth below an evolutionary strategy for fission power. With this approach, nuclear power plants would continue for decades to operate on once-through fuel cycles, with a shift after the year 2000 from current light water reactors (LWRs) to more uranium-efficient burner reactors—which in the remainder of this paper will be called advanced converter reactors (ACRs). Candidate ACRs include the heavy water reactor (HWR) which has already been developed by Canada, the high temperature gas cooled reactor (HTGR) which has been brought to an advanced state of development in both the USA and FR Germany, and an advanced light water reactor.

Unlike the fast neutron breeder reactors, ACRs can operate effectively on once-through fuel cycles as well as on highly uranium-efficient, proliferation-resistant closed fuel cycles. It is possible therefore to deploy ACRs without making a commitment to the extra complexities and proliferation vulnerabilities associated with a closed fuel cycle. The option of shifting to a highly uranium-efficient closed fuel cycle would always exist, however. Such a shift could be made if (a) very tight uranium supply constraints appeared to be developing, and (b) it became possible in the meantime to institute adequate technical and institutional arrangements to safeguard fuel recycling facilities against the diversion of weapons-usable materials.

An ACR operating on a closed isotopically denatured uranium—thorium fuel cycle would be two to five times more uranium-efficient than a light water reactor operating on the current once-through fuel cycle, and would as a result compete economically with the plutonium breeder up to very high uranium prices. Because of the large fissile inventories required to start up a breeder economy, the U-235 requirements of an efficient advanced converter reactor system operating on a closed fuel cycle and a

---

[2] Natural uranium contains 0.7 per cent U-235 and 99.3 per cent U-238. The U-235 isotope is chain-reacting ('fissile'). The U-238 isotope is not, but can be transmuted into chain-reacting isotopes of plutonium by neutron capture. Th-232, the principal naturally occurring isotope of thorium, is 'fertile' like U-238. A burner or converter reactor, while it usually produces a significant amount of new fissile material, is a net consumer of fissile material and, therefore, in the absence of a source of artificially produced fissile isotopes, requires a supply of U-235. A breeder reactor is a net producer of fissile material, and therefore, after its initial inventory of fissile material has been established, a breeder reactor economy can grow at a few per cent annual rate without an external supply of fissile material.

breeder reactor system would be comparable for the 50- to 75-year period during which either system is being established.

For these reasons, any economic incentive for a shift to plutonium breeder technology recedes to a time well beyond the middle of the next century. During the intervening period, it appears likely that the neutronics of ACRs could be improved to the point where they would become nearly as resource-conserving as the breeder. It is possible, therefore, that the evolutionary strategy could be extended indefinitely.

In the next two sections we discuss the rationale for making proliferation resistance a design criterion for nuclear energy systems and the value of isotopic denaturing of reactor fuel in achieving such proliferation resistance. Then we discuss the uranium efficiency and economics of power generation for the alternative reactor systems considered. Finally we describe in greater detail our evolutionary strategy for nuclear power development—first in terms of its applicability to the United States, and then for the world. To test the margin of safety in the evolutionary strategy, we will assume high nuclear power growth and limited availability of high-grade uranium resources.

II. Latent proliferation

No set of technical constraints on the nuclear fuel cycle can prevent a nation determined to acquire nuclear weapons from doing so. Nevertheless, to date most nations have deliberately refrained from launching programmes dedicated to nuclear weapon production. They have been deterred presumably in part at least by internal opposition, by the potential for weakening alliance relationships with the great powers, and by the potential for igniting regional nuclear arms races.

Many of these same nations are interested, however, in acquiring the technology for nuclear fuel reprocessing. A reason given has been that this technology would be required by the plutonium breeder reactor. Indeed, until recently it was widely expected that most new reactors coming into operation by the year 2000 would be breeder reactors.

With reprocessing plants comes the ability to recover nuclear weapons-usable material from spent reactor fuel, however. Civilian nuclear power programmes around the world which are aimed at plutonium recycle are therefore 'legitimizing' an activity which was previously conducted only in the course of nuclear weapon production. Because such an approach to a nuclear weapon capability is inherently ambiguous and does not force a nation to signal or even decide in advance its actual intentions, it is termed latent proliferation [3−5]. Latent proliferation reduces the stability of a nation's non-nuclear status because it reduces the time and resources required to move from the decision point to the possession of nuclear weapons.

An important objective of the non-proliferation effort therefore is to try to encourage the adoption of civilian nuclear technologies in which the gap between nuclear power and the production of nuclear weapon materials is kept as wide as possible.

Recently, it has been suggested that the plutonium breeder fuel cycle could be made more proliferation-resistant through a fuel-reprocessing arrangement in which the recovered plutonium would be kept mixed with a substantial fraction of the short-lived radioactive fission products present in the spent fuel. This procedure (termed the Civex process by its authors) would indeed make it more difficult to chemically process the fresh fuel to recover the plutonium. As has been pointed out by Walter Marshall, one of the authors of the Civex proposal, however, it would be decades after the deployment of a plutonium fuel cycle before the Civex process could become fully operational. In particular, he points out that the process would not be practical for LWR recycle fuel or for the initial loads of liquid metal fast breeder reactor (LMFBR) fuel. This is because the plutonium in this fuel would be obtained from many years-old LWR spent fuel in which the short-lived fission products used in Civex would have decayed away [6, 7]. Furthermore, the high economic cost of the Civex process and the occupational hazards relative to working with decontaminated plutonium would provide a ready excuse for a nation to abandon the process.

In any case, even with Civex, each 1-GW(e) breeder reactor (1 GW(e) = 10^6 electrical kilowatts) would have delivered to it annually in fresh fuel over 1 000 kg of fissile plutonium—over 100 times the quantity required to produce a nuclear weapon.[3] Because this plutonium is chemically separable from radioactive contaminants in the fuel, this strategy would be much less effective than isotopic denaturing in making the nuclear fuel cycle proliferation-resistant.

III. Isotopically denatured fuel cycles

Fuel cycles in which the fuel is effectively 'isotopically denatured' are those where the fissile material in nuclear reactor fuel is mixed with non-fissile isotopes of the same element to the point where the fissile material is so

---

[3] Another fuel cycle with this same problem would involve the use of uranium containing a high proportion of fissile isotopes ('highly enriched uranium') in reactor fuel. Fortunately there is not much pressure currently to deploy such a fuel cycle commercially—although both the high temperature gas-cooled reactor and the light water breeder reactor prototypes and many small research reactors are currently fuelled with 'weapons grade' uranium. All of these systems can be fuelled with isotopically denatured uranium—in almost all cases with an acceptable performance penalty.

Table 1.1. Uranium requirements and fissile plutonium discharge rates

1 GW(e) reactor, operating at a 65 per cent average capacity factor.

Fuel cycle type	Reactor type	Fissile enrichment of fresh fuel (per cent in heavy metal)	Core discharge burn-up (MWd/kg HM)	Uranium requirements[a] (short tons U_3O_8) Inventory[b]	Annual consumption	Fissile plutonium discharged annually (kg)
Low enriched Uranium-once through	LWR (PWR)[c]	3.2	33	242	143	141
	LWR (PWR)[d]	4.4	55	286	120	98
	HWR (CANDU)[e]	0.7 (natural U)	7.5	142	135	295
	HWR[e,f]	1.2	20.6	142	88	127
Denatured uranium–thorium[c] Once-through Recycle	HTGR[g]	8.5	125	242	91	22
	LWR[e]	4.8	35.6	706	86	52
	HWR[e]	1.7	16	610	30	26
	HTGR[g]	4.5	65	294	59	39
	LMFBR[h]	14.8	50	1755	87	350
Plutonium recycle	LMFBR[i]	13.6	75	—	—	1029[j]

Notes:

[a] These are the U_3O_8 requirements from the uranium mill. The tails stream at the enrichment plant is assumed to contain 0.1 per cent U-235. For enrichment priced at $75/SWU this tails assay becomes more economical than the 0.2 per cent ordinarily assumed when the price of U_3O_8 reaches $60–$70/lb. Reprocessing losses in recycle systems are assumed to be 2 per cent.

[b] The inventory given here is the U_3O_8 equivalent of the fissile material in the system in excess of three years' (n = 3) make-up requirements (M) for reactors which are refuelled annually (LWR, LMFBR), two years' (n = 2) make-up requirements for reactors which are refuelled continuously (HWR), and two and one half years' (n = 2.5) make-up requirements for reactors

which are refuelled semi-annually (US HTGR). The fissile inventory in the reactor is approximately $I_R = \frac{1}{2}(L + D) T_R$, where L and D are the equilibrium annual fissile load and discharge rates respectively and T_R is the fissile residence time (in years) of the fuel in the reactor. We assume that the fissile inventory outside the reactor in the fuel cycles is approximately equivalent to $2L$. The total fissile inventory as defined above is therefore $I_T = I_R + 2L - nM$.

The fissile inventory is then converted to an equivalent U_3O_8 inventory by treating all fissile atoms as if they were U-235 atoms and allowing for losses of U-235 to the enrichment plant tails. Note that for once-through fuel cycles $L = M$.

c The 'denatured uranium' here is a mixture of fissile uranium isotopes and U-238, such that the percentage of the fissile isotopes in the mixture is: 12 per cent when the fissile isotope is U-233, 20 per cent when the fissile isotope is U-235, or the weighted average when both are present. For the denatured recycle systems it is assumed that plutonium is not recycled.

d See reference [43].

e See reference [44]. The HWR numbers have been put on a common basis corresponding to a reactor with 30.5 per cent thermal–electric conversion efficiency.

f See reference [14b].

g See references [18, 19, 45].

h See reference [46a]. A linear combination of two oxide-fuelled reactors (one of which is fuelled with denatured U-233 and one of which is fuelled with denatured U-235) is taken such that the net production rate of U-233 is zero. The numbers have been changed to correspond to an LMFBR with 40 per cent thermal–electric efficiency (from 35.5 per cent).

i For an oxide-fuelled LMFBR with an in-core plutonium fissile inventory of 3 190 kg, a fuel residence time of 3.6 years and a breeding ratio of 1.28 (based on reference [46]).

dilute that it cannot sustain a fast (explosive) chain reaction in a small critical mass. Weapons-usable material can be recovered from such denatured fuel only by use of isotopic separation techniques which today require considerable technical resources and time.[4] In contrast, in plutonium-recycle systems, weapons-usable material can be extracted chemically from fresh reactor fuel. The isotopically denatured fuel cycles we have considered are shown in table 1.1.

Isotopic denaturing is a major proliferation-resistant characteristic of the once-through fuel cycle of today's water-moderated reactors, which we believe should be given up only reluctantly in considering alternative nuclear systems. In the water-moderated reactor once-through fuel cycles shown in table 1.1, the fissile U-235 in the fresh fuel is diluted with non-fissile U-238 to a concentration of less than five per cent—well below the U-235 level required to sustain a fast fission chain reaction. Although it was originally assumed that high temperature gas-cooled reactors would be fuelled with a mixture of highly enriched uranium and thorium, they may also be operated effectively on a once-through isotopically denatured fuel cycle, as is shown in table 1.1.

Denatured uranium fuel cycles have no chemically separable weapons-usable material in fresh fuel, but some plutonium is inevitably present in the spent fuel as a result of neutron capture on the U-238 denaturant. During the first few months after the fuel has been discharged from the reactor, the fuel is so intensely radioactive that the recovery of the plutonium would be very difficult, but the plutonium becomes increasingly accessible with remote chemical-processing techniques as the fuel cools. It is therefore critical that spent fuel from reactors operating on once-through fuel cycles be removed as soon as practicable after discharge from the many individual power-plant sites around the world to a few centralized depots where it could be stored under tight international control.

If and when it becomes worthwhile to realize the uranium savings achievable with recycle and it becomes institutionally and technically possible to safeguard the associated fuel cycle facilities, it would still be possible to preserve much of the proliferation resistance of the fresh fuel for slow neutron reactors by recycling only isotopically denatured fissile isotopes. This could be accomplished by operating these reactors on the denatured uranium–thorium fuel cycle.[5] In this system, the artificial fissile isotope U-233 would be produced and recycled instead of plutonium.

---

[4] This is true despite recent advances in centrifuge and laser isotope enrichment technologies. The development of these technologies does raise significant proliferation policy issues, however. See for example references [8, 9].

[5] See reference [10]. A mixture of fissile uranium isotopes and U-238 is considered to be 'denatured' for nuclear weapon purposes if the percentage of fissile isotopes is less than the fissile isotope weighted average of 12 per cent (for pure U-233) and 20 per cent (for pure U-235). Although a fast neutron chain reaction can be sustained at somewhat lower concentrations of the fissile isotopes in such a mixture, the corresponding critical mass becomes so large that weapon designers consider it impractical for nuclear explosives purposes. (See, for example, reference [11]).

Uranium-233 is formed as a result of neutron capture on a relatively abundant fertile isotope, Th-232. The U-233 would be isotopically denatured for nuclear weapon purposes (as U-235 is currently in low-enriched uranium fuel) by an admixture of the principal natural uranium isotope U-238. No corresponding denaturing isotopes exist for plutonium.

In a denatured uranium–thorium fuel cycle, fresh reactor fuel would be a mixture of denatured uranium and thorium. For systems involving reprocessing (see figure 1.1(b)), only the uranium and thorium would be recycled (supplemented with thorium and 'make-up' enriched uranium)—not the by-product plutonium. Make-up uranium from the enrichment plant would be highly enriched to compensate for the fact that the recycled uranium would be fractionally more depleted in fissile isotopes than in U-238. This make-up uranium would therefore be weapon-grade material until it was denatured by mixing with the recycled uranium.

In terms of proliferation resistance, this fuel cycle would require a lower level of international surveillance than the plutonium fuel cycle between the fuel-fabrication facility and the 50–100 dispersed nuclear power reactors which it would serve, but it would share the vulnerability of the plutonium fuel cycle to diversion of plutonium at the reprocessing plant and the vulnerability of fuel cycles based on highly enriched uranium to diversion of this material at and between the uranium enrichment and the fuel-fabrication plants. It is likely that these 'sensitive' facilities would have to be operated under direct international or at least multinational control.[6] This would be difficult to arrange, but it should be easier to do so than internationalizing the far more numerous power reactors—as would be necessary in order to achieve a comparable degree of security in a plutonium economy. Ten to twenty 'commercial-scale' fuel cycle centres (a fuel cycle centre would include enrichment, fuel-reprocessing and fuel-fabrication plants) would suffice to service a world nuclear capacity of 1 000 power reactors. It is more likely that a high level of international control would be accepted at these fuel cycle centres than at the reactors, because these centres would not be tightly coupled to national power systems.

In table 1.1 we show the annual fissile plutonium discharge rates for various fuel cycles. It will be seen that in all cases the discharge rates for reactors operating on isotopically denatured fresh fuel are much lower than for an LMFBR operated on a plutonium fuel cycle. However, reactors operating on low-enriched uranium once-through fuel cycles still discharge annually a great deal of plutonium. This arises because the only fertile material present in these fuel cycles is U-238.

In denatured uranium–thorium fuel cycles, plutonium production is reduced because thorium is substituted for U-238 until only enough U-238

---

[6] See references [12, 13]. The multinational (FR Germany, the Netherlands and the United Kingdom) arrangements for management of the European URENCO uranium centrifuge enrichment facilities appear to be institutionally quite close to those required for an international fuel cycle facility. These institutional arrangements grew in part out of safeguards concerns.

remains to dilute the fissile uranium isotopes to the upper concentration limit specified for isotopic denaturing. High fuel burn-up[7] can also be effective in reducing the plutonium discharge rate. This is because a large fraction of the by-product plutonium is fissioned in the fuel before it is discharged. Both high burn-up and the fact that a large fraction of the fertile material is thorium are responsible for the relatively low fissile plutonium discharge rate for the once-through HTGR fuel cycle shown in table 1.1. The total rates of discharge of fissile plutonium in both this case and that of the HWR operating in the denatured recycle mode are comparable to the expected one to two per cent loss rate from the plutonium breeder fuel cycle.

IV.   Uranium efficiency

Isotopic denaturing was our first criterion for candidate fuel cycle alternatives. Uranium efficiency is our second, so that most of the alternative fuel cycles shown in table 1.1 have annual uranium make-up requirements substantially less than the LWR operated on today's fuel cycle. (Uranium inventories are also shown in table 1.1. These represent one-time capital investments which can be passed on to start up replacement reactors when a plant is retired.)

As may be seen from a comparison of the first five entries in table 1.1, each of which represents an isotopically denatured once-through fuel cycle, substantially improved uranium utilization efficiency is possible with current reactor types without the recycle of chemically separable weapons-usable material.

1. The first entry corresponds to the most prevalent system today—the LWR operating on a once-through low-enriched uranium fuel cycle.[8]

2. The second entry indicates the projected results of the first stage of a programme aimed at increasing the uranium efficiency of the LWRs operating once-through fuel cycles—a 15 per cent saving associated with an increase in the fuel burn-up.

3. The third entry shows that the Canadian heavy water reactor

---

[7] Burn-up is measured in units of MWd/kgHM (megawatt days of fission energy released per kilogram of heavy metal [uranium plus thorium plus plutonium] in the fuel). Plutonium from high burn-up fuels contains large percentages of isotopes with large critical masses and high spontaneous fission rates (Pu-240 and Pu-242). Although this does not denature the plutonium, it does increase the critical mass and significantly decrease the probable explosive yield of nuclear weapons made with the mixture.

[8] LWRs are currently operated at a considerably lower uranium utilization efficiency than is indicated by the design value in table 1.1. One reason is that US enrichment plants are currently operated with the depleted uranium 'tails' still containing 0.25 per cent U-235 versus the 0.1 per cent assumed in table 1.1. Another is that fuel often does not reach its design burn-up.

(CANDU) operating on its currently used natural uranium once-through fuel cycle uses slightly less uranium than the LWR with the present design burn-up.

4. The uranium efficiency of the CANDU on a once-through fuel cycle could be significantly improved if it were fuelled with slightly enriched uranium.[9] As will be seen from the fourth entry in table 1.1, the uranium requirements of this ACR would then be only about 60 per cent of those of today's LWR operating on a once-through fuel cycle. This is approximately the same reduction in uranium resource requirements as could be achieved through the recycle of both the residual U-235 and the fissile plutonium in low-enriched uranium LWR fuel.

5. The fifth entry shows that the uranium requirements of another advanced converter reactor, the HTGR, operating on a once-through denatured uranium–thorium fuel cycle, could be about as low as those of the HWR on the slightly enriched uranium fuel cycle.

A major determinant of the uranium efficiency of a fission power system is its 'conversion ratio', that is, the number of new fissile atoms produced in its fuel per fissile atom consumed. Indeed, the good uranium efficiency of the HWR stems directly from its high conversion ratio. This high conversion ratio stems in turn from the relatively low wastage of neutrons in the HWR.[10] Once the necessary number of neutrons have been committed to maintain the chain reaction, every additional neutron which is saved from loss via leakage out of the reactor core or capture in the reactor structure or coolant can be used for the production of new fissile atoms. The deuterium in heavy water captures far fewer neutrons than the ordinary hydrogen in light water. The neutron losses in the Canadian HWR are also reduced greatly by a design which allows continuous refuelling.[11] As a result the large changes in reactivity which occur in the LWR core between annual refuellings are eliminated, fewer artificial neutron absorbers are needed to control the reactivity of the HWR

---

[9] The uranium efficiency of a once-through fuel cycle can be measured in terms of the number of fissions in the fuel per U-235 atom purchased at the uranium mill. Enriching the fuel entails some losses of U-235 atoms at the enrichment plant, but up to a point it also allows an increase of the number of fissions per initial fissile atom (fifa) in the fuel. In the case of the CANDU, the U-235 savings associated with increased fuel fifa more than offset the increased losses at the enrichment plant up to an enrichment of approximately 1.2 per cent. See, for example, reference [14].

[10] Another variable which affects the conversion ratio is the energy spectrum of the neutrons which sustain the chain reaction. The high conversion ratio ('breeding ratio') of the LMFBR, for example, stems from the high neutron energies characteristic of a 'fast neutron' reactor. A greater number of neutrons is released on the average by fast neutron-induced fissions than by fissions at the low neutron energies characteristic of thermal neutron reactors.

[11] From the point of view of proliferation resistance, the on-line refuelling capability of the CANDU is a disadvantage because it makes possible the diversion of spent fuel at any time—not just during the annual refuelling. Also, since the reactor can be operated on natural uranium fuel, the operator is not as susceptible to threats of fuel cut-offs as a penalty for safeguards violations. For a description of a proposed safeguards system for CANDU reactors, see reference [15].

core, and the neutron loss term is correspondingly reduced.[12]

The high uranium efficiency of the HTGR on the once-through uranium−thorium fuel cycle stems not from the high conversion ratio of the reactor but rather from its high thermal−electric conversion efficiency (40 per cent versus 33 per cent for the LWR) and the high burn-up achievable with its fuel.

Uranium efficiency improvements beyond those achievable in once-through fuel cycles can be obtained by reprocessing and fissile recycle. Examples shown towards the bottom of table 1.1 illustrate the savings predicted by current calculations for isotopically denatured recycle in LWRs and HWRs without major design modifications.[13] It appears that the uranium savings realizable with denatured uranium recycle in HTGRs would be much less than for the HWRs.[14] As shown in table 1.1, the

---

[12] Other methods than the use of neutron 'poisons' exist to compensate for slow changes in the core reactivity. In the light water breeder reactor the geometry of the core is adjustable [16]. In the 'spectral shift' reactor concept the neutron energy spectrum is modified by making the moderator a variable mixture of light and heavy water [17].

[13] The uranium−thorium fuel cycle has a higher conversion ratio than the plutonium−uranium fuel cycle in slow neutron reactors because U-233, the fissile isotope which is produced by neutron absorption on thorium, releases approximately 0.2−0.6 more neutrons per thermal neutron absorbed than Pu-239, the fissile isotope which is produced by neutron capture on U-238. This situation is reversed in a fast neutron rector, where Pu-239 releases more neutrons per neutron captured than U-233. In either case the extra neutrons are valuable because they can be used to breed more fissile material.

A characteristic of these uranium−thorium recycle systems is that their in-reactor inventory of fissile uranium is equivalent to many years' consumptive requirements. As a result, if the inventory were discarded at the end of the reactor lifetime, the uranium requirements of these systems would be considerably increased. Inexplicably, in most analyses of the viability of a long-term dependence on converter reactors operating on a uranium−thorium fuel cycle it is assumed that these fissile inventories must be discarded when a plant is retired, while in analysing the economics of the breeder reactor, the same analyses assume that the corresponding plutonium inventories can be passed on from reactor to reactor.

One issue which has been raised in this connection relates to the build-up of U-236 in recycled uranium. Uranium-236 is a neutron poison and its presence therefore results in a lowered conversion ratio and increased uranium requirements. The uranium feed requirements shown in table 1.1 for recycle fuel cycles correspond to the requirements after the fissile uranium has been recycled five times when the U-236 poisoning effect has already reached a significant fraction of its equilibrium value. At equilibrium the rates of U-236 creation and destruction become equal. We have estimated the U-236 poisoning effect at equilibrium using effective cross-sections furnished by Dr Y. Chang of Argonne National Laboratory and find that it would increase the uranium feed requirements of the recycle fuel cycles included in table 1.1 by only 10−20 per cent. This does not appear to be a large enough effect to justify discarding the fissile inventories. There would be only small net effects from detailed corrections for the U-236 poisoning effect in the cumulative uranium requirements shown in figure 1.5 for long-term operation with uranium recycle.

[14] See references [18, 19]. Because of the different neutron energy spectrum, the fissile enrichment and hence the proportion of U-238 denaturant in HTGR fuel is higher than that in HWR fuel. Furthermore U-238 competes more effectively with Th-232 for neutrons in the HTGR than in the HWR. As a result of these two effects, a larger fraction of the bred material is plutonium in an HTGR. Unfortunately the number of neutrons released by fissile plutonium following neutron capture is particularly low in an HTGR neutron spectrum and consequently the conversion ratio of the HTGR is more seriously degraded by denaturing than is that of the HWR.

The fraction of neutrons absorbed by U-238 can be minimized through fuel element redesign. The numbers shown in table 1.1 for the HTGR with denatured uranium recycle were obtained from such a reoptimization, assuming semi-annual refuelling.

performance of an LMFBR restricted to denatured uranium recycle is degraded by the denaturing requirement to the point where it requires about as much annual make-up uranium as the more efficient once-through fuel cycles considered here.[15]

Since neutron-absorbing fission products build up in the fuel with increasing fuel burn-up, the conversion ratio can be increased by reducing the time spent by the fuel in the reactor.[16] With current reactor types, the uranium savings would not offset the extra costs of more frequent fuel reprocessing and refabrication until very high uranium prices were reached, but for nations concerned about reducing their uranium import requirements, more frequent recycle could be a practical option. Indeed, on a non-denatured uranium−thorium fuel cycle, the conversion ratios of both the HWR and the HTGR can be pushed to 1.0 (a break-even breeder) at recycling costs that may be economically practical [21, 22].

V. Economics

At least until late 1975, the US Atomic Energy Commission (AEC) and its successor agency, the Energy Research and Development Administration (ERDA), were basing their civilian nuclear energy research and development (R & D) programme on the belief that, even if the price of uranium did not rise, the plutonium breeder would have a decisive economic advantage over the LWR operated on a once-through fuel cycle. This belief stemmed from the expectations that the capital cost of the LMFBR would decline to

---

[15] The numbers in table 1.1 show clearly that, in shifting an LMFBR onto an isotopically denatured uranium−thorium fuel cycle, the objective of achieving high uranium utilization efficiency and proliferation resistance would be in basic conflict. When the discharged plutonium is not recycled, the LMFBR is effectively reduced from a fissile isotope breeder to a burner reactor, with make-up U-235 requirements comparable to the make-up needed by an HWR operating on a once-through fuel cycle. At the same time, the hundreds of kilograms of fissile plutonium discharged annually from an LMFBR operated on a denatured fuel cycle would represent such a large energy resource that it is hardly credible that it would not be used. Indeed, in virtually all scenarios combining the LMFBR and denatured fuel cycles, it is assumed that the plutonium will be recycled in special safeguarded reactors, with the result that a significant fraction of the nuclear power system is put back on the plutonium fuel cycle.

The LMFBR would produce so much plutonium on the denatured uranium−thorium fuel cycle because most of the fertile material in the LMFBR core would be the U-238 denaturant. This is due to the high concentration of fissile uranium required in an LMFBR core. In order to sustain the fission chain reaction the concentration of fissile atoms must be sufficient so that, on the average, one of the typically two or three neutrons released per fission will be captured by another fissile atom. For neutrons in a 'slow' neutron reactor (the neutron energies are typically of the order of 0.1 electron volts in the LWR and HWR cores) the fission cross-sections of fissile atoms are about two orders of magnitude larger than the capture cross-sections of the fertile atoms, thus making it possible to sustain the chain reaction in mixtures with only a few per cent fissile atoms. For 'fast' neutrons (typically of the order of 10^5 electron volts in an LMFBR core) the fissile atom cross-sections are only about one order of magnitude greater than the fertile atom cross-sections, with the result that a much higher concentration of fissile atoms is required.

[16] The proposed molten salt thermal neutron breeder reactor would achieve a conversion ratio greater than one in part because neutron poisons would be removed continuously [20].

the same level as the LWR within 15 years of its introduction, and that its fuel cycle cost would be very much smaller than that for the LWR in 1974.[17]

More recently, however, both the LMFBR fuel cycle and capital cost estimates have been revised upward by the US Department of Energy (DOE), the successor agency to ERDA. The capital cost of a commercial LMFBR is now expected to be 25–75 per cent greater than the capital cost of an LWR with the same generating capacity, and the expected costs for reprocessing and refabricating breeder fuel have risen to a level where, at today's uranium prices, they would roughly offset the savings in uranium and enrichment costs which the breeder would make possible.

Even if the LMFBR fuel cycle costs were negligible, the capital cost disadvantage is so great that the breeder could probably not compete with the LWR until uranium prices rose very much higher than they are today. For an LMFBR costing 25–75 per cent more than the LWR, the capital charge differential would be 0.7–2 times the *total* LWR fuel cycle cost for uranium at recent prices (approximately \$40/lb U_3O_8 for new contracts— one pound U_3O_8 = 0.38 kg U). Including estimated LMFBR fuel cycle costs, we find, as shown in figure 1.2, that LWR-generated electricity would be cheaper than electricity from an LMFBR system,[18] at least until the price of U_3O_8 reached \$60/lb, and may be cheaper for U_3O_8 prices as high as \$180/lb, because of the uncertainties in the LMFBR capital cost. (In these economic calculations, the 1977 DOE fuel cycle cost numbers shown in the appendix have been used.[19] These numbers are very uncertain, but the contribution of this uncertainty to the range of uncertainty of LMFBR electricity prices is considerably smaller than the range of uncertainty associated with the LMFBR capital cost estimate.)

By the time uranium prices reach this range, however, other reactor–fuel cycle combinations are likely to become economically competitive with the LWR; because they are more uranium-efficient than the LWR, they would remain economically competitive with the LMFBR up to even higher uranium prices. Advanced light water reactors, heavy water reactors, and the high temperature gas-cooled reactors in particular all lie between the LWR and the LMFBR in terms of uranium efficiency and probably capital

---

[17] For the breeder cost numbers, see reference [23a]. For the LWR fuel cycle cost numbers, see reference [24].

[18] Our LMFBR *system* includes for each 2.4 GW(e) of LMFBR capacity one GW(e) of LWR capacity which is fuelled by the net surplus of plutonium produced by the LMFBR.

[19] Unless otherwise documented, the cost figures shown in the appendix are obtained from reference [25]. One change which we have made is to raise the capital costs for an LWR from \$625 to \$800/kW(e), based on the analysis in reference [26]. Other nuclear power plant capital costs have been increased by the ratio of these two costs.

The principal reason for the high LMFBR capital cost is the use of sodium instead of water as a coolant. Because sodium is extremely reactive chemically, it is necessary to design special features into the system to keep the sodium isolated from air (a hermetically sealed refuelling machine and an inert gas atmosphere surrounding the reactor plumbing system) and water and to limit the consequences of those sodium leaks which will inevitably occur (secondary sodium loops to carry the heat from the radioactive primary sodium coolant to the steam generators, guard vessels, containment purge systems, complex steam generator designs, and so on).

24

Figure 1.2. Levelized cost of delivered electricity as a function of uranium price

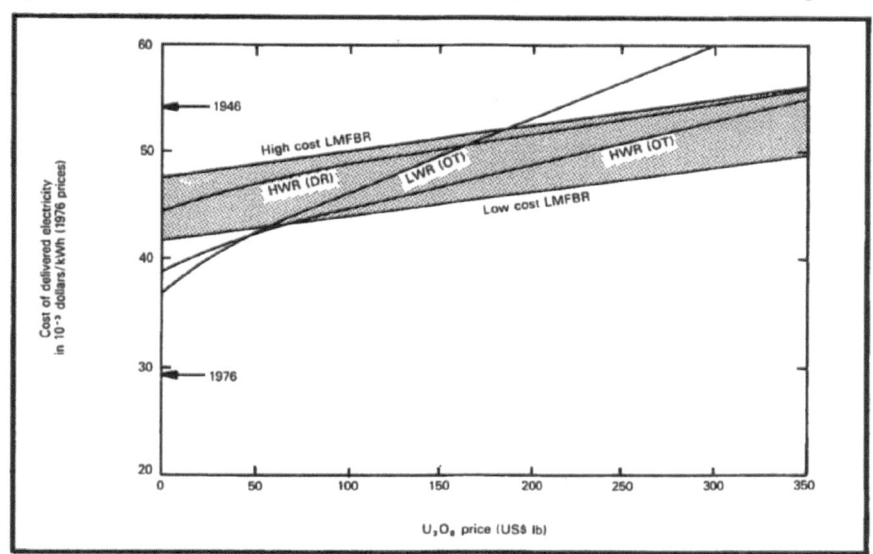

Note: Here the estimated price of delivered electricity (in 1976 dollars) levelized over the life of the power-plant is shown for alternative reactor systems, as a function of the price of U_3O_8. To offer an historical perspective, US average electricity prices (in 1976 dollars) for the years 1946 and 1976 are indicated by arrows on the left-hand scale.

Two once-through (OT) fuel cycles for reactors operating on low-enriched uranium are shown here: the LWR (OT) and the HWR (OT). The HWR (DR) system involves denatured uranium−thorium recycle. A range of electricity prices is shown for an LMFBR/LWR system as a shaded band.

The cost band shown for the LMFBR system is actually the cost for a non-growing system of LMFBRs and the plutonium-burning LWRs that could be supported with excess LWR. The low (high) side of the shaded band corresponds to an LMFBR capital cost 25 per cent (75 per cent) greater than that of the LWR.

For the LWR and HWR fuel cycles the cost of electricity corresponds to 0.2 per cent enrichment tails below $60/lb U_3O_8 and 0.1 per cent tails above $60/lb (hence the break in the slope of these curves at this uranium price).

For simplicity we have set the price of make-up uranium equal to the price of inventory uranium in constructing these curves. It should be noted, however, that: (a) the levelized cost for make-up uranium over the lifetime of a reactor will probably be somewhat higher than the cost of its initial uranium inventory, and (b) once the fissile inventory of a reactor system has been purchased the rate of increase of the cost of electricity with uranium price will be lower than shown. In the case of the LMFBR the cost of electricity would stabilize. For the HWR operating on the denatured fuel cycle the slope of the cost curve would be reduced more than 55 per cent, whereas for the once-through fuel cycles the slope would be reduced by only 10 per cent. These differences represent the different relative contributions of inventories and make-up fuel to the cost of electricity for different fuel cycles. See table 1.2.

cost, and are therefore candidates for systems which would generate power at prices competitive with the LMFBR over a large intermediate range of uranium prices. For specificity the potential economic role of advanced converter reactors will be illustrated using the example of the Canadian HWR, because there is already considerable commercial experience with this reactor. Within the uncertainties of the economic and technical data, it appears that similar cases could be made for the HTGR or an advanced LWR.

Table 1.2. Levelized cost of delivered electricity from alternative fuel cycles[a]

	PWR Once-through	PWR denatured recycle	PWR Pu cycle[b]	HWR once-through	HWR denatured re-cycle	LMFBR[b] (stand alone)
Capital[c]	14.88	14.88	14.88	15.48	15.48	(18.60–26.04)
Heavy water[d]	–	–	–	2.19	2.19	–
U₃O₈[e]						
Inventory[f]	0.0065C_{uo}	0.0185C_{uo}	–	0.0038C_{uo}	0.0163C_{uo}	–
Make-up[g]	0.0609C_u	0.0366C_u	–	0.0372C_u	–	–
Conversion	0.09	0.07	–	0.05	0.04	–
Enrichment[e]	2.24	2.76	–	0.59	1.48	–
Plutonium						
Inventory[h]	–	–	–	–	–	0.0632C_{po}
Make-up[i]	–	–	0.0765C_p	–	–	–
Credit[j]	–	–	–	–	–	-0.0214C_p
Thorium	–	0.12	–	–	1.45	0.98
Reprocessing[k]	–	0.69	0.69	–	3.37	2.00
Fabrication	0.48	2.43	1.84	0.47	–	–
Spent fuel storage	0.40	–	–	0.65	–	–
Operation and maintenance[l]	2.30	2.30	2.30	2.30	2.30	2.30
Total cost of electricity including transmission and distribution costs and losses[m]	37.81 +0.0071C_{uo} + 0.0669C_u	40.95 +0.0203C_{uo} + 0.0403C_u	37.06 +0.0749C_{po} +0.0841C_p	39.28 +0.0042C_{uo} +0.0409C_u	44.55 +0.0179C_{po} +0.0140C_u	(41.64–49.81) -0.0235C_p

Notes:

[a] The costs given here are in units of 10^{-3} dollars/kWh (1976 dollars) levelized over the nominal 30-year life of the power plant, based on the parameters given in appendix (see page 45). The effective cost of money (corrected for inflation) is assumed to be 3.44 per cent/year, which would be appropriate for investor-owned utilities with a 50:50 mix of debt/equity financing and a 50 per cent combined federal-state income tax rate. Straight

line depreciation is assumed for both the physical plant and the fuel cycle investments. For fuel cycle changes, depreciation occurs only during the fuel burn-up period.

[b] The cost of electricity from an equilibrium LMFBR system is equal to the average cost of electricity from the system consisting of one plutonium-fuelled LWR plus the 2.4 'stand alone' LMFBRs of the same capacity which are needed to provide the fissile plutonium make-up for the LWR.

[c] The capital charge is based on the capital costs and the capacity factors given in the appendix and a capital charge rate (corrected for inflation) of 10.6 per cent per year. The capital charge rate includes a 3.1 per cent annual charge on the initial investment for property taxes, insurance, and capital equipment replacements over the life of the power plant.

[d] The initial heavy water inventory is treated as a non-depreciating asset which is sold for its original price when the plant is retired. The annual make-up is assumed to be 2 per cent of the initial inventory. (See the appendix.)

[e] For 0.1 per cent enrichment tails. With the enrichment cost assumed here (see the appendix), 0.1 per cent enrichment tails would be more economic than 0.2 per cent tails when the U$_3$O$_8$ price exceeds $60–70/lb.

[f] The amount of U$_3$O$_8$ equivalent to the fissile uranium inventory is treated as a non-depreciating asset which is sold (transferred to a replacement reactor) for its initial price (expressed as 'C$_{uo}$' 1976 dollars per pound) when the plant is retired. In the economic calculations the inventory for each fuel cycle is that given in table 1.1, with one adjustment: the HWR inventories given in table 1.1 must be reduced to reflect the fact that the HWR is assumed to have an average capacity factor of 75 per cent (versus 65 per cent for other reactor types).

[g] The cost of make-up is expressed as a function of the price of U$_3$O$_8$ (in 'C$_u$' 1976 dollars per pound) levelized over the life of the power plant.

[h] The plutonium inventory is treated as a non-depreciating asset and is expressed as a function of 'C$_{po}$' the price of fissile plutonium at the time of reactor startup in 1976$ per gram. The total inventory includes both the core inventory and the out-of-reactor inventory. The out-of-reactor plutonium inventory of the LMFBR is assumed to be equivalent to the plutonium in two annual fuel reloads. The relationship between C$_{po}$ and C$_{uo}$ is given in the appendix.

[i] The cost of make-up plutonium is expressed as a function of 'C$_p$,' the price of fissile plutonium in 1976 dollars per gram levelized over the life of the power-plant. The relationship between C$_p$ and C$_u$ is given in the appendix.

[j] It is assumed that half of the excess plutonium is produced in the radial blanket (for which it is assumed the residence time is five years, with one fifth of the radial blanket discharged per year) and that half is produced in the axial blanket (half is discharged every T$_R$/2 years, where T$_R$ is the residence time in the reactor for the core and the axial blanket). Here we consider a core burn-up of 75 MWd/kg so that T$_R$ = 3.6 years. It is assumed that the excess fissile plutonium is sold at a price of C$_p$ (in 1976 dollars per gram).

[k] The reprocessing cost given here includes the cost of long-term waste disposal (including the shipment of the waste to the waste repository).

[l] The actual reported operation and maintenance cost for 40 operating LWRs in 1976 was 2.3×10^{-3} dollars/kWh. (The cost of transmission and distribution averaged 11.5×10^{-3} dollars/kWh in 1972, according to reference [48]. If this cost increased at the same rate as general inflation, it would have been 15.4×10^{-3} dollars/kWh in 1976.)

Including its heavy water inventory and allowing for the higher average capacity factor that is achieved with on-line refuelling, the heavy water reactor is estimated to have an approximately 20 per cent higher capital charge per kilowatt hour generated than the LWR. By the time the price of U_3O_8 reaches $60/lb, the lower costs of uranium and enrichment operating on a slightly enriched uranium once-through fuel cycle would offset this capital cost disadvantage relative to the LWR. This once-through HWR system would then generate electricity at prices within the range of uncertainty calculated for the LMFBR system up to U_3O_8 prices of over $350/lb.

Over this enormous U_3O_8 price range, the rise in the price of HWR-generated electricity would only be approximately one-half of the price *decrease* shown for electricity in the USA during the 30-year period 1946—76. To the consumer of electricity, therefore, the economic implications of the choices being discussed here are hardly earthshaking.

Although not economically justified until very high uranium prices are reached, considerably greater uranium efficiency could be achieved by shifting advanced converter reactors to fuel cycles involving the recycle of denatured uranium. In figure 1.2, it may be seen that the estimated price of electricity from an HWR operating with denatured uranium recycle lies within the range of uncertainty of corresponding prices for electric energy from an LMFBR system over the full range of uranium prices shown. The cost of electricity from this HWR system rises only slightly more rapidly with the rising cost of uranium than does the cost of electricity from the LMFBR system. (The cost of electricity from the LMFBR system rises with the price of uranium because of the investment required to purchase its large fissile inventory increases with the price of uranium.)[20]

Of course, with full deployment of a breeder economy, uranium requirements would eventually go down to essentially zero and the price of uranium would stop climbing. The uranium requirements of ACRs operating with uranium recycle would be so low, however, that the price of uranium would rise only very slowly once these systems were established— about one-fifth as rapidly with time as with the LWR on a once-through fuel cycle in the case of the HWR, for the same rate of increase of uranium prices with consumption. Because this price would have to be paid for only 0.2 times as much uranium per kilowatt hour, the rate of increase in the price of electricity generated by an HWR operating with denatured uranium recycle due to uranium price increases would be much less—only $0.2 \times 0.2 = 0.04$ times the rate for an LWR operating on a once-through fuel cycle. Even for an ACR on a once-through fuel cycle requiring 0.6 times as much uranium as an LWR once-through system, the rate of increase in the price of ACR-generated electricity would only be $0.6 \times 0.6 = 0.36$ times the rate for an LWR. Thus, in summary, an ACR both would be relatively

---

[20] The plutonium-fuelled LMFBR and the plutonium-fuelled LWR each have inventories of approximately 5 000 kg of fissile plutonium (including the inventory outside the reactor in the fuel cycle). For the relationship between the market price of this plutonium and the price of U_3O_8, see the appendix.

insensitive to high uranium prices and, by reducing the rate of uranium consumption, would slow down the rate of uranium price increases.

Thus it would only be when uranium prices were extremely high that the cost savings associated with the very low uranium requirements of the LMFBR could give it a decisive economic advantage over advanced converter reactors, and even here it appears that the evolutionary strategy could be taken one step further by improving the neutron economy of the current advanced converter reactors to the point where they too become independent of an external source of fissile material.[21]

The timetable for the evolution to increasingly uranium efficient fission systems depends on how rapidly uranium prices escalate. Assuming that on the average uranium is purchased by utilities at competitive market prices, the rate of price escalation depends upon both the rate at which nuclear electricity production grows and the uranium resources situation—subjects to which we now turn.

VI. A high nuclear growth scenario

In order to test the viability of an evolutionary strategy for even a large US economy heavily dependent on fission, we describe in this section a scenario in which, by the year 2020, nuclear energy provides one-third of the primary energy consumed by a US economy with a gross national product (GNP) 2.5 times larger than in 1975. For simplicity, we assume that all of this fission energy is used to generate electricity.

The nuclear growth scenario which we derive is shown in figure 1.3. Nuclear power grows through 300 GW(e) installed capacity in the year 2000 to 640 GW(e) by 2020, after which it levels out. The electricity energy generated annually by the 640-GW(e) nuclear capacity corresponds to approximately twice the *total* electric energy generated in the USA in 1975.

This is certainly a large fission economy by any of today's yardsticks, but it is small in comparison with the nuclear power growth projections which were being used to justify the LMFBR development programme only a few years ago. As recently as 1975, ERDA was projecting a rapidly growing US nuclear economy exceeding 3 700 GW(e) capacity by the year 2025.

At that time, however, the art of projecting the growth of electric power consumption corresponded to very little more than projecting past exponential growth rates into the future. The ERDA 'reference case'

---

[21] See references [21, 22], and the report on the light water breeder reactor cited in reference [16]. Note that all of the conversion ratios calculated in these references are for *non-denatured* thorium—U-233 fuel cycles. Denaturing reduces these conversion ratios, particularly in the case of HTGR.

Figure 1.3. A high nuclear growth scenario

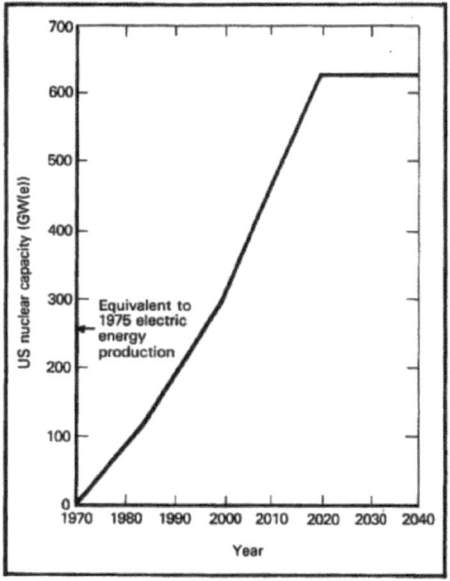

Note: If fission energy comes to play a major role in the future US energy supply, its growth might look approximately as shown. Our results are not sensitive to the detailed form of the growth curve, but would be undermined if electric demand continued indefinitely to double every 10–15 years—as was assumed until recently by most official studies. The capacity shown is for a nuclear system operating at an average 65 per cent capacity factor and at 33 per cent thermal–electric conversion efficiency. The plateau which is reached at 2020 corresponds to a rate of electrical energy production from nuclear power plants almost twice as great as total US electric energy production in 1975.

assumed a doubling of US electricity consumption approximately every 13 years for the next 50 years with no indication of a plateau even at the end of that period [23b].

In the past few years, however, it has become clear that there are many trends which are tending to damp out such rapid growth in the consumption of electric power. These include the end of the post-World War II 'baby boom', an approach to saturation in the degree of electrification of the US energy economy, and the transition from an era of declining to one of rising electricity prices. These effects are taken into account in our nuclear growth scenario.

The primary energy variable in this scenario is the total energy E consumed in the economy. We relate this to the level of economic activity, as measured by the (deflated) GNP, through an 'energy efficiency' factor η:

$$E = \text{GNP}/\eta \qquad (1.1)$$

Between 1920 and 1950, η increased rapidly (approximately 1.6 per cent per year), probably because of the technological revolution which accompanied the replacement of the direct combustion of coal by the more convenient energy forms: petroleum, natural gas and electricity. Between

1950 and 1970, η fluctuated about an almost constant level. Energy-efficiency improvements no longer came automatically with the introduction of new technologies and, since energy prices were generally declining (measured in constant dollars) there was little economic incentive to make energy-efficiency improvements.

Since 1970, however, η has increased at an average rate of 1.3 per cent per year, reflecting deliberate improvements in the efficiency of energy use in response to increasing energy prices and expectations of constrained energy supply in the future. If, as seems probable, US energy prices (in constant dollars) double over the next 25–35 years, it is likely that the average rate of increase in η will be at least one per cent per year [27–29]. We will make this an assumption in our scenario, although it is less than half of the average rate of improvement which could be achieved if the barriers to implementing economically justified energy-efficiency improvements could be overcome [30].

Until recently, future US GNP growth was projected simply by extrapolating the exponential growth of the post-World War II decades. It now appears, however, that if the US fertility rate continues near the relatively low level which has been experienced over the past several years, the US GNP will grow much more slowly in the future, even if per capital economic growth continues at the average rate experienced over the past decades.

The effect on GNP growth of population trends may be most easily discussed through a representation of the GNP as a produce of two factors —employment (measured in terms of full-time equivalent employees, L) and GNP per worker (the average productivity, P):

$$GNP = L \times P. \tag{1.2}$$

Using this representation, approximately half of the average 3.5 per cent annual rate of GNP growth experienced during the period 1950–75 can be associated with the increased number of (full-time equivalent) workers born in the 'baby boom' which ended in the late-1960s. Under the assumption that the US fertility rate stops declining and stabilizes at 1.9 children per woman (slightly higher than the level of 1973–77) and that immigration persists at the average rate of 400 000 per year, the growth in employment even in a 'full employment' economy (4–5 per cent unemployment) would average only one per cent per year over the period 1975–2010. This dramatically slower labour force growth appears virtually certain in this period. Even if the fertility rate should rise to 2.1, the working age population would not be growing between 2010 and 2030 and would grow at only 0.3 per cent per year thereafter.[22]

With respect to productivity, we assume that the average 1950–75 rates of productivity growth for the goods and service sectors of the economy will continue for the next several decades. The overall average annual rate of

---

[22] The 1.9 fertility rate corresponds to the 'high' population scenario of E.L. Allen *et al.* [27]. The 2.1 fertility rate corresponds to the Series II projection of the US Census Bureau.

growth in US productivity would then be 1.5 per cent in the period 1975–2010, an upswing from the very low level of 1.2 per cent per year average growth rate for the period 1965–75.[23]

As a result of these trends in employment and productivity, the US economy would grow more slowly in the future than in the past, averaging 2.5 per cent per year over the period 1975–2010. Per capita GNP in this period would grow at approximately the same rate as the period 1950–75, however.

Combining the assumption of a one per cent average annual improvement in the energy efficiency of the overall economy with the above assumptions for the trend of the growth of the US labour force and its productivity, total annual US energy use would increase at an average annual rate of 1.5 per cent (or by a factor of 1.7) over the period 1975–2010. The rate of increase toward the end of this period would be less as the labour force levels out. We assume in our nuclear growth scenario, however, that the nuclear power sector continues to grow rapidly after 2010 until it levels out in about 2020 at an asymptotic level of approximately 640 GW(e). This plateau could come about either as a result of saturation of energy use and electrification, with any further slow growth of the GNP largely offset by increases in the energy efficiency of the economy, or as a result of the penetration of new energy sources such as solar energy or fusion in a growing energy economy. At least one of these eventualities seems likely. In any case, our conclusions regarding an evolutionary strategy for nuclear power would not be significantly affected by continued slow growth in the nuclear electric sector after 2020.

Perhaps the most important conclusion to be drawn from this analysis is that, even with continued economic growth, a relatively modest rate of improvement in energy efficiency, and a large-scale commitment to nuclear energy, it appears quite plausible that US nuclear generating capacity will approach an asymptote below 1 000 GW(e). This conclusion is in striking contrast to previous projections of indefinitely continued exponential growth and has profound implications for decisions relating to alternative nuclear technologies.

---

[23] The 1.5 per cent rate would be slightly less than the 1.8 per cent average growth rate during the period 1950–75. Our projection is based upon the assumption that the greater growth in productivity in the goods sector will continue to be slightly more than offset by the more rapid growth in service sector employment. Here the service sector is defined as services, government, communications, finance, insurance and real estate, as these industries are defined by the Office of Business Economics, US Department of Commerce. The goods sector is everything else. During the period 1950–75 the productivity growth rates for these sectors were 2.4 per cent per year for goods and 0.9 per cent per year for services. During the same period the fraction of the gross product originating in the goods-producing sectors declined at an average annual rate of slightly less than 0.2 per cent (see reference [31]).

VII.   Uranium prices

The problem of the uranium supply for US nuclear power plants is usually put in terms of the amount of time it will take a given projected nuclear capacity to consume an assumed US resource base of a few million tons U_3O_8. However, this formulation is not correct. The uranium resource base of the United States cannot be represented by a single number. As with other minerals, it is an increasing function of price. At higher prices it becomes economic to mine lower-grade and less accessible ores.

This point is particularly germane to our evolutionary strategy. The traditional strategy for developing fission power technology was to go directly from the 'uranium-guzzling' LWR to the LMFBR, which requires hardly any make-up uranium at all. With such a strategy it may make sense to ignore uranium resources recoverable above a certain price because at such a uranium price the LWR fuel costs would be high enough for it to be economical to switch to the breeder. With an evolutionary strategy based on ACRs, however, it would be possible to reduce uranium requirements to lower and lower levels, with the result that these systems are likely to remain economically competitive with the LMFBR even if fuelled from uranium resources considerably more costly than those which have usually been included in estimates of the US uranium resource base. Our purpose in this section, therefore, is to lay out the basis for a conservative (in this case 'reasonable lower bound') estimate of the cumulative availability of US uranium as a function of price.

The 1977 uranium supply curve estimate by the Supply Evaluation Branch of the Division of Uranium Resources and Enrichment of DOE is shown in figure 1.4.[24] Below a price of $100/lb, this estimate includes principally the U_3O_8 which the DOE estimates is recoverable in the USA at a forward cost of $50/lb or less. (Forward costs, which do not include the costs of exploration, land acquisition, royalties or profits, are lower than prices by about a factor of two in this price range.) The official 1978 DOE estimate of resources in this category was 4.4 million short tons U_3O_8 (1 short ton $U_3O_8 = 0.77\,Mg\,U$) [33].

For U_3O_8 prices above $100/lb it was assumed in deriving the DOE supply curve that it would be possible to begin to recover uranium from the black marine 'Chattanooga' shale. This shale lies relatively near the surface under much of central Tennessee and adjacent areas of Kentucky and Alabama. A recent US Bureau of Mines compilation of data for an area of $7\,000\,km^2$ finds that the uppermost (Gassaway) stratum of the deposit

---

[24] The estimate includes the production constraints which would be associated with a nuclear economy growing to $450\,GW(e)$ in the year 2000. It also includes 0.34 million tons of by-product U_3O_8 which are assumed to be producible by 2040. The price represents the cost of production of the highest cost resource being mined at the corresponding cumulative production level plus a 15 per cent rate of return on investment [32]. The estimates in reference [25] are based on Combs's analysis.

Figure 1.4. Uranium availability as a function of price

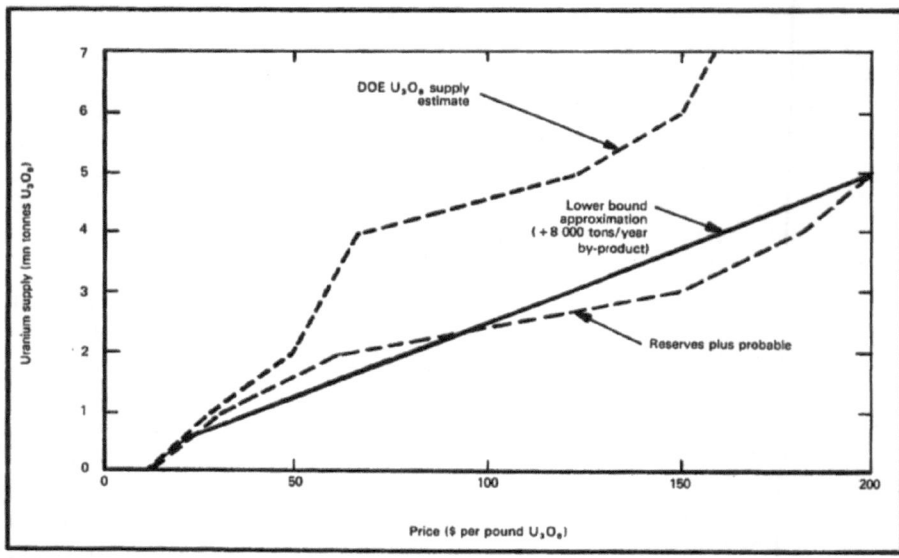

Note: The upper dashed line shows the 1977 Department of Energy estimate of US uranium resources as a function of equilibrium price. The lower dashed line includes only identified reserves and assumed resources in the 'probable' category (see text). Both curves include contributions from Chattanooga shale beginning above $100/lb U_3O_8. The solid line represents our estimated lower bound on the availability of US non-by-product uranium as a function of price. In addition we assume that the recovery of uranium as a by-product of phosphate-processing rises to a rate of 8 tons U_3O_8 per year by the year 2000.

averages 60–70 p.p.m. U_3O_8 in a layer 3–5 metres thick, corresponding to a resource of 4–5 million short tons U_3O_8 [34].

The Supply Evaluation Branch has in addition generated a "small U_3O_8 supply case" also shown in figure 1.4. This differs from the first supply extrapolation mainly by including only reserves and estimated 'probable' resources in unexplored extensions of known deposits or deposits which are expected to be discoverable within specific trends or areas of mineralization in known uranium districts. The small supply case is also somewhat more pessimistic regarding the costs of exploiting the Chattanooga shale. We adopt this low supply case as a 'lower bound' estimate, approximating it by a straight line, as shown in figure 1.4. On this resource curve the availability of uranium rises to 5 million short tons U_3O_8 at a price of $200/lb U_3O_8.

Uranium is also available in substantial quantities as a by-product associated with the recovery of other minerals. Probably the most significant source of by-product uranium in the future will be from phosphate recovery. Uranium is commonly present at concentrations of 50–200 p.p.m. in marine phosphorite rock, the source of most of the world's phosphate fertilizer [35]. While the cost of recovering uranium from phosphate rock as the primary product is likely to be at least as high as that for recovering uranium from Chattanooga shale, extraction of uranium as a by-product of phosphate recovery is relatively inexpensive, since the uranium goes into solution when the phosphate is extracted as phosphoric acid. As a result of the rise of uranium prices in 1974–75, by-product uranium recovery operations have begun in Florida, and it is estimated that the United States could be producing 8 000 short tons of by-product U_3O_8 annually by the end of the century [6]. In our analysis below, we assume that by-product U_3O_8 will be available at a rate which rises linearly with time starting in 1980 until it stabilizes at a level of approximately 8 000 short tons per year in 2000. The cumulative production of by-product U_3O_8 would amount to 0.32 million short tons by 2030 and 0.72 million short tons by 2080.

VIII. A timetable for the USA

We are now in a position to address the question posed initially—whether the evolutionary strategy we have put forward represents a credible alternative to an LMFBR strategy. We first consider the situation in the United States.

The critical components of the evolutionary strategy are first to rely as far as is practical on once-through fuel cycles based on contemporary reactor types, with a shift after the year 2000 to more uranium-efficient advanced converters, and second, to maintain the option of shifting these same ACRs to denatured uranium recycle fuel cycles if, for some reason, the uranium supply becomes constrained.

Figure 1.5, based on our high nuclear growth scenario, shows in simplified form how such a strategy compares in terms of uranium resource consumption with the two extreme strategies ordinarily discussed: continued reliance on unimproved LWRs operating on a once-through fuel cycle, and a shift to plutonium breeder reactors beginning in the year 2000 at the maximum rate allowed by the availability of plutonium for start-up inventories.[25]

Figure 1.5 shows the cumulative uranium resource requirements as a function of time for the LWR once-through and LMFBR strategies, and between these curves the uranium requirements for the evolutionary strategy based on an advanced converter reactor with and without denatured recycle (using the uranium requirements of the HWR given in table 1.1). In

---

[25] The LMFBR in this scenario has the characteristics given in the appendix for a fuel burn-up of 75 MWd/kg heavy metal. We assume that all plutonium generated in LWRs is saved to start breeders until the equilibrium breeder generating capacity is reached. We assume that after 2000 all uranium recovered from LWR fuel is recycled in LWRs. In equilibrium, the LMFBR capacity totals 392 GW(e), the LWR capacity fuelled by LMFBR generated plutonium totals 163 GW(e), and the by-product uranium-fuelled LWR capacity (assuming both uranium and plutonium are recycled) totals 85 GW(e).

Figure 1.5.   Cumulative uranium requirements for different fission strategies

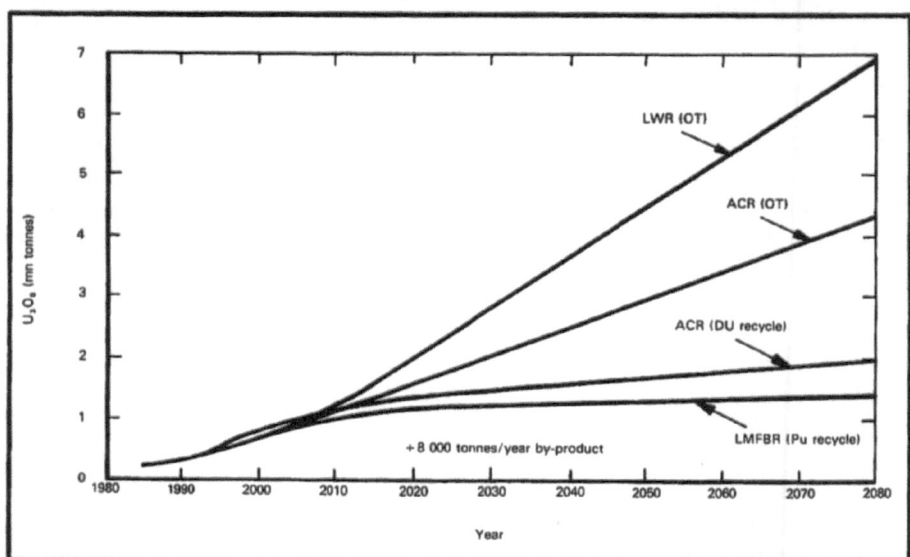

Note: Cumulative non-by-product US uranium requirements are shown for four alternative strategies, given the growth in nuclear-generated electricity pictured in figure 1.3. The top curve (LWR [OT]) corresponds to continued operation of light water reactors on a low-enriched uranium once-through (OT) fuel cycle. The bottom curve (LMFBR [Pu recycle]) corresponds to the introduction of a mixed-oxide-fuelled plutonium breeder reactor beginning in the year 2000.

For the middle two curves, all new reactors introduced after the year 2000 are advanced converter reactors (ACRs). All reactors are assumed to be replaced at the end of a 30-year life. The curve labelled ACR (OT) corresponds to the continued operation of all nuclear reactors on a low-enriched uranium once-through fuel cycle indefinitely. The curve labelled ACR (DU recycle) shows how cumulative uranium requirements could be reduced substantially by shifting the LWR to denatured uranium–thorium recycle beginning in 1995 and introducing the ACR on this fuel cycle beginning in 2000.

The uranium requirements assumed for the ACR in these two scenarios are respectively: (*a*) those of an HWR operated on the 1.2 per cent enriched once-through uranium cycle and (*b*) those of an HWR operated in the recycle mode with denatured uranium–thorium fuel. All uranium inventory and annual make-up requirements for these alternative scenarios are those listed in table 0.1 except that the uranium requirements of the once-through fuel cycles have been increased by assuming a tails assay at the enrichment plant of 0.2 per cent U-235 before the year 2000.

the curves associated with the evolutionary strategy, it is assumed that all new reactors coming on-line after the year 2000 are ACRs and, in the case of recycle systems (either ACR or LMFBR), that all reactors are put on closed fuel cycles beginning approximately in the year 2000. These assumptions are, of course, only for computational simplicity—the actual transition would be spread over a period of at least a decade.

The most striking aspect of these curves is that, as a result of a shift to ACRs after the year 2000, it would be possible even with a continued reliance on once-through fuel cycles to keep the cumulative non-by-product uranium requirements over the next 100 years down to the 4.4 million tons of U_3O_8 which the DOE estimates is available in US high-grade uranium deposits. Even with the lower bound uranium supply curve shown in figure 1.4, the price of U_3O_8 in 2080 for this once-through scenario would be only about \$180/lb and the corresponding cost of the delivered electricity from a once-through ACR would still be within the mid-range of the electricity cost uncertainty band for the LMFBR system.

The phasing-in of ACRs and the introduction of denatured uranium recycle after the turn of the century would result in a reduction of cumulative uranium requirements during the next century nearly as great as with the deployment of LMFBR. Although introduction of recycle would not be economically justified at such an early date, the cost of electricity generated by the recycle ACR system would remain within the range of uncertainty calculated for the plutonium breeder for hundreds of years, assuming a nuclear economy of the size considered here.

It would be much easier to establish a closed fuel cycle for ACRs than for LMFBR, because the ACRs have the flexibility to operate economically on a once-through fuel cycle. For this reason, it would be possible, for example, to wait until 50–100 GW(e) of ACR capacity had been deployed and economies of scale in fuel-cycle facilities could be fully exploited before closing the ACR fuel cycle. Similarly the ACR system could better tolerate delays or failures in reprocessing and refabrication facilities than the LMFBR system. The orchestration of the simultaneous deployment of a system of LMFBRs and their associated fuel cycle would be so complex and susceptible to 'common mode' failure that it would almost certainly require very substantial government participation in all aspects of the operation.

Each of the candidate ACR technologies based on currently developed reactor types has special advantages and disadvantages; an improved LWR would probably require the least changes in the nuclear industry, and, by permitting conversion of already deployed reactors to improve their uranium utilization efficiency, it could affect the greatest uranium savings in the shortest time.

The HWR probably provides the clearest path at present to a more uranium-efficient system both on once-through and recycle fuel cycles. It also requires the least enrichment of the fresh fuel in the case of the once-through fuel cycle—which might lead to a more proliferation-resistant enrichment industry. As a reactor, however, the HWR may be the least proliferation-resistant of the ACRs. It can be operated on natural uranium to produce large quantities of plutonium, and its on-line refuelling capability may require the application of more stringent international safeguards.

The HTGR can be about as uranium efficient as the HWR on a once-through fuel cycle, and it has the proliferation-resistant advantage of a relatively low fissile plutonium discharge rate—about 20 kg per year for a denatured uranium–thorium once-through fuel cycle. But the uranium requirements of the HTGR are reduced much less dramatically with denatured uranium recycle, than are those of the HWR.

IX.   Nuclear energy for the world

It is often argued that, while the United States may be able to postpone the commercialization of a breeder reactor, other industrialized nations less well endowed with indigenous resources of uranium and fossil fuel cannot afford to do so.

In figure 1.5 we presented a scenario, however, in which the introduction of uranium-efficient advanced converter reactors operating with denatured uranium recycle in the year 2000 instead of the breeder would result in little change in cumulative US uranium requirements before the middle of the next century. This result is not peculiar to the United States—indeed the LMFBR would have an advantage in the United States in the year 2000 relative to many other nations, since the United States is building up in the spent fuel from LWRs a very large inventory of plutonium for starting up breeders. It appears therefore that the introduction of an LMFBR instead of an ACR would not necessarily reduce the uranium requirements of any nation for many decades.

The advanced converter reactor system which competes most favourably with the LMFBR in terms of uranium conservation, however, involves reprocessing, which, for non-proliferation reasons it would be desirable to postpone—at least until arrangements have been made to internationalize and safeguard sensitive fuel cycle facilities. We therefore now consider: (a) whether the world's uranium supply–demand balance is sufficiently favourable to allow a continued dependence on a once-through fuel cycle for the next decades; and (b) the special concerns in uranium-poor nations about the security of their fuel supply.

For the purposes of a discussion of the global uranium supply–demand balance, it is convenient to divide the nations of the world into three groups: those with 'centrally planned' economies (principally the USSR, Eastern Europe, and China); those in the 'Third World' (Latin America, the Middle East, Africa, East Asia and South Asia); and the members of the Organization for Economic Cooperation and Development (OECD) (North America, Western Europe, Japan, New Zealand and Australia).

We know little about the uranium supply–demand situation in the centrally planned nations, except that they do not currently trade in uranium with the other two economic groups. We assume here that they will not become uranium importers on a scale large enough to worsen significantly the uranium supply–demand balance in the remainder of the world.

Current the Third World countries taken together account for only a few per cent of total world nuclear capacity and uranium demand.[26] They

[26] Total nuclear generating capacity in the Third World (operating or scheduled to operate in 1978) is 3.7 GW(e). Total capacity including that under construction or on order is 20.6 GW(e). The Third World countries contributing to the latter total are Argentina 1 GW(e), Brazil 3 GW(e), India 1.7 GW(e), Iran 4.2 GW(e), Korea 1.8 GW(e), Mexico 1.3 GW(e), Pakistan 0.1 GW(e), the Philippines 0.6 GW(e), South Africa 1.8 GW(e), and Taiwan 4.9 GW(e) [37].

account for over 20 per cent of the identified high-grade uranium resources outside the centrally planned nations, however, and it is likely that their share will increase, since generally they have not been intensively explored for uranium [38]. In addition most of the world's phosphate resource and therefore its potential resource of by-product uranium is in the Third World.[27] It seems likely, therefore, that the Third World will not be a net importer of uranium for many decades.

In considering the uranium supply–demand balance in the OECD region, we will assume that in the next century the average per capita nuclear electricity consumption in the OECD is the same as in the United States. Then, with the same nuclear energy system, the total OECD uranium demand would be approximately three times that of the United States.

On the supply side, the OECD uranium reserves available at forward costs of less than \$50/lb U_3O_8 are approximately 2.3 times greater than those of the United States. This ratio is less than the demand ratio. In the past, the level of uranium exploration effort has been relatively high in the United States, however, and the supply ratio is climbing higher as exploration efforts are intensified in Canada and Australia.[28]

In any case, in view of the large margin of safety we found in the exploration of our strategy to the United States, it would appear that the uranium supply–demand situation would allow the same evolutionary fission energy strategy to work for the OECD and indeed for the world. Specifically, on the basis of uranium supply–demand considerations it appears to be possible to postpone for decades any commitment to fuel cycles involving reprocessing and for even longer any commitment to the plutonium breeder.

The possibility of balancing uranium supply and demand for the world as a whole may not at first sight appear to be sufficiently reassuring for some uranium-poor nations, however. In particular Japan and OECD Europe, if they continue to rely on the once-through fuel cycle, would have to be willing to import about 50 per cent more uranium over the next 50 years than they would with the breeder, while other nations would have to be willing to export this additional uranium. Japan and OECD Europe are today heavily dependent upon petroleum imports and are understandably concerned about the new dependence upon imported uranium which they are creating.

---

[27] See reference [39]. About 83 per cent of the world's phosphate resources are estimated to lie in the Third World—mostly (70 per cent) in Morocco (see table 2, page 820). The non-US demand for phosphate rock in the year 2000 is projected to be more than 200 million tons per year (compared to 80 million tons in the year 1973—see table 10, p. 833). Assuming that this phosphate rock contains uranium in concentrations averaging 100 p.p.m. [35], these phosphate-mining operations could support about 200 GW(e) of HWR capacity operating on a once-through fuel cycle.

[28] As this report goes to press, the discovery of an extremely high-grade deposit under Midwest Lake in Saskatchewan, Canada, was announced. Reserves in this single deposit are estimated to total 50 000–100 000 tons of U_3O_8, or more. This may prove to be one of the largest high-grade uranium deposits in the world. Other high-grade ore discoveries are expected in this area. See reference [40].

The analogy between uranium dependence and petroleum dependence is a poor one, however—simply because uranium is so inexpensive. For the HWR operating on the 1.2 per cent enriched once-through fuel cycle, for example, the cost of U_3O_8 would have to rise to over \$500/lb before the fuel cost associated with the released fission energy would be as great as that for oil at \$13/barrel! Uranium imports represent a much smaller balance-of-payments issue than petroleum, therefore, and importing nations can guard against uranium supply disruptions at relatively low cost by stockpiling several years' requirements. (The carrying charge associated with a 10-year stockpile of enriched uranium for an HWR operating on a once-through fuel cycle would add only 1.5 per cent to the cost of a kilowatt hour of electricity at \$60/lb U_3O_8.) As an indication of how easy it is to stockpile uranium we note that, as a result of delays in its nuclear power-plant construction programme, France will have extra enriched fuel in its stockpile in 1985 equivalent to four years' consumption.[29]

For the uranium-poor countries, therefore, the flexibility of the evolutionary strategy should be emphasized. So long as these countries remain convinced of a secure supply of uranium, the operation of ACRs on a once-through fuel cycle would be a low-cost nuclear alternative to the breeder. If, at some time, these nations become concerned about their uranium supply, the same ACRs could be shifted to a recycle system which is nearly as resource-conserving, more proliferation-resistant, and approximately as economical as an LMFBR system.

X. Summary

The LMFBR has dominated the imagination (and research budgets) of the nuclear establishments of the industrialized countries for more than a decade, to the point where no other system has appeared practical or worthy of serious attention.

Recently, however, this reactor has become controversial, as the proliferation vulnerabilities of its fuel cycle have become increasingly evident and as its economics have begun to look less favourable.

In contrast, the evolutionary strategy which we have described involves both simpler and more flexible advanced converter reactors which, unlike the LMFBR, can be operated economically on isotopically denatured fuel

---

[29] France will have an estimated surplus of 12.4 million SWU worth of enriched uranium in 1985 [41] corresponding to approximately 20 000 short tons of U_3O_8 feed for a plant producing 3.2 per cent enriched uranium with an operating tails assay of 0.2 per cent. France will have a total estimated nuclear capacity of about 35 GW(e) in 1985 [42], of which approximately 4 GW(e) will be either plutonium fuelled or fuelled by natural uranium. At 65 per cent load factor 20 000 tons U_3O_8 would fuel 30 GW(e) for four years.

cycles in either the once-through or the recycle mode. Adoption of the evolutionary strategy would therefore enable the world to postpone for decades decisions regarding fuel reprocessing while the true world nuclear power growth and uranium resource situations are clarified and the possibilities for a strengthened international safeguards régime are determined. If uranium supplies are found to be limited and arrangements can be agreed upon which make fuel cycles involving reprocessing sufficiently proliferation-resistant, then the converter reactors could be shifted to denatured uranium recycle operation.

No strategy for nuclear power development can completely sever the connection between fission energy technology and nuclear weapons. For this reason an urgent objective of the larger energy research and development programme should be to determine if solar or other long-term energy sources are practical alternatives to fission. Here too, the evolutionary strategy with its dependence on less complex and interdependent technologies has an advantage in that it would make less difficult a disengagement from fission power if such a course should ultimately be decided upon.

An interesting aspect of the evolutionary strategy is that it would bring to the fore uranium-efficient technologies which have been developed in several nations but which have been overshadowed by the various breeder reactor development programmes. These technologies include the heavy water reactor technology developed by Canada, the high temperature gas-cooled reactor technology developed in FR Germany and the United States, and light water reactor modifications being studied in the United States.

Thus the evolutionary strategy would build on the familiar and encourage nuclear experts to focus during the next few years on making more efficient, safe, and secure the nuclear systems on which nations now rely instead of dissipating their talents devising safeguards for new systems which are inherently more vulnerable to the diversion of weapon-usable materials.

References

1. Nye, J., 'Nonproliferation: a long-term strategy', *Foreign Affairs*, Vol. 56, No. 3, April 1978, pp. 601–23.
2. US Congress, Office of Technology Assessment, *Nuclear Proliferation and Safeguards*, Appendix VI (US Government Printing Office, Washington, D.C., 1977).
3. Feiveson, H. A., *Latent Proliferation: The International Security Implication of Civilian Nuclear Power* (Ph.D. Thesis, Woodrow Wilson School of Public and International Affairs, Princeton University, 1972).
4. Wohlstetter, A. *et al.*, *Moving Toward Life in a Nuclear Armed Crowd?* Report to the US Arms Control and Disarmament Agency (Pan Heuristics Inc., Los Angeles, 1976).
5. Wohlstetter, A., 'Spreading the bomb without quite breaking the rules', *Foreign Policy*, Winter 1976–77, p. 88.

6. Marshall, W., 'Nuclear power and non-proliferation', Speech at the Uranium Institute, London, 12 July 1978.

7. Levinson, M. and Zebroski, E., 'A fast breeder system concept, a diversion resistant fuel cycle', Paper presented at the Fifth Energy Technology Conference, Washington, D.C., 27 February 1978.

8. Krass, A. S., 'Laser enrichment of uranium: the proliferation connection', *Science*, No. 196, 1977, p. 721.

9. Technical Support Organization, Brookhaven National Laboratory, *Preliminary Safeguards Analysis of Denatured Thorium Fuel Cycles* (10 November 1976, unpublished).

10. Feiveson, H. A. and Taylor, T. B., 'Security implications of alternative nuclear futures', *Bulletin of the Atomic Scientists,* December 1976, p. 14.

11. Taylor, T. B., 'Nuclear safeguards', *Annual Review of Nuclear Science,* 1975, p. 407.

12. *Regional Nuclear Fuel Cycle Centers* (IAEA, Vienna, 1977).

13. Chayes, A. and Bennett Lewis, W., eds., *International Arrangements for Nuclear fuel Reprocessing* (Ballinger, Cambridge, Mass., 1977).

14. Till, C. E., and Chang, Y. I., *CANDU Physics and Fuel Cycle Analysis*, Report No. RSS-TM-2, Argonne National Laboratory, 1977.
   (a) —, figure 9.
   (b) —, table XII.

15. Waligura, A. *et al.*, *Safeguarding On-Power Fueled Reactors—Instrumentation and Techniques*, Chalk River Nuclear Laboratories, May 1977.

16. 'Light water breeder reactor technology', *Final Environmental Statement on the Light Water Breeder Program*, ERDA Report No. 1541 (ERDA, Washington, D.C., 1976), secton II, E.

17. Edlund, M. C., 'Development in spectral shift reactors', *Proceedings of the Third UN Conference on the Peaceful Uses of Nuclear Energy,* Vol. 6 (United Nations, New York, 1964), p. 314.

18. Turner, R. F., *Optimization of MEU/Th Cycles for HTGR*, Memorandum, General Atomic Co., 26 July 1978.

19. Turner, R. F., private communication, 25 August 1978.

20. Perry, A. M. and Weinberg, A. M., 'Thermal breeder reactors', *Annual Review of Nuclear Science,* 1972, p. 317.

21. Slater, J. B., *An Overview of the Potential of the CANDU Reactor as a Thermal Breeder*, Chalk River Nuclear Laboratories, Canada, 1977.

22. Rütten, H. J., Lee, C. E. and Teuchert, E., *The Pebble-Bed HTR as a Net Breeding System*, Report No. 1521, Jülich Nuclear Research Center, FR Germany, 1978.

23. *Final Environmental Statement on the Liquid Metal Fast Breeder Reactor Program,* ERDA Report No. 1535 (ERDA, Washington, D.C., 1975).
   (a) —, figure III F-19 and p. III F-77;
   (b) —, table III F-7.

24. *The Nuclear Industry*, US Atomic Energy Commission Report No. WASH 1174-74 (USAEC, Washington, 1974), p. 20.

25. Haffner, D. R. *et al.*, *An Evaluated Uniform Data Base for Use in Nuclear Energy Systems Studies,* Hanford Engineering Development Laboratory, 18 October 1977.

26. Rudasil, C. L., *Coal and Nuclear Generating Costs,* Report No. EPRI PS-445-SR, Electric Power Research Institute, 1977.

27. Allen, E. L. *et al.*, *U.S. Energy and Economic Growth, 1975–2000*, Report No. ORAU/IEA-76-7, Institute for Energy Analysis, Oak Ridge, Tenn., 1976.
28. Energy Research Group, University of Cambridge, 'World energy demand to 2020', in *World Energy Resources, 1985–2020* (IPC Science and Technology Press, New York, 1977), p. 213.
29. 'US energy demand: some low energy future', *Science*, No. 200, 1978, p. 142.
30. Ross, M. H. and Williams, R. H., *Energy and Economic Growth*, Study prepared for the Subcommittee on Energy of the US Congress Joint Economic Committee (US Government Printing Office, Washington, D.C., 1977).
31. *Survey of Current Business*, National Income and Product Accounts, US Department of Commerce.
32. Combs, G. F. Jr., Supply Evaluation Branch, Division of Uranium Resources and Enrichment, US Department of Energy, Private communication, 22 June 1978.
33. US Department of Energy, *Statistical Data of the Uranium Industry*, Report No. GJO-100(78), Grand Junction, Colorado, 1978.
34. US Bureau of Mines, *Uranium from Chattanooga Shale*, Informational Circular No. 8700, 1976.
35. Cathcart, J. B., *Uranium in Phosphate Rock,* Report No. 75–321, U.S.G.S., Open File, 1975.
36. Klemenic, J. and Blanchfield, D., 'Production Capability and Supply', Contribution to the Uranium Industry Seminar, US Department of Energy, Grand Junction, October 1977.
37. 'Power reactors 1977', *Nuclear Engineering International*, Supplement, April 1977.
38. OECD Nuclear Energy Agency and the International Atomic Energy Agency, *Uranium: Resources, Production, and Demand* (OECD, Paris, 1977).
39. Stowasser, W. F., 'Phosphate', *Mineral Facts and Problems*, US Department of Interior, Bureau of Mines Bulletin No. 667, 1975, pp. 819–34.
40. Bennett, A. and Katzenstein, D., 'Canadian uranium find of venture led by Exxon unit buoys analysts, traders', *Wall Street Journal*, 12 September 1978, p. 4.
41. *Nucleonics Week*, 1 June 1978, p. 1.
42. *Nucleonics Week*, 20 July 1978, p. 6.
43. Till, C. E. and Chang, Y. I., *Once-Through Fuel Cycles*, Report No. RSS-TM-13, Argonne National Laboratory, 1978, table I.
44. Till, C. E., *Fuel Cycle Options and Fueling Modes*, Argonne National Laboratory, 9 January 1978, unpublished.
45. Teuchert, E., *Once-Through Cycles in the Pebble Bed HTR,* Jülich Nuclear Research Center, FR Germany, 1977.
46. Chang, Y. I. *et al., Alternative Fuel Cycle Options,* Report No. ANL 77–70, Argonne National Laboratory, 1977.
    (a) —, tables V and VI.
    (b) —, tables IV, V, and VIa.
47. Energy Research and Development Administration, 'Examination of 1976 nuclear generating costs', *Update: Nuclear Power Program Information and Data*, Report prepared by Division of Nuclear Research and Applications, June 1977, pp. 28–35.
48. Baughman, M. L. and Bottaro, D. J., *Electric Power Transmission and Distribution Systems: Costs and Their Allocation*, Report of the Center for Energy Studies, University of Texas at Austin, July 1975.

49. Till, C. E. *et al.*, 'A Survey of Considerations Involved in Introducing CANDU Reactors into the US', Report No. RSS—TM-1, Argonne National Laboratory, 1977.
50. Banal, M. *et al.*, 'Work starts on Super Phénix at the Creys-Malville Site', *Nuclear Engineering International*, May 1977.
51. Kasten, P. R. *et al.*, *Assessment of the Thorium Fuel Cycle in Power Reactors*, Report No. ORNL/TM-5565, Oak Ridge National Laboratory, 1977, table 3.

Appendix 1A

Assumptions used in the economic calculations[a]

	Reactors		
	LWR	HWR	LMFBR
Physical parameters			
Average capacity factor (per cent)[b]	65	75[c]	65
Thermal−electric conversion efficiency (per cent)	33[d]	30.5[d]	40[e]
Mass of heavy metal (10^3 kg/ GW(e)) for UO_2-based fuels)			
Core	74−80[f]	113−126[f]	26[e]
Axial blanket	–	–	20
Radial blanket	–	–	20
Fissile plutonium (LMFBR)	–	–	3.2[e]
Breeding ration (LMFBR)			1.28[e]
Heavy water[g] (10^3 kg/GW(e))			
Inventory	–	800[c]	–
Annual make-up	–	16	–
Cost (1976 dollars)			
Capital costs[h] ($/KW(e))	800	960[c]	1 000−1 400
Fuel cycle service costs ($ per kg HM-oxide fuel)[i]			
Reprocessing (incl. spent fuel transport cost)	180	175	340
Spent fuel disposal, incl. transport (for once-through cycles)	115	110	–

			Core	Axial blanket	Radial blanket
Fabrication					
Low-enriched uranium	100	60	–	30	170
Plutonium−uranium mixture	380	–	1 110	–	–
Denatured U-233−thorium mixture	620	390	–	–	–
Denatured U-235−thorium mixture	160	100	–	–	–

45

Other costs

Heavy water ($/kg)[j]	200
Enrichment ($/SWU)[k]	75
Value of plutonium for LMFBR/LWR system ($ gram)[l]	$5.6 + 0.34C_u$
Conversion of U_3O_8 to UF_6 for enrichment ($/kg U)	3.5
Transportation and disposal of reprocessing wastes (10^{-3} kWh(e))[m]	0.2
Thorium ($/lb Th O_2)	40

Time of purchase of fuel cycle services relative to time of charge (discharge) into (from) reactor (years)

Uranium from mill	charge −2.0
Conversion of U_3O_8 to UF_6	charge −1.5
Enrichment	charge −1.0
Fabrication	charge −0.5
Reprocessing	discharge +1.0
Spent fuel disposal	discharge +1.0
Reprocessed waste disposal	discharge +10.0
Purchase of plutonium inventory (LMFBR)	charge −1.0
Sale of excess plutonium (LMFBR)	discharge +1.0

Notes:

[a] Unless otherwise indicated, all cost data given here were obtained from reference [25]. The information of this data base was obtained from the results of detailed studies conducted by ERDA's (now DoE's) national laboratories and contractors, and was evaluated before its inclusion in the data base in a cooperative effort by the Oak Ridge National Laboratory, Combustion Engineering, United Engineers and Constructors, Argonne National Laboratory, Hanford Engineering Development Laboratory, and the Nuclear Research and Applications and the Uranium Resource and Evaluation divisions of ERDA.

[b] Historically LWR capacity factors have averaged about 60 per cent and CANDU capacity factors (4 unit Pickering station) have averaged about 75 per cent [49]. One advantage of the CANDU over both the LWR and LMFBR is that it is refuelled without being shut down.

[c] Recently Combustion Engineering (private communication, Norman Shapiro) has carried out an assessment of the HWR slightly modified from the CANDU design. This study concluded that an HWR plant for the US would cost 8.7 per cent more than the LWR (we assume 20 per cent more) and that the heavy water inventory would be 735 kg/GW(e) (we assume 800 kg/GW(e)). The Combustion Engineering capital cost estimate was predicated on what they felt was a reasonable application of licensing criteria, where criteria and practices seem to apply equally well the HWR and LWR; however consideration was given to the unique attributes of the HWR when applying these criteria. Combustion Engineering also estimates that on line refuelling provides a 2−3 percentage points' advantage in the capacity factor for the HWR (we assume 10 points). The net effect of the different Combustion Engineering assumptions is a busbar cost for the HWR 0.15×10^{-3} dollars/kWh higher than that given in table 1.2.

[d] See reference [46].

[e] The LMFBR design assumed in this analysis is similar to that described in reference [46b], for an oxide-fuelled system, corrected for a higher thermal efficiency. (The efficiency assumed in the Argonne analysis is 35.5 per cent. We assume that 92 per cent of the thermal power is released in the core and that the mixture of fissile plutonium isotopes in the core corresponds to that in plutonium discharged from a uranium-fuelled LWR. The breeding ratio we assume (1.28) is somewhat higher than that 1.18 breeding ratio expected for the Super Phénix. See reference [50].

[f] See reference [44]. The lower value of the heavy metal inventory is appropriate for uranium−thorium fuel, while the higher value is appropriate for uranium fuel.

[g] See reference [51].

[h] The capital cost of an LWR is based on reference [26], in which it is estimated that a new plant would cost $649−934/kW. Capital costs for other reactor types are estimated as multiples of the LWR cost, assuming the same relative costs as those given in reference [25].

[i] The fuel cycle costs used in this analysis are based on reference [25]. For recycle fuel cycles that report projects a high initial cost and a lower value after a 10-year period as a result principally of economies of scale. In this analysis we have assumed a weighted average of these

46

two values, with the initial cost having a weight 0.2 and the cost after 10 years having a weight of 0.8.

^{j} The 1975 price was $110–120/kg (see reference [49]).

^{k} This is the price for 'Fixed Commitment contracts' announced by ERDA on 30 September 1977. In reference [25] it is projected that with advanced technology the cost will fall by the year 2000 to $55/SWU.

^{l} The plutonium price is the break-even price for fissile plutonium extracted from the fuel of a uranium-fuelled PWR used as make-up fuel for another PWR when the price of uranium is C_u dollars per lb U_3O_8, the cost of a separative work unit is $75, and the cost of fabricating mixed Pu–U oxide fuel is $280/kg heavy metal more than that for low-enriched uranium fuel. The amounts of make-up uranium and plutonium required for the two PWRs is based on reference [44], for an enrichment tails assay of 0.1 per cent.

^{m} In reference [25] these costs are given as $55/kg heavy metal for the LWR, or $1.67/MWd of fuel burn-up. We assume the latter rate can be applied to all fuel cycles.

Chapter 2. Enrichment

Paper 2. Practical suggestions for the improvement of proliferation resistance within the enriched uranium fuel cycle

J. H. COATES and B. BARRÉ

Square-bracketed numbers, thus [1], refer to the list of references on page 59

I. Introduction

Concern about nuclear proliferation began to be voiced soon after the end of World War II, and has since then grown in parallel with the development of nuclear energy. Until recently, attempts to solve the problem have mainly consisted of the introduction of increasingly stringent regulations in the field of nuclear activities. Among the most outstanding of these measures are probably the NPT dispositions and the London Club's tacit rules on the export of nuclear technology.

Without decrying the use of such institutional measures, another possibility has recently developed, which consists in concentrating on the development and application of more proliferation-resistant technologies. Of course, this task may not prove easy, especially since these technologies must be as efficient and competitive as their more proliferation-prone counterparts. However, even if this new approach does not prove entirely satisfactory by itself, we maintain that in combination with the appropriate regulations, it may bring about an adequate solution without hampering the development of peaceful nuclear applications.

There are two major examples which illustrate the efforts made by the Commissariat à l'Energie Atomique (CEA) in order to apply this new line of thinking. One is the chemical enrichment process, the existence of which was disclosed by M. Giraud at the 1977 Salzburg IAEA International Conference, and the other is the medium enrichment fuel for research reactors which avoids the use of weapon-grade material.

In both cases, good results are expected in spite of the additional constraint which lies in avoiding proliferation risks.

II. Chemical enrichment

Historical background of chemical exchange compared to other processes

It may be instructive to recall the reasons which guided past choices regarding the development of enrichment technologies. The first of these choices was made by the United States at the end of World War II in order to select, from electromagnetic, thermal diffusion, gaseous diffusion, centrifuge and chemical exchange, the process which would allow the fastest weapon-grade production. Having reluctantly abandoned centrifuge because of the lack of resistant bowl materials, the final choice was made in favour of gaseous diffusion. As far as chemical enrichment was concerned, this was discarded presumably because of its low enrichment factor, high uranium hold-up and long start-up period which were unacceptable in war-time conditions.

These early choices proved to be most persistent. They gave at least a 20-year lead to gaseous diffusion, the consequences of which are still perceptible. In France, where a promising chemical process had been discovered in 1968, the projects regarding new commercial units which were contemplated in the early 1970s were all based on gaseous diffusion which, it is fair to add, was then far more mature than the new chemical process. If the option of calling upon chemical enrichment was kept open as late as 1975, it should be pointed out that this was due not so much to non-proliferation concerns, which were to some extent inherent in a nuclear weapon state, as to fear of not being able to gather the large partnerships which were needed in order to launch a large competitive diffusion plant. Once this uncertainty vanished, due to the 1973 oil crisis, and also to the smooth running of the project, there was no point in taking a possible risk of leading the EURODIF partners astray by announcing the existence of a new process.

This is why the Salzburg announcement was delayed until 1977, when it was felt that more attention should be given to the proliferation risks of uranium enrichment.

During the last year of the Ford presidency, signs came from the United States that a major change of opinion regarding proliferation risks was taking place in that country. Abandoning the previous 'Atoms for Peace' policy, President Carter was to confirm this new tendency by taking a number of spectacular steps against breeders and reprocessing plants. However, it was felt in France that even stronger concern should be given to uranium enrichment, which is by far the shortest and most efficient route to weapon capacity. Hence the Salzburg announcement.

Proliferation resistance of the chemical process

Slow kinetics and criticality limitations are the two main intrinsic features of the process, which do not allow the attainment of high U-235 assays and therefore make the chemical enrichment proliferation-proof.

Depending on the assay to be attained, but assuming this to be of high-grade military value, the time needed to bring the cascade equilibrium would be in the order of one generation, that is, more than two decades. Of course this has been calculated on the basis of industrial components which have been tested hitherto and one can imagine that 'military' versions of these components could be developed. In this case, however, it is likely that more serious difficulties would have to be overcome in other areas where upgrading meets limitations. This is the case with problems of criticality which can be solved by adequate geometry but which bring constraints on the process itself as well as on its ancillaries. It would also be the case with process regulation which becomes more elaborate as the length of the cascade increases.

Moreover, the difficulties of attempting to establish a military plant would be even greater in trying to misuse a commercial plant.

But the strongest argument in favour of the non-proliferant character of the chemical process is perhaps not of a technical but of a psychological nature. Even after having studied the process for 10 years, French technicians cannot yet declare themselves confident that high U-235 upgrading can be achieved at all. In fact, such certitude could only be obtained if the attempt were endeavoured and if it proved successful. But then who would decide to make such an attempt? Certainly not a nuclear weapon state, for the cost would be disastrous compared to gaseous diffusion, for instance. Thus, it could only be a non-nuclear weapon state endeavouring to attain military capacity. But is it reasonable to suppose that such a state would be ready to undertake a 20-year venture without having any certainty that the goal aimed at would be attained at all? In other words, would one take the risk of being obliged to start again from scratch after having lost 20 years? Obviously, there is not the slightest chance of such a decision being taken given the fact that, unfortunately, several other processes offer no such uncertainty regarding feasibility.

For all these reasons, it can be stated that the chemical exchange process is perfectly secure, even though, in theory, it cannot be demonstrated that its proliferation resistance is 100 per cent.

At most, it could only be used as a low-grade pre-enrichment unit which only brings a small advantage along the path of proliferation, since it does not prevent the use of other processes that are more adapted to highly enriched uranium production. Given the existence of these other processes, one may assume that they may be used for the total enrichment of natural material. One could argue, of course, that pre-enrichment might help to reduce the costs of the operation, or facilitate clandestine activity, but these do not seem to be very convincing arguments.

Competitiveness of the chemical exchange process

Despite its high resistance to proliferation, the French chemical exchange process seems to stand up to competition with other processes in a remarkable way. This is due to a number of factors. First, the isotopic effect obtained is roughly twice as high as those previously identified. This means that, other things being equal, the specific separative work delivered by the process is four times cheaper than the best of the other known processes in this category. The second reason is that the process is relatively simple—fairly similar to those found in petrochemistry, for example. Finally, considerable improvements have been made by the CEA to the basic components, such as the contactors. In spite of certain drawbacks, particularly high uranium retention, the overall economy of the process appears to be quite close to that of gaseous diffusion in the French version. Moreover, it is well adapted to small or medium-scale applications (in the range of 0.3 to three million SWU/year) and its energy consumption is quite low (roughly four times less than gaseous diffusion).

Possible utilization of the chemical enrichment process

The industrial capability of many countries is increasing at such a rate that the technological bolt which prevented the spreading of uranium enrichment is now giving way. In addition, the uncertainties regarding the availability of energy provide another incentive to countries to develop their own means of access to reliable energy sources. In other words, there is an increasing number of countries potentially having both the will and the means of realizing their own enrichment plants. Offering participation in multinational centres may avoid to some extent the proliferation of national plants, but it may not satisfy the requirements of all the states interested in building an enrichment plant. It is in this context that the French process may play a role. Chemical enrichment would allow the possibility of national plants without increasing the risks of proliferation. This is the proposal which, after having been approved by the French government, was made by M. Giraud at Salzburg in 1977.

It is obvious that this initiative is not related to French enrichment needs, which will largely be satisfied by the EURODIF and COREDIF plants. On the other hand, it is clear that the CEA is not trying to make special commercial profits out of the new process, since it is ready to carry out the development of the technique within a multinational framework, and thus share the rights of utilization with foreign partners.

Current prospects

The CEA has already investigated and chosen the process conditions adapted to industrial utilization. It has also developed, constructed and tested full-size prototypes of the main components. The time has now come to demonstrate the feasibility and attractiveness of the process by constructing, starting up and operating a small production unit. This operation should take roughly five years, which means that by the mid-1980s full confirmation of the process could be obtained, allowing commercial projects to be undertaken from then on. It is contemplated that the demonstration phase would be open to a number of foreign partners who might share the French views regarding proliferation hazards, and bring substantial support in trying to reach a breakthrough beneficial to the community. Discussions are under way in order to try to bring this new approach to proliferation resistance into application.

III. Enriched uranium fuels

Enrichment levels in reactor fuels

Natural uranium, with its 0.711 per cent U-235 isotopic content, barely offers the possibility of sustaining chain reactions with thermal neutrons. It must be used with efficient and non-capturing moderators, heavy water and graphite. Graphite, which is slightly less efficient than heavy water, requires the use of metallic uranium as a fuel. In both cases, however, excess reactivity is very low, which leads to an almost continuous renewal of the fuel, through on-load refuelling machines, which are very intricate and expensive devices requiring heavy shielding.

On-load refuelling devices allow removal of fuel elements with very low burn-up without having to shut down the reactor, and the low initial U-235 content of the fuel in combination with short irradiation give birth to almost pure plutonium-239 mixed with very limited amounts of radioactive fission products. Therefore natural uranium fuelled reactors constitute a high proliferation hazard,[1] as is clearly shown by the fact that they have been, and probably still are, the main source of military plutonium in the nuclear weapon states.

The high capture cross-section of hydrogen makes it impossible to use natural uranium fuel in light water moderated reactors. The level of enrichment selected—around 3 per cent U-235—gives enough reactivity margin to permit long operating periods between refuelling operations, and eliminates the need for on-load fuelling machines. The light water reactors (LWRs) must be shut down, and the pressure vessel opened, before any fuel element can be removed. Hence, removal of under-irradiated fuel elements for military-grade plutonium recovery would be lengthy, expensive in terms of reactor immobilization, and rather obtrusive. This is why LWRs, which constitute the majority of nuclear power reactors in operation or under construction, present a low and acceptable level of proliferation risk when under international controls and safeguards. Were they to be fitted with rapid refuelling devices, this opinion would probably need to be revised.

---

[1] These comprise the French UNGGs (natural uranium gas/graphite reactors), the British Magnox, neither of them in commercial use any longer, and the Canadian Heavy Water Reactors (CANDUs).

It is generally admitted that any uranium whose U-235 isotopic content is below 20 per cent is unfit, without further enrichment, for explosive purposes. Of course, 20 per cent enriched feed uranium would provide quite a short cut to a highly enriched product in an enrichment plant, but in the absence of such a plant, or if only the chemical treatment method is used, 20 per cent enriched uranium is proliferation-safe. This applies even more to 3 per cent enriched uranium.

There are two kinds of reactor currently using, or planning to use, fuel elements containing uranium enriched above the 'safe' level: high-temperature reactors (HTRs), and research and irradiation reactors. Suggestions for dealing with the specific problems associated with each of these two types of reactor will now be given.

High-temperature reactors

High-temperature reactors, through a combination of very finely divided fuel, all-refractory core (graphite and ceramics) and helium cooling, offer the possibility of obtaining temperatures in the range of 700−1000°C at the core outlet. They may be the key to opening nuclear energy to fields other than electricity production and district heating. This could include coal gasification, hydrogen production, steel making, and various other chemical processes which call for high temperatures.

This type of reactor has been under development, both in Europe and in the United States, since the early 1960s. Small-scale prototypes have been built by the Organization for Economic Cooperation and Development (OECD) countries (Dragon reactor, Winfrith), the United States (Peach Bottom 1) and FR Germany (AVR, Jülich). Premature attempts at commercialization by General Atomic coupled with the none-too-successful start-up of the 300 MW(e) Fort Saint-Vrain plant, and delays in the 300 MW(e) German thorium high-temperature reactor (THTR) have dashed previous hopes of an early deployment of HTRs, but this kind of reactor still offers hopeful prospects for the future.

HTRs are very flexible, so far as the fuel cycle is concerned, due to the nature of the fuel that they use. Each fuel element consists of a microsphere of oxide of carbide fuel, coated with several layers of pyrolytic carbon and/or silicon carbide, the external dimension being of the order of 1 mm (see figure 2.1). These coated particles are embedded in either graphite prismatic blocks with coolant channels (US design) or graphite spherical pebbles (West German design), and assembled or heaped together to constitute the core stack. The cooling geometry, defined by the block shape or the pebble size, is then independent of the specific fuel loading, which means that the moderating ratio, particle diameter and fissile/fertile materials choice can be made almost at will without any modification in design.

Fuel economy has been the basis for the current choice of the thorium cycle in HTRs, since in-pile U-233 formation permits a good neutron

Figure 2.1. Cross-section of General Atomic coated particles

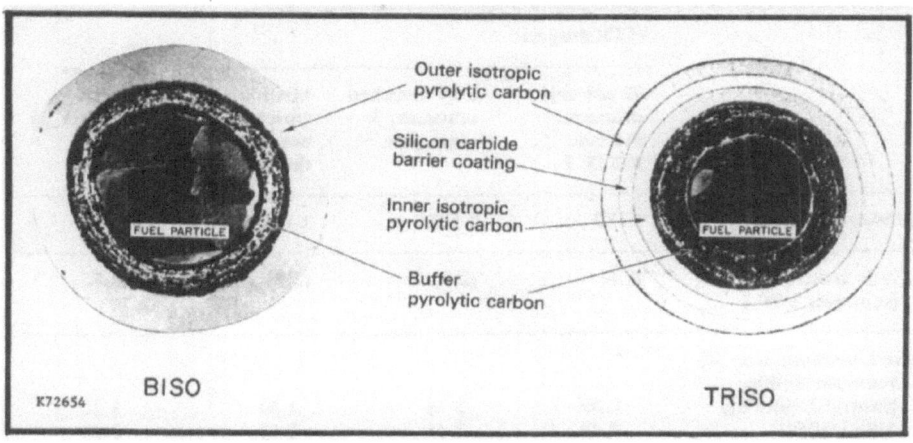

^{a} BISO is the type of coating consisting of a double layer applied to fertile thorium oxide particles.
^{b} TRISO is the type of coating consisting of several layers applied to fissile uranium carbon particles.

Source: From an internal General Atomic report, No. A 14067, September 1976.

balance in the HTR spectrum, and they have a much higher conversion ratio than LWRs. But natural thorium does not possess any fissile isotope: thorium fuel requires a fissile make-up, which is usually 93 per cent enriched uranium. Each fuel element thus comprises a mixture of uranium carbide and thorium oxide particles, embedded in the graphite matrix.

Once it has been completely fabricated, and even without any irradiation, an HTR fuel element does not constitute a serious diversion hazard; a typical prismatic block would contain only 400 grams of U-235, and recovering this uranium from the block would prove no minor task (as anybody having designed an HTR head-end for a reprocessing plant would tell you). But choosing the thorium cycle for HTRs *inevitably implies having an enrichment plant designed for highly enriched products*—a very sensitive item, indeed—and manipulating vast amounts of 93 per cent enriched uranium in the fuel-fabrication plant. These imply very serious proliferation dangers.

Fortunately, the inherent flexibility of HTR fuel permits the use of 6−12 per cent enriched uranium, instead of a U-235/thorium mixture, with only a slight penalty in terms of conversion ratio and fuel economy. This cycle used to be the basis of the development studies and experiments carried out between 1966 and 1972 by the OECD Dragon team.

Moreover, plutonium production in the low-enriched uranium fuelled HTR is noticeably less than in a LWR, and the irradiated graphite blocks themselves, with proper canning, would not seem to be an unrealistic form

Table 2.1.   Enrichment levels in HTR fuel cycles

	HTR projects			Reactors PWR[a] (BUGEY-2)
	93 per cent uranium/ thorium RHTF 1	Low-enriched uranium: 3-year-cycle	Medium-enriched uranium/ thorium	
Power (MW(e))	1 170	1 170	1 170	928
Cycle cost, open (centimes/kWh)	[2.8]	[2.9]	[2.8]	2.8
Requirements over 20 years per GW(e)				
Natural U (mn kg,	2.30	2.74	2.50	3.22
Million SWU	2.47	2.41	2.43	2.09
Plutonium production				
kg fissile Pu/reload	1.6	97	38	164
Fissile Pu/total percentage	20	58	55	69
kg U-233/Reload	170	..	86	..

[a] Pressurized water reactor

Source: From an internal CEA report, October 1977.

of disposal for the irradiated fuel (this being a tentative view, and subject, of course, to further study).

Intermediate solutions (mixing 20 per cent enriched uranium and thorium), as shown in table 2.1, do not seem to offer commanding advantages over the pure low-enriched uranium cycle.

Our practical suggestion is then very clear and simple: HTR development should be carried out only along the low-enriched uranium cycle, which is compatible with both prismatic blocks and pebbles.

Research and irradiation (pool type) reactors

According to the IAEA list of power and research reactors [1], there are something over 300 research reactors spread throughout the world. Of these reactors, around 125 use highly enriched (more than 80 per cent) uranium fuel, and many are situated in non-nuclear weapon states.[2] Highly enriched uranium is a very sensitive material and it should be noted that most of these reactors are situated at universities, research centres, and similar

---

[2] For example, Argentina, Australia, Austria, Belgium, Colombia, Denmark, FR Germany, Israel, Italy, Japan, Netherlands, Pakistan, South Africa, Spain, Sweden, Thailand and Turkey.

institutions, all of which are subject to much less control than power plants usually are. These reactors fall into one of three categories:

1. A very few sophisticated reactors, aiming at fundamental research (such as, for instance, the Franco–West German–British High Flux reactor at Grenoble) have specifications which can only be obtained through the use of 93 per cent enriched uranium, at least for the time being.

2. Most of the low- or zero-power reactors for teaching or training purposes have requirements which could be accommodated by using 20 per cent enriched uranium.

3. Reactors for technological irradiation and/or isotope production have characteristics of flux power density and a useful volume which, in most cases, have resulted in the choice of almost the same design: an open pool, light water moderated and cooled reactor, using materials testing reactor (MTR)-type flat plate metallic fuel elements, where 93 per cent enriched uranium is co-laminated between two aluminium sheets (the U/Al technique).

Converting type (3) reactors to low-enriched fuel cycle is not an easy task, as they require (a) power densities of the order of, or greater than, $300 \, MW/m^3$; (b) small cores, to limit the total reactor power; (c) a high number of thin plates (less than 1.5 mm) per fuel element; and (d) high reactivity margin to accommodate for neutron leakage out of the core or to irradiation specimens: this implies a high fissile density per unit core volume.

The co-laminated U/Al technique has a practical limitation in that the uranium percentage weight in the fuel meat cannot reasonably exceed around 45 per cent, as shown in figure 2.2. This, in turn, means that one cannot go below 40 per cent enrichment without severely impairing reactor performance.

Since early 1976, a French team has been working on adapting to research reactors the Caramel[3] fuel type which had previously been designed for small and medium-size power reactors (Chaudière Avancé de Série (CAS).

To overcome the enrichment limit of MTR technology, Caramel fuels use uranium oxide, whose density is $10.2 \, g/cm^3$ instead of a maximum of 2.8 for the U/Al.

As can be seen in figure 2.3, each plate is an assembly of zircalloy leak-tight casings wrapping 'toffees' of sintered uranium oxide, with two 0.4-mm zircalloy cover plates.

Original power reactor-type Caramels were parallelepipeds of square 2×2 cm faces and 4 mm thickness. To increase the power density, in order to meet the research reactor requirements, this thickness has been decreased to about 1.5 mm.

To demonstrate the validity of this design, CEA has decided to adapt its

---

[3] Caramel is the French word for toffee.

Figure 2.2. Enrichment versus uranium percentage in materials testing reactor fuels for a typical fissile volume loading

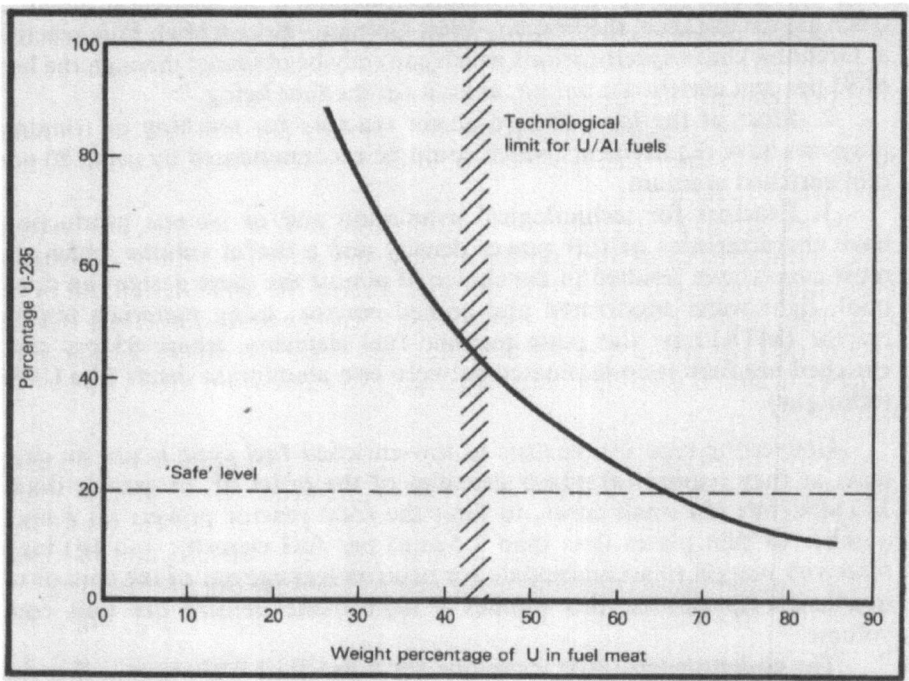

Source: Document 1 submitted by Working Group 8, Australia, to the INFCE.

Figure 2.3. Caramel uranium oxide plates

Table 2.2. Comparison of U/Al and Caramel fuels inside the Osiris reactor

		Co-laminated U/Al	Caramel
Power (MW(th))		70	70
Power density (MW/m^3)		390	345
Coolant flow (m^3/h)		4 250	5 700
No. fuel assemblies		39	44
No. plates/assembly		24	17
Cycle length (days)		21	28
		(1/4 core)	(1/4 core)
Total uranium (kg)		15.5	370
Enrichment (per cent)		93	7
Average neutron fluxes	Thermal	2.5×10^{14} n/cm^2/s	$\sim 2 \times 10^{14}$
	Fast (> 0.18 MeV)	4×10^{14} n/cm^2/s	$\sim 3.5 \times 10^{14}$

most powerful irradiation reactor, Osiris, to a low-enriched fuel cycle, using Caramel fuel elements. To date, all the fuel has been fabricated, and operation with the new core is planned for March 1979.

It has been necessary to refit the primary pumps, to increase the coolant mass flow by 30 per cent, and the start-up configuration has been fixed as shown in table 2.2. As a result, the total power of the reactor remains unchanged, but the neutron fluxes are slightly lowered, this probably being subject to later improvement.

Hitherto the fabrication has proceeded only on a pilot scale,[4] but one promising result has been that the fabrication cost of such pilot scale production has been only 30 per cent higher than the industrial-scale U/Al fabrication cost.

Building new reactors adapted to this fuel will be much easier than the retrofit exercise we have just described, and reactors which are less demanding in terms of power density than Osiris will accommodate even thicker 'Caramels'. We should, therefore, by the end of 1979, have a full-scale demonstration of the reliability and irradiation performance of the Caramel fuel for research reactors, and we should then be in a position to make actual firm proposals to reduce immediately to much less than 20 per cent the enrichment necessary for the research reactor fuels, with very few exceptions.

Reference

1.   *Power and Research Reactors in Member States* (IAEA, Vienna, annual).

---

[4] 2 500 plates had been fabricated by late 1978.

Paper 3. Jet nozzle and vortex tube enrichment technologies

P. BOSKMA

Square-bracketed numbers, thus [1], *refer to the list of references on pages 70–1.*

I. Introduction

This paper deals with two uranium enrichment technologies based on aerodynamic principles, the jet nozzle process and the so-called advanced vortex tube process. Pilot plants are in operation in FR Germany and South Africa, respectively, and the technology is completely in the hands of non-nuclear weapon states. Together with centrifuge technology, already at the threshold of commercial application, these technologies constitute further steps in the process of proliferation of enrichment facilities and technologies outside the territories of the nuclear weapon states.

The paper discusses the principles and characteristics of the technologies that are dealt with in the open literature, and evaluates their position in the field of enrichment technologies and their impact on the proliferation of nuclear weapon capabilities.

II. The technologies

Enriching uranium to U-235 involves separating isotopes of the same chemical element. A large number of processes are possible which are essentially based on the difference in mass of the isotopes, the difference in spectra and related phenomena such as diffusion velocities, centrifugal forces in rotating streams and curvature of ion beams in magnetic fields. The prospects for their use as viable technologies largely depend on the characteristics of the process such as their separation factor and separative work capacity per separating element, their power consumption and their process characteristics for application as a mass-scale technology. The facts that details on enrichment technologies have hitherto been withdrawn and that costs for developing the technology are high prohibit easy access to a uranium-based nuclear weapon capability and the use of highly enriched uranium as a trigger for the hydrogen bomb.

The five nuclear weapon states all developed the gaseous diffusion technology for their military programmes. Even now the major enrichment facilities are of this type. At the end of the 1960s, however, a tripartite undertaking by FR Germany, the Netherlands and the United Kingdom started construction of pilot plants using centrifuge separation. A decision on commercial facilities with a total capacity of 2 000 tons SWU/year in Capenhurst (UK) and Almelo (Netherlands) was taken at the end of 1977. Centrifuge technology is considered to be an attractive alternative to gaseous diffusion, mainly because of its high separation factor, low energy consumption (about 10 per cent) and flexibility for adjustment to market conditions. Production of the centrifuge requires advanced materials and sophisticated technology, however. Research on centrifuge technology is now under way in many industrialized countries and recently the USA decided to install centrifuge technology in its projected 8.8-million SWU/year enrichment plant in Portsmouth.

Many more processes are at the research and development stage. Apart from the aerodynamic processes discussed below, chemical exchange methods, plasma-based processes and photo-excitation methods using lasers should be mentioned. Laser enrichment technology especially is considered to be very promising. Because of its very high separation factor, it may be able to produce highly enriched uranium in a very few stages, making the technology especially attractive for military applications.

The jet nozzle technology

In the jet nozzle process, developed by Becker and co-workers, a stream of about 5 per cent UF_6 in 95 per cent H_2 flows along a curved fixed wall (see figure 3.1). Isotope separation is achieved because of the centrifugal forces generated by the deflection. Prototypes of cascades have been in operation at the Kernforschungszentrum in Karlsruhe (FR Germany) since 1967, and separation factors of 1.013–1.015 have been attained [1, 2]. Since 1970 the West Germany company STEAG has been involved in both technological development and commercial implementation of this process. In 1975, within the context of West German–Brazilian nuclear cooperation, the Brazilian company NUCLEBRAS and the West German company INTERATOM joined the effort with the construction of a demonstration plant with a capacity of 200 tons SWU/year as its primary objective [3]. Nozzle-separation elements are produced either mechanically by the Messerschmitt–Bölkow–Blohm company in Munich or by the stacking of photo-etched metal foils by Siemens. Thus, the technology has so far remained under West German control.

In 1975 the process developers estimated the total energy requirement of a 5 000-ton SWU/year enrichment plant at about 2 500 megawatts [4]. Becker claimed in 1977 that "the specific power consumption of the separation nozzle process corresponds approximately to that of the existing US

Figure 3.1. Cross-section of the separation nozzle system used in the commercial implementation of the process

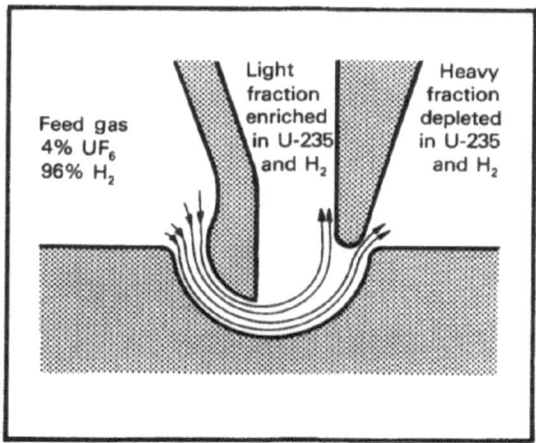

Source: See reference [3b]

gaseous diffusion plants'' and that "there is no doubt that a further significant reduction of that figure is to be expected" [3a]. Other authors give figures between 2 500 and 3 500 kWh/kg SWU for specific power consumption [2, 5, 6].

As to their policies for the future, the developers of the jet nozzle technology are especially interested in cooperation with uranium-producing countries which, because they possess alternative sources of energy, have a favourable electricity price level and will soon be in a position to export their uranium in an enriched form. Local industry should be permitted to participate in these projects so as to obtain direct access to the technology [4].

Apart from the Brazilian project, contacts have been reported with India [7]. But quite a number of African and South American countries, such as Zaire and Argentina, could fit into such a policy.

The advanced vortex tube process

The advanced vortex tube process, developed by the Uranium Enrichment Corporation of South Africa (the UCOR process), is another aerodynamic separation process. It uses as the separating element a stationary walled centrifuge with UF_6 in hydrogen as a process fluid [8]. Process pressures are well above atmospheric. The separation has a high degree of asymmetry with respect to the UF_6 flow in enriched and depleted streams. To deal with the small UF_6 cut, a cascade technique has been developed called the helikon technique [9]: an axial flow compressor simultaneously transmits several streams of different isotope composition without significant mixing,

Figure 3.2. Flow path through part of the helikon module

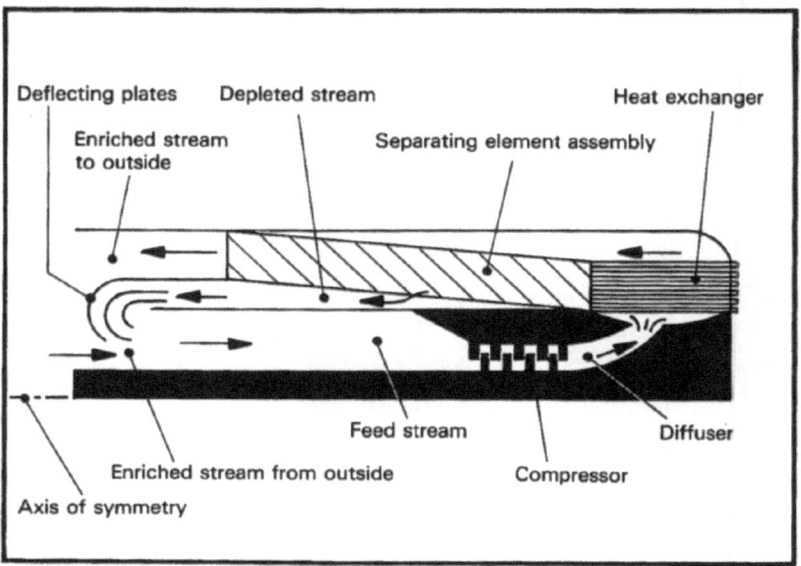

Source: See reference [9a].

thus making the process more efficient. Separation elements and the helikon technique together constitute the modules of the UCOR process (see figure 3.2). Separation factors up to 1.030 have been reached [8]. Modules with a separative work capacity of 10 tons SWU/year have been achieved and those of 80–90 tons SWU/year are considered realistic prospects. Varying degrees of enrichment up to a maximum of several times the separation factor over the separating element can be realized. The process has a low uranium inventory, resulting in cascade-equilibrium times of the order of 16 hours for enrichment to 3 per cent. The process is thus suited for producing highly enriched uranium.

The process developers claim that specific power consumption for a commercial plant is of the order of 3 300 kWh/kg SWU. According to 1977 estimates, the minimum capacity for economic viability under South African conditions could be "appreciably lower" than 5 000 tons SWU/year [9].

Due to low process temperatures and relatively low UF_6 partial pressures, extensive use can be made of normal uncoated construction steels. The process developers claim further that the UCOR process is, in comparison with diffusion and centrifuge technology, "the least exacting of the enrichment processes so far developed as far as manufacturing and materials specifications are concerned" [8a].

A pilot plant in Valindaba with a six-ton SWU/year module began operation in 1975. Decisions are expected to be made by 1978–79 for a commercial plant of about 5 000 tons SWU/year for full operation around

1986. UCOR claims that it will be in a position for delivery during the second half of the 1980s.

During the 1970s several contacts have been reported between the jet nozzle group in FR Germany and the UCOR group [10, 11], among which was a contract making a cost comparison of the two techniques. Contacts with Israeli and Dutch scientists have also been reported. South Africa is willing to share its know-how with "friendly countries" and in particular to undertake the establishment of a large-scale enrichment plant in South Africa as an international venture [8b].

Other aerodynamic processes

Research on several other aerodynamic processes has been reported. Among them are projects on the separation probe, crossed beams, velocity slip and the jet membrane (Muntz—Hamel) process. Separation factors of 1.1 have been claimed for the velocity slip process [12]. French estimates for specific power consumption for the most promising process are 2 500—3 000 kWh/SWU [13], while a US prognosis believes the power costs to be higher than those of the gaseous diffusion process [14]. Equilibrium times are estimated by the French study at about two days for 3 per cent enrichment and four weeks for enrichment to 90 per cent. So far as is known, none of those processes is yet out of the laboratory phase.

Characterization relative to present technologies

For an evaluation of the jet nozzle and UCOR technologies, a review of the present field of enrichment technologies and the supply and demand situation for enrichment services is relevant.

Table 3.1 shows some characteristics of the gaseous diffusion and centrifuge technologies on the one hand, and the jet nozzle and UCOR technologies on the other. Although neither of the two aerodynamic processes is much more attractive from the technological point of view, they do not appear to be very inferior to gaseous diffusion which is still the dominant technology in this field. Expected lower capital investment and medium-size plant capacity for economies of scale could, under certain local conditions, compensate for their substantial power consumption. Another attractive feature could be their somewhat lower level of technological and industrial engineering, especially for the UCOR process. Moreover, the uranium-enrichment field is so strongly dominated by political considerations and secrecy that less promising technologies may certainly have a future, for example, because of the desire for national independence with regard to energy policies. This holds especially for the enrichment market which is so strongly dominated by the nuclear weapon states.

Supply and demand prospects for enrichment are very speculative,

Table 3.1. Some characteristic data of enrichment technologies

Substantial errors are possible: (a) due to the secrecy of the technologies; (b) because of an eventual bias of the estimates as made by the process developers; and (c) because of extrapolations from pilot plant to commercial facilities.

Characteristics	Gaseous diffusion	Centri-fuge	Jet-nozzle	Advanced vortex tube	Reference	Year
Separation factor	1.004	1.2–1.5	1.010–1.015		[6]	1977
			1.013–1.015		[1, 2]	1975, 1972
				1.030	[8]	1975
Specific power	2 050–2 400	300–400			[19]	1974
consumption	2 400	200	2 400		[5]	1974
(kWh/kg SWU)			4 400		[4]	1975
	2 500	250	2 500–3 500		[6]	1977
				3 300	[9]	1977
Minimum	9 000	1 000	2 500		[6]	1977
economic capa-city (tons SWU/year)				<5 000[a]	[9]	1977
Specific invest-	145[c]	145[c]	150[c]		[5]	1974
ment costs[b]		210[d]			[20]	1975
$/kg SWU/year)	300	330			[21]	1975
	280[e]				[22]	1975
				~200[f]	[8]	1975
Construction time (years)	6	4	5	5–6	[8, 9]	1975, 1977

[a] 'Appreciably lower' than 5 000 tons SWU/year.
[b] In money units of the date of the reference.
[c] Exchange rate $ = DM3.
[d] URENCO estimate: $1 thousand million for 10 000 tons SWU/year; exchange rate £1 = $2.
[e] EURODIF estimate: 15 thousand million French francs for 10 700 tons SWU/year; exchange rate $1 = 5 francs.
[f] UCOR estimate: 1974 money values.

depending largely on the growth of nuclear power stations, reactor types to be used, plutonium recycling and tails assays. Earlier estimates for the growth of nuclear power have generally been far too high. Alertness on the part of public opinion and some governments, especially in industrialized capitalist countries, with regard to waste disposal, reactor accidents and nuclear proliferation have delayed many programmes and will probably continue to do so. Recent prognoses of the demand for enrichment services by the US Energy Research and Development Administration (ERDA) in 1976 [14a] and the US Office of Technology Assessment in 1977 [15] suggest that existing and planned gaseous diffusion and centrifuge facilities can probably supply sufficient enrichment capacity up to the end of the 1980s and, through stockpiling, sufficient enrichment services up to about the middle of the 1990s. Nevertheless, in this field too, political considerations of independence and national industrial development may be of sufficient weight to initiate enrichment facilities.

In conclusion, it can be said that the jet nozzle and advanced vortex tube processes, although not very promising from the technological point of view, could provide an attractive alternative for countries with specific local conditions and political–economic objectives, even apart from military considerations. This would hold especially for countries with uranium resources of their own, a medium level of technological and industrial development and relatively low energy prices.

III. Problems of proliferation of enrichment technologies and facilities

Uranium enrichment is one of the so-called sensitive technologies of the nuclear fuel cycle because it directly leads to weapon-grade materials. Moreover, it is the only field of nuclear technology, apart from weapon design, where there is considerable secrecy. The development of nuclear power has stimulated extensive research on enrichment technologies and is leading to a proliferation of enrichment know-how and facilities to several countries. URENCO, a tripartite centrifuge technology group under FR Germany, the Netherlands and the United Kingdom, has established enrichment facilities in Capenhurst (UK) and Almelo (Netherlands), the latter being the first enrichment plant outside the territory of the nuclear weapon states. Preparations are under way for a URENCO plant on West German territory (probably at Gronau). EURODIF has a gaseous diffusion plant under construction of 10 800 tons SWU/year at Tricastin (France), with capital participation from Belgium, Iran, Italy and Spain. Thus a closer access to enrichment technology and facilities is spreading in Europe and the Middle East, where, apart from Iran's participation in EURODIF, Israel is known to be active in the field of laser enrichment.

From the point of view of weapon capabilities, pilot or demonstration plants with capacities of 20–100 tons SWU/year are already relevant. Such plants could produce 100–500 kg of 90 per cent enriched uranium per year, sufficient for about 5–20 nuclear weapons. Roughly speaking, a 100-ton SWU/year capacity is sufficient for fuelling a nuclear power plant of 1 000 MW(e).

The jet nozzle and advanced vortex tube processes could provide an independent road to enrichment for industrially and technologically less developed countries, especially if local industry is permitted direct access to the technologies, as advocated by the developers of the jet nozzle process. This holds especially for countries which want to keep open their option on nuclear explosives, either because of perceived threats to their security or because of the desire for international status and prestige.

The proliferation policies of the countries involved

In this context it is relevant to analyse the non-proliferation commitments of the countries involved in these enrichment processes.

The Federal Republic of Germany declared in 1954, when becoming a member of NATO, that it undertook not to manufacture nuclear weapons on its territory [16]. It is a party to the Non-Proliferation Treaty (NPT) and is thus obliged not to transfer control over such weapons or explosive devices directly or indirectly or to manufacture them anywhere. At the Review Conference of the NPT in 1975, it declared its intention to share its technological know-how only with countries which have decided to renounce the manufacture or acquisition of nuclear explosive devices [17].

However, FR Germany is selling its jet nozzle technology to Brazil, a country that insists on the right to carry out nuclear explosions for peaceful purposes "including explosions which involve devices similar to those used in nuclear weapons"[18]. Moreover, Brazil is not a party to the NPT and does not feel bound to the Treaty of Tlatelolco for the prohibition of nuclear weapons in Latin America, so long as the attached protocols have not been signed.

Brazil has promised that the imported technological know-how will not be used for the manufacture of nuclear explosive devices and that the transfer of technology is under International Atomic Energy Agency (IAEA) safeguards. However, safeguards on the transfer of know-how are deemed to be very weak and, at best, of very limited duration. The continuous technological change and growing information in the open literature can hardly prevent indigenous developments being claimed by the receiving country as autonomous national developments. Adherence to the NPT or at least full-scope safeguards are minimum conditions for any effective security system. Moreover, the projected enrichment capacity of 200 tons SWU/year makes the plant militarily relevant.

The first deal in jet nozzle technology is thus very disturbing from the point of view of non-proliferation. If the conditions of this agreement become a precedent for future deals in jet nozzle plants, this technology may become a very serious contribution to proliferation on the part of FR Germany. And the situation will worsen if Brazil, in the more distant future and on the basis of its present commitments of non-proliferation, becomes an exporter of the technology.

South Africa is not a party to the NPT, although it has stated that it shares the objectives of the treaty. Its objections to joining have been the so-called risks for industrial espionage and the extra costs due to the safeguards system which, in their view, may damage the competitive position of South African industry. South Africa should definitely be regarded as a near-nuclear weapon country. There have even been reports of preparations for a nuclear explosive test, which were denied by the South African government. Recently South Africa has stated a willingness to reconsider its position with regard to the NPT, following US political pressure. South Africa is rather isolated in the international arena, due to its apartheid régime. Racial conflicts in Southern Africa are growing and many observers consider it a region where there is a serious risk of major war. It is difficult to imagine how a nuclear weapon capability could be of use to internal South African problems. But it could certainly have a deterrent function against neighbouring countries and intervention from other states, whether or not such threats were realistic.

Little is known about the sharing of the UCOR process with other countries, apart from the fact that in 1975 South Africa reported "negotiations at an advanced stage" with unnamed partners. It is not to be expected that South Africa, which has no IAEA safeguards on its own enrichment facility, would require safeguards in the context of such agreements. Thus the UCOR process also provides a serious problem for non-proliferation. International cooperation in a plant on South African territory in the form of capital would give closer access for countries to an uncontrolled enrichment technology.

In conclusion, it can be said that so far, the policies of the countries involved in jet nozzle and advanced vortex tube enrichment technology represent a danger for proliferation. They open the option for countries not party to the NPT to obtain enrichment technology, possibly with insufficient or even no safeguards. Such practice could easily lead to a further spread of nuclear explosive capabilities.

IV. Conclusions

The much greater awareness of the dangers for nuclear proliferation involved in the spread of nuclear energy technology has not yet led to an effective non-proliferation régime. The control of technology transfer and emerging new technologies in a national context create especially serious problems. This is particularly the case for sensitive technologies such as uranium enrichment.

The jet nozzle and advanced vortex tube processes are relevant examples. Although technologically not so attractive, they can provide independent national enrichment alternatives, which is an advantage for countries with uranium resources of their own. Thus many more countries could arrive at a national option on weapon-grade material. A strict non-proliferation policy would be helped by a moratorium on the development and export of the discussed technologies. Present policies with regard to those technologies are certainly insufficient from the standpoint of non-proliferation.

References

1. Becker, E. W. *et al.*, 'Physics and development potential of the separation nozzle process', Paper presented at the International Conference on Uranium Isotope Separation, London, 5–7 March 1975, figure 2.
2. Mohrhauer, H., 'Stand der Urananreicherung in Europa', *Atomwirtschaft*, Vol. 17, June 1972, p. 300.
3. Becker, E. W., 'The separation nozzle process for uranium isotope enrichment', *Nuclear Power and its Fuel Cycle*, Proceedings of an International Conference, Salzburg, 2–13 May 1977, Vol. 3 (IAEA, Vienna, 1977).
   (a) —, p. 170.
   (b) —, p. 163.
4. Geppert, H. *et al.*, 'The industrial implementation of the separation nozzle process', Paper presented at the International Conference on Uranium Isotope Separation, London, 5–7 March 1975, table 3.
5. Braun, P. *et al.*, 'Zum Wirtschaftlichkeitsvergleich von Urananreicherungs-verfahren', *Atomwirtschaft*, Vol. 19, November 1974, p. 536.
6. Hildenbrand, G., 'Nuclear energy, nuclear exports and the nonproliferation of nuclear weapons', Paper presented at the US Atomic Industry Forum Conference on International Commerce, and Safeguards for Civil Nuclear Power, March 1977.
7. 'India buys German route to enriched uranium', *New Scientist*, Vol. 53, 9 March 1972, p. 546.
8. Roux, A. J. A. and Grant, W. L., 'Uranium enrichment in South Africa', *Progress in Nuclear Energy: Nuclear-Energy Maturity*, Proceedings of the European Nuclear Conference, Paris, 21–25 April 1975, Vol. 12 (Pergamon Press, Oxford, 1975).
   (a) —, p. 40.
   (b) —, p. 43.
9. Roux, A. J. A. *et al.*, 'Development and progress of the South African enrichment project', *Nuclear Power and its Fuel Cycle*, Proceedings of an International Conference, Salzburg, 2–13 May 1977, Vol. 3 (IAEA, Vienna, 1977).
   (a) —, p. 176.
10. Muller, M., 'The enriching politics of South Africa's uranium', *New Scientist*, 2 May 1974, p. 252.
11. 'South Africa's process may not be so new after all', *Science*, Vol. 183, 29 March 1974, p. 1271.
12. Anderson, J. B. and Davidovits, P., 'Isotope separation in a "seeded beam"', *Science*, Vol. 187, 21 February 1975, p. 642.
13. Fréjaques, C. *et al.*, 'Evolution des procédés de séparation des isotopes de l'uranium en France', *Nuclear Power and its Fuel Cycle*, Proceedings of an International Conference, Salzburg, 2–13 May 1977, Vol. 3 (IAEA, Vienna, 1977), pp. 203–13.
14. Vanstrum, P. R. and Levin, S. A., 'New processes for uranium isotope separation', *Nuclear Power and its Fuel Cycle*, Proceedings of an International Conference, Salzburg, 2–13 May 1977, Vol. 3 (IAEA, Vienna, 1977).
   (a) —, p. 219.
15. *Nuclear Proliferation Fact Book*, Committee on International Relations, US Congress (US Government Printing Office, Washington, D.C., September 1977), p. 174.

16. Paris Agreements of 23 October 1954: Protocol III on the control of armaments, annex I, 23 October 1954.
17. NPT Review Conference Document NPT/CONF/SR.3.
18. *World Armaments and Disarmament, SIPRI Yearbook 1976* (Almqvist & Wiksell, Stockholm, 1976, Stockholm International Peace Research Institute), p. 369.
19. 'Der Markt für Urananreicherung in West-Europa', *Jahrbuch der Atomwirtschaft*, 1974, p. B45.
20. Allday, C. and Kehoe, R. B., 'Urenco–Centec progress and plans', Paper presented at the International Conference on Uranium Isotope Separation, London, 5–7 March 1975.
21. O'Donell, A. J., 'Uranium enrichment associates, a private initiative', Paper presented at the International Conference on Uranium Isotope Separation, London, 5–7 March 1975.
22. 'Eurodiff-Anlagenkosten verdoppelt', *Atomwirtschaft,* Vol. 20, February 1975, p. 57.

Paper 4. Laser separation of isotopes

K. L. KOMPA

Square-bracketed numbers, thus [1], *refer to the list of references on pages 89–90.*

I. Basic requirements and classification of methods

There is no single universal method best suited to the separation of any isotope. Rather a choice has to be made among various techniques depending on the element, the isotopic purity required, and the scale of the operation. The present discussion centres around the principal possibilities, the determining parameters, and the first experimental results in uranium laser isotope separation.[1] Some of the more important separation principles are listed in table 4.1.

Table 4.1. Methods of isotope separation

1. Electromagnetic
2. Thermal diffusion
3. Gaseous diffusion
4. Mass and sweep diffusion
5. Gas centrifuge
6. Separation nozzle
7. Chemical exchange
8. Chromatographic and ion exchange
9. Distillation and exchange distillation
10. Electrolysis and electromigration
11. Photochemical and laser

Source: Reference [2].

In a rough classification of these methods, the electromagnetic method, combining general applicability with high separation efficiency, has been called the most versatile means for the production of research quantities of isotopes. Methods 3–6 are considered to be the most economical methods for the separation of heavy elements, particularly uranium, although they have only small single-step separation factors. Method 9 is best suited for the separation of isotopes of light element [3].

---

[1] Besides uranium enrichment, the separation of deuterium from hydrogen by lasers may also have, at least indirectly, some influence on proliferation risks associated with heavy water reactor concepts. Results for heavy water production by laser means have been reviewed recently [1] and are therefore ignored in the present discussion.

Table 4.2. Classification of the principal laser isotope separation methods

Initial state of system	Laser excitation	Primary response of system	Separation step	Final state of system
Atomic vapour	U.v./visible lasers	(a) Momentum transfer	Geometrical deflection	Atomic vapour
		(b) Electronic excitation	Chemical reaction	Chemical product
	I.r. and u.v./visible lasers	(c) Ionization through sequence of excitation steps	Electrical collection of ions	Pure element
Gaseous molecules	U.v./visible laser(s)	(d) Electronic excitation	Chemical reaction	Chemical product
		(e) Excitation to predissociative electronic state	Dissociation	Fragments
	I.r. laser(s)	(f) Excitation to long-lived vibrational state	Chemical reaction	Chemical product
		(g) Excitation to dissociative vibrational levels	Dissociation (isomerization)	Fragments (isomer)
	I.r. and u.v./visible lasers	(h) Direct photo dissociation after initial selective vibrational excitation	Dissociation	Fragments

Photochemical separation (method 11) is not new and had some success with the separation of chlorine, mercury, and carbon isotopes some decades ago (compare references [4—8]). A strong interest in this method, however, developed only after high-power laser sources had become available at many wavelengths in the infra-red, visible, and ultra-violet spectral regions. The general optimism for this new approach is based on the expectation that high single-stage separation efficiencies are compatible with large-scale operation and that the method will be applicable to nearly all elements. This, however, is still a long-term prospect. The field is now undergoing a rapid and diversified development in most industrial countries. Some of the experimental approaches chosen can be discussed with reference to table 4.2. This list restricts itself to experiments in (mostly low pressure) gases, despite some possibilities for separation schemes in condensed phases and a recent successful demonstration of U-235/238 separation in a solid matrix [9]. Most workers in the field would agree, however, that isotope separation will eventually be practical only in the gas phase. The three schemes which have received most attention so far as

Figure 4.1.   Classification of laser isotope separation methods (see table 4.2)

(a) Deflection of mechanical trajectory.

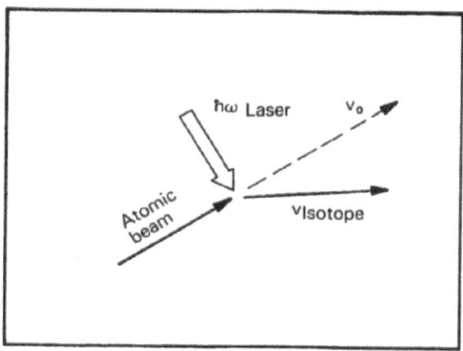

(b), (c).   Electronic excitation followed by ionization or chemical reaction

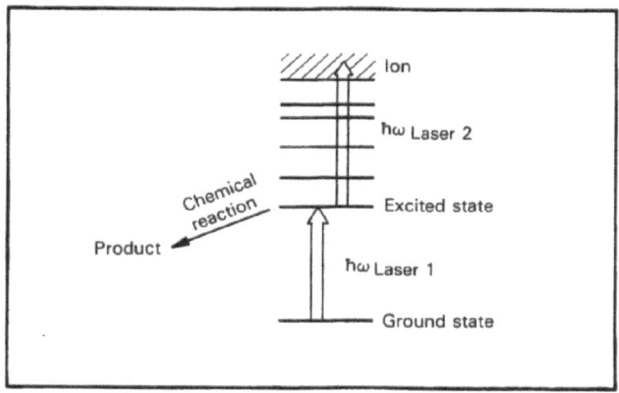

Figure 4.1. (continued)

(d), (e), Laser excitation of a molecule AB to an electronic state AB* followed by dissociation AB* → A + B or a chemical reaction AB* + C → A + BC.

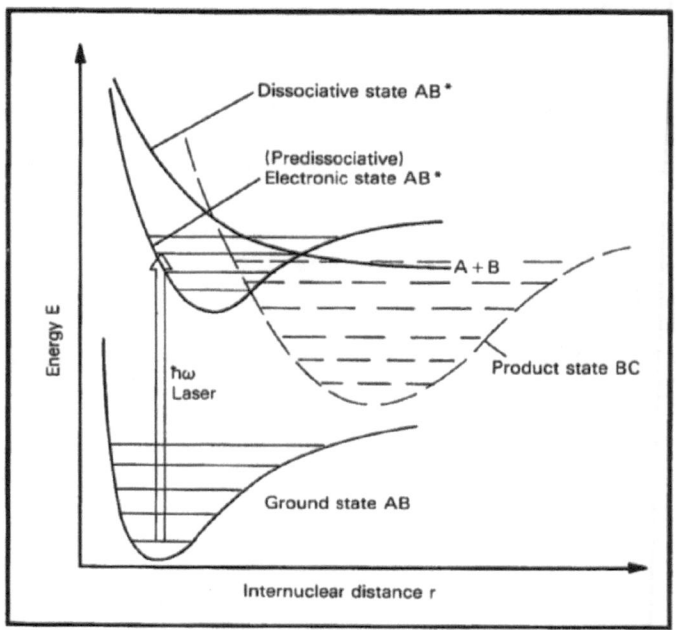

uranium separation is concerned are shown in boxes in table 4.2. These include the selective ionization of uranium atoms followed by deposition on electrical collector plates and the selective dissociation of uranium hexafluoride, UF_6 either by infra-red radiation alone or by the combined action of infra-red and ultra-violet radiation. The details of the laser separation principles listed under (a)–(h) in table 4.2 are schematically explained in figure 4.1. It becomes apparent that many separation principles exist because:

1. The chemical reactivity of atoms or molecules is usually increased by excitation.

2. The ionization energy of an excited atom or molecule is smaller than that of unexcited ones.

3. The dissociation energy of an excited molecule is smaller than that of an unexcited one.

4. Predissociation occurs when an excited molecule passes spontaneously into a dissociative state.

5. The excitation of a molecule may result in isomerization. The isomer, because of its different internal structure, has different chemical properties.

Figure 4.1.   (continued)

(*f*), (*g*), (*h*).   Excitation to a low-lying vibrational state followed by photodissociation, further vibrational excitation up to dissociation in the electronic ground state, or a chemical reaction of the excited molecules to form the product BC.

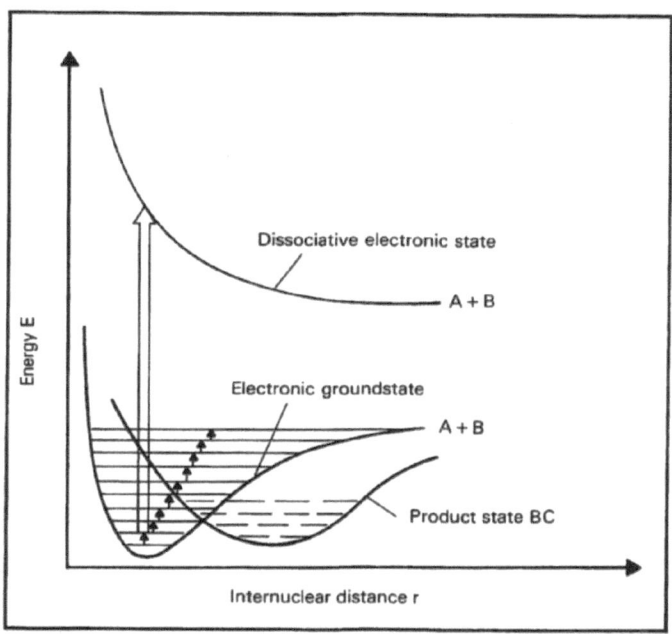

6. The recoil of an atom or molecule occurs when it absorbs a photon with its momentum $h\omega/c$. This gives a very small but observable particle photodeflection.

7. Excited atoms and molecules may have a higher polarizability, a different symmetry of wave function, and so on. This may cause changes in cross-sections for scattering by other particles, in its motion in external fields, and so on. Many possible photochemical and photophysical methods of laser separation follow from this classification [10].

Once a suitable separation principle has been selected, which other parameters determine the separation efficiency? So far this discussion has been idealized, as spectral selectivity has been assumed to be infinite. The following limiting conditions also have to be met to make the separation practical:

1. Energy transfer and relaxation must not ruin the initially selective excitation.

2. Cross-reactions and reversibility of the separative reaction steps which both lead to isotopic scrambling have to be avoided.

3. A laser of the right wavelength, frequency bandwidth, energy, spatial and temporal properties must be available.

4. The overall systems technology must not become too complex, expensive and energy-consuming.

Some brief comments on these problems may be given in the following way: without a detailed discussion of the causes of isotopic shifts in absorption spectroscopy of atoms and molecules, such shifts can be said to be generally large enough to permit selective interaction with existing lasers. For molecules, however, one is not just dealing with single absorption lines, but with bands which often consist of many densely spaced lines. Even if the individual line positions of the isotopic molecules in question are sufficiently separable, there is often considerable band overlap obscuring the isotopic shift. The principal method of countering this problem is to work at very low pressures and to cool the molecules to very low temperatures. Often, especially for compounds of heavy elements, this is only possible in a gas dynamic way in supersonic molecular beams. The rather complicated technology of such arrangements provides additional benefits with regard to the problems listed above under 1 and 2, but tends to act against achievement of requirement 4.

So far as the availability of laser sources is concerned, it is clear that all necessary parameters can be achieved from the physics point of view. Compatibility with technological requirements may still prove to be difficult. These arguments will become clearer in the following discussion of practical cases.

Laboratory separation experiments using photochemical and laser methods are surveyed in table 4.3.

II. Results for light elements

Every isotope separation concept is easier for light elements, since the corresponding molecular compounds are more volatile, show simpler spectral features and have larger isotopic shifts. The molecule which has been most extensively studied with regard to laser excitation and isotopically selective dissociation is sulphur hexafluoride, SF_6. It is instructive to summarize some of the details here as the separation works very well in this case and some conclusions can be drawn about factors which are also important in the separation of heavier elements.

SF_6 is a non-toxic gaseous substance which has six observable infra-red absorptions at 940, 769.4, 639.5, 614, 522 and 344 cm^{-1}. The highest energy absorption band at 940 cm^{-1} corresponds to the ν_3 mode which is used for the laser excitation. The spectral absorption is shown in figure 4.2. Strong

Table 4.3. Catalogue of isotope separation experiments

Photoionization in atomic beams (compare concept (c) in table 4.2 and figure 4.1)
Uranium-235/238; Calcium-40

Laser-induced chemical reaction (concepts (d) and (f) in table 4.2 and figure 4.1)
Nitrogen-15; Chlorine-35, 37

Selective photodissociation (concept (e))
Deuterium; Carbon-12, 13; Nitrogen-14, 15; Bromine-81

Multiple photon absorption by infra-red laser (concept (g))
Boron-10, 11; Sulphur-32, 33, 34; Deuterium; Carbon-13; Silicon-29, 30; Tellurium;
Osmium; Chlorine-35, 37; Uranium-235; Molybdenum

Selective two-step photodissociation (concept (h))
Nitrogen-15; Chlorine-35, 37; Boron-10; Deuterium;[a] Tritium;[a] Oxygen-16, 18;
Titanium-50;[a] Uranium-235

---

[a] Proposed.

Source: Reference [11a].

absorption is exhibited for the CO_2 laser which is one of the simplest and most common infra-red gas lasers. The overlap between the various CO_2 laser emission lines with the absorption of either 32 SF_6 or 34 SF_6 is indicated in the figure. When the CO_2 laser is tuned to either one of the absorption bands and the power level is adjusted to at least 20 MW/cm² or about 1 J/cm², isotopically selective dissociation can be accomplished in low-pressure SF_6 gas. Figure 4.3 shows the dependence of the dissociation yield on laser energy fluence. It is seen that yields close to 100 per cent can be obtained for moderate laser energies in a totally selective fashion. The difference between the two curves of figure 4.3 resembles the higher dissociation efficiency in a molecular beam experiment. However, even for the lower plot, which corresponds to dissociation in a static gas volume, the efficiency is still high. In the primary dissociation process, one fluorine atom is split off leaving a molecular fragment which is chemically totally different from the parent molecule, in this way facilitating the removal of the separated product. The separation factor in such a process can be defined as follows. If $[N_A]_0$, $[N_b]_0$ and $[N_A]_f$, $[N_B]_f$ are the initial and final concentrations of two isotopic species to be separated, the coefficient of separation is

$$K(A/B) = ([N_A]_f/[N_B]_f)/([N_A]_0/[N_B]_0).$$

With no separation present this coefficient would be

$$K(A/B) = 1.$$

For the SF_6 dissociation, the final concentration ratio can easily be made to be

$$[N_A]_f/[N_B]_f \approx 100.$$

Starting out with a natural abundance of S-32/S-34 $= [N_A]/[N_B]_0 \approx 25$, one gets

Figure 4.2. Absorption in the v_3 band of 32 SF$_6$ and 34 SF$_6$ (natural mixture 32 SF$_6$ ≈ 95 per cent, 34 SF$_6$ ≈ 4 per cent) under low resolution in comparison with the range of CO$_2$ laser emission lines around 10.6 μm

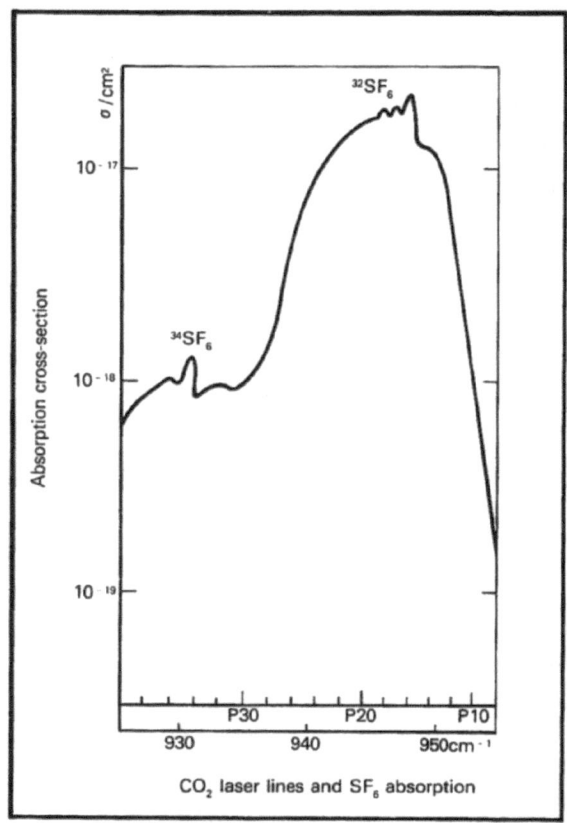

Source: See references [15, 16].

$$K(A/B) \approx 4.$$

The results of figure 4.3 gave rise to a laboratory-scale separation experiment [10, 12–14] for the two sulphur isotopes S-32 and S-34 which is schematically shown in figure 4.4. With a 15 J per pulse CO$_2$ laser, about 10 000 shots are needed to produce one gram of 99 per cent pure S-32. With the table-top experiment sketched in figure 4.4, this corresponds to a few hours' work and requires 1 kWh of electrical input to the CO$_2$ laser. A more detailed consideration of the economics of such a simple separation scheme is given in the last section of this survey.

Although SF$_6$ is the best studied example of isotope separation by multiple photon infra-red excitation, the next best studied case is probably uranium hexafluoride UF$_6$. For the purpose of the present discussion, we will therefore not go into detail about the many interesting results for other

Figure 4.3. Plots of SF$_6$ dissociation probability versus laser energy flux

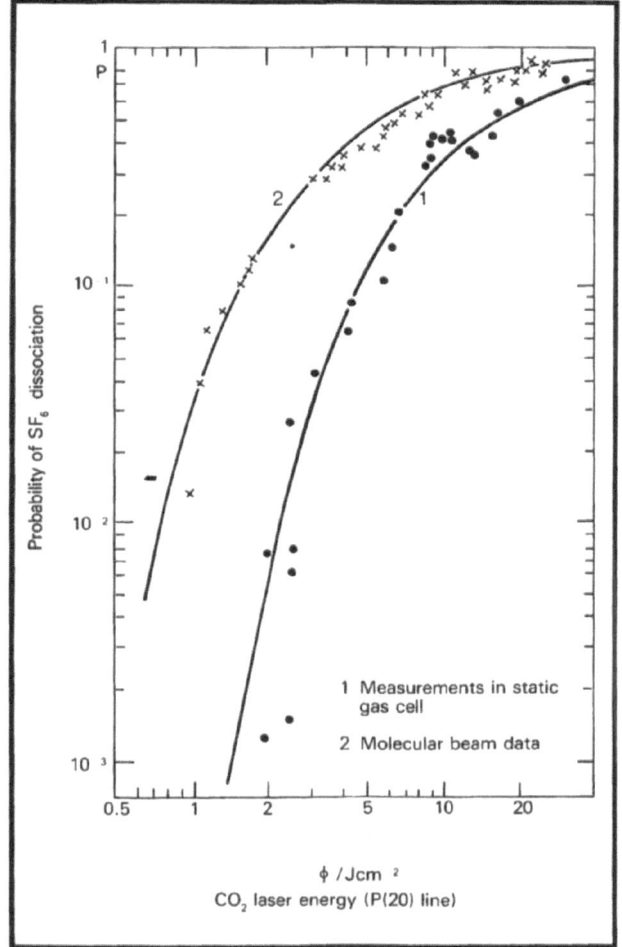

Note: It is seen that dissociation yields approach 100 per cent even for quite moderate laser energies.

Source: References [13, 14].

light elements. For a general overview, the reader is referred to table 4.3 and the literature quoted there.

III. Results for heavy elements

Despite the structural similarity between SF$_6$ and UF$_6$, there are some important changes to be noticed in changing to UF$_6$:

81

Figure 4.4. Elements of a table-top sulphur isotope separation experiment

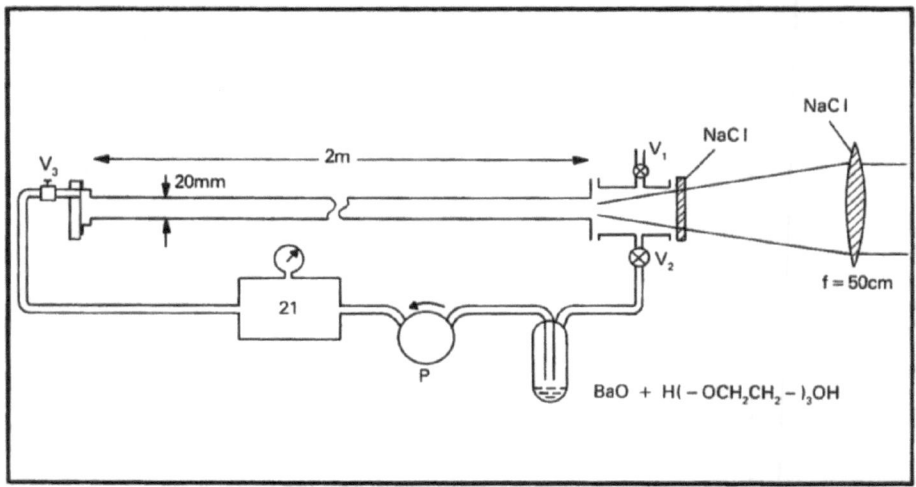

Note: The irradiation and dissociation of either one of the isotopic forms of SF_6 takes place in a long internally gold-plated absorption tube. The laser beam enters through the tube throat via sodium chloride optics. The gas mixture is circulated with the aid of a compression stage P, a buffer volume of two litres, and is re-expanded into the absorption cell by valve V_3. The chemical deposition of the isotopically pure dissociation fragments is achieved in the trap with the barium oxide suspension. (See also figure 4.9.)

Source: See references [15, 16].

1. UF_6 also has six infra-red absorptions which are now at 666.6, 626, 535, 200, 189 and 144 cm^{-1}. The absorption corresponding to the O_3 mode, which is again best suited for isotopically selective excitation, is at 626 cm^{-1} $\approx 16\,\mu$m. There is no powerful and technologically simple tunable laser available at present for this wavelength [13a]. Much of the work towards laser uranium enrichment is therefore still concerned with laser development.

2. While the vibrational isotope shift between S-32 and S-34 SF_6 is ≈ 17 cm^{-1}, due to the much smaller reduced mass effect between U-235 and U-238 UF_6 this is now of the order of 1 cm^{-1}. No usable isotopic spectral difference has so far been found in the u.v. absorption of UF_6.

3. While the vibrational isotope shift is quite sufficient in principle to permit selective interaction of one isotopic species with a laser field, the spectral structure is totally blurred at room temperature due to thermal populations in all low-lying vibrational levels and corresponding hot band absorptions.

4. If an isotopically selective dissociation were to be achieved, the dissociation product would separate itself from the gas mixture, since UF_6 is a solid compound.

The statements made under 2 and 3 are borne out by the plots of figures 4.5(*a*) and (*b*), 4.6 and 4.7. It is seen that the spectra can be recorded at very low temperatures where all the relevant absorption features are fully

Figure 4.5.

(*a*) **Absorption spectrum of expansion-cooled UF₆ compared with room-temperature spectrum [9, 17]**

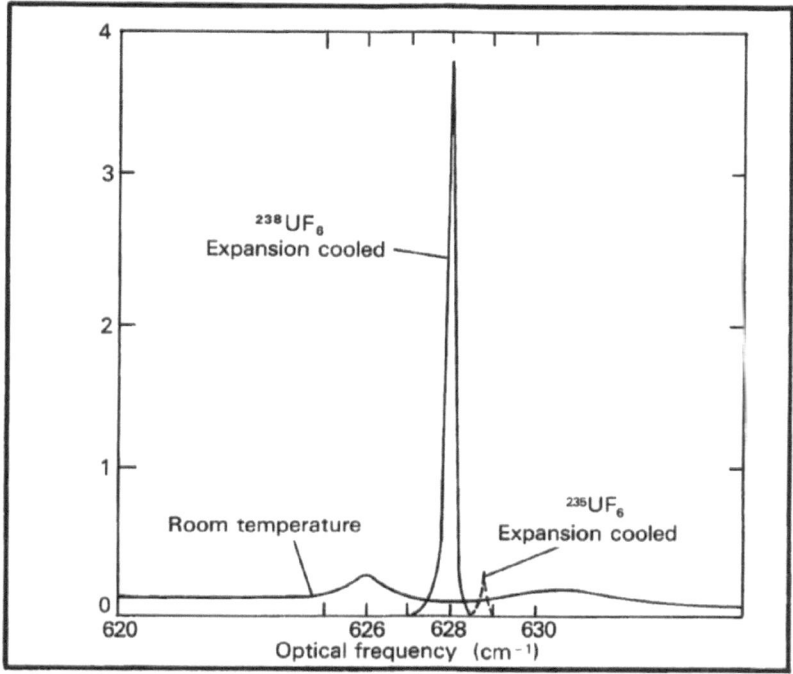

(*b*) **Spectrum of 238 UF₆ at room temperature with ground-state PQR structure of v_3 band; spectrum of 235 UF₆ is shown at the bottom for comparison [9, 17]**

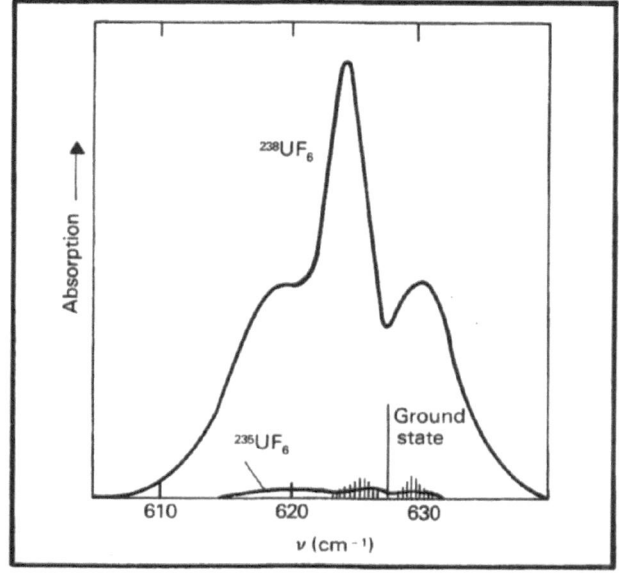

Figure 4.6. **Fractional population of the 10 lowest-lying vibrational levels of UF_6 with temperature, indicating the blurring of the ground-state absorption at room temperature**

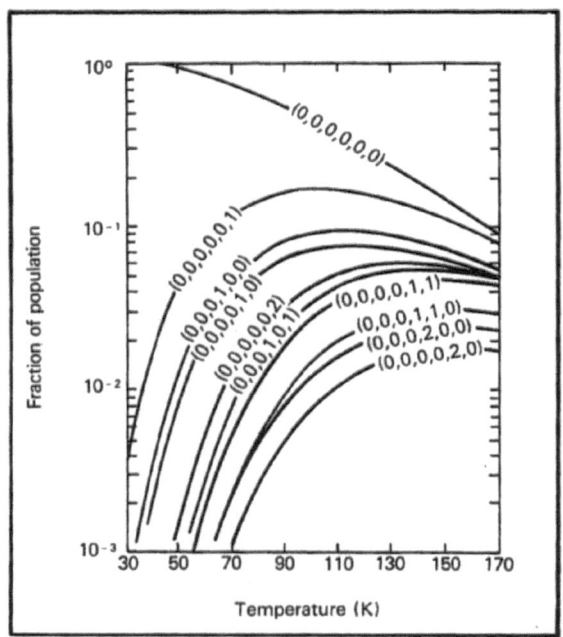

Source: Reference [9].

resolved. The cooling to these low temperatures is not possible in a static gas, as UF_6 would solidify, but can be realized in supersonic molecular beams. Figure 4.8 shows a shaped expansion nozzle to be used in such experiments. The achievable temperature is related to the pressure P and the coefficient γ, the ratio of heat capacities at constant pressure and constant volume ($\gamma = C_p/C_v$) by

$$T/T_0 = (P/P_0)^{(\gamma - 1)/\gamma}.$$

Temperatures of very few degrees have been used in this way.

After preparing the UF_6 molecules as indicated, selective laser action becomes possible. A wide variety of concepts have been applied to generate the right laser frequency. Some of the most promising approaches use schemes where the frequency of a CO_2 or other infra-red laser is shifted to longer wavelengths ($\sim 16\,\mu m$) by some molecular transformer. In technical terms this is just an accessory part to the infra-red laser. Complications arise, however, from the high precision and stability requirements. The present status is such that several laser sources exist which can be used in laboratory experiments but which are not yet suitable for large-scale

Figure 4.7.   Low-resolution u.v. absorption spectrum of UF$_6$ showing no resolved isotope shift

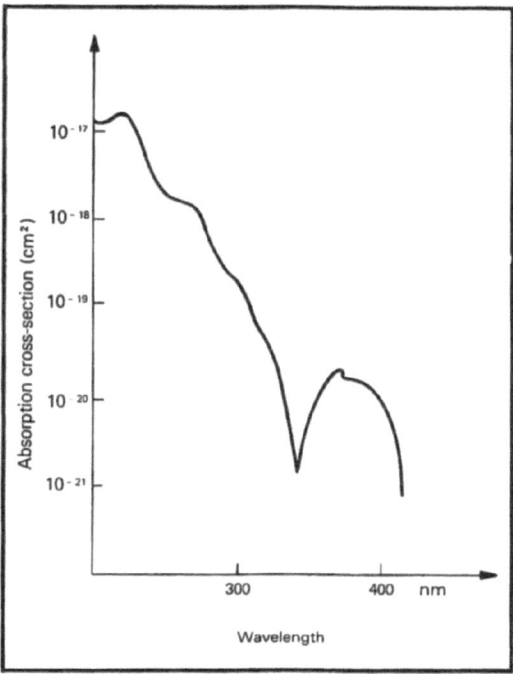

Source: See reference [18].

Figure 4.8.   Expansion nozzle scheme for generation of supersonic cold UF$_6$ beams for isotope separation experiments

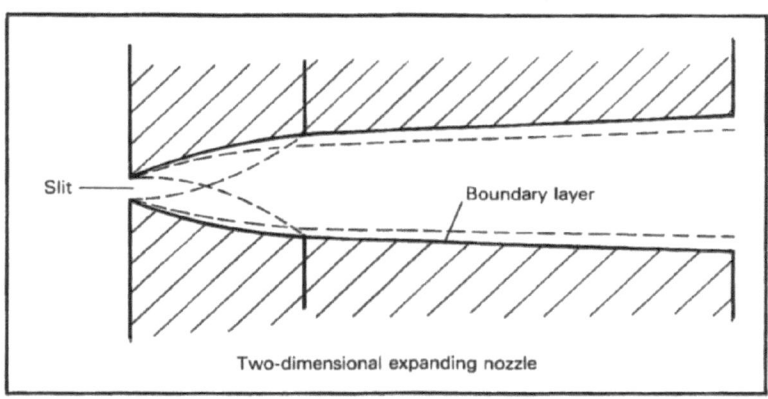

operations. It must be emphasized, however, that this technology is within the reach of many laboratories.

A report has been given recently [19] on the dissociation of UF$_6$ by multiple infra-red photon dissociation using a combination of two infra-red lasers, one at 16 μm, the other a CO$_2$ laser. In this work no enrichment was

Figure 4.9.   Photograph of equipment described in figure 4.4

specified. There are claims, however, that some minor enrichment has been achieved by this technique even at relatively high temperatures. It must be questioned whether this enrichment is already technologically meaningful.

Another scheme which is pursued mainly by the Los Alamos Scientific Laboratory in the USA uses photodissociation around 300 nm (3 000 Å) (figure 4.7) by an ultra-violet laser after selective preparation of 235 UF_6 by 16 μm irradiation of a cold molecular beam (concept (h) of table 4.2). Substantial enrichment is reported to have been achieved as early as 1976 but no detailed figures have been published. U.v. lasers for this purpose can easily be made available. Direct isotopically selective u.v. photodissociation (concept (e) of table 4.2) has not had any success, although the possible simplification of the spectral structure in comparison to figure 4.8 by super-

Figure 4.10.   Close-up of ultra-violet laser

Figure 4.11. A typical ultra-violet laser used in laboratory isotope separation experiments

sonic cooling does not seem to have been investigated in detail.

As table 4.2 shows, the uranium is used in all these schemes either in molecular or atomic form. In fact the atomic process of 2- or 3-step photo-ionization was the earliest concept and is actively used, for instance, in the US Lawrence Livermore Laboratory. Three milligrams of reactor-grade uranium, starting from natural uranium, are reported to have been produced several years ago. For a more detailed discussion and a comparison of the molecular and atomic approaches, the reader is referred to reference [10]. Industrial activity in this area, by AVCO EXXON Nuclear Isotopes Inc., is reported to be comparatively close to large-scale testing [20].

The following conclusions can be drawn in summary:

1. Uranium isotope separation has been demonstrated by at least three principles: 2-photon ionization of uranium vapour, u.v. photodissociation of that fraction of UF_6 molecules which was excited before by 16 μm radiation, and multiple-photon dissociation by the action of two infra-red lasers.

2. The enrichment achieved must be assumed to be still rather small, although no technical details have been published.

3. Laser excitation sources suitable for the present phase of the work, which is still predominantly of an academic nature, exist and an extrapolation to large-scale operations is not unreasonable.

4. A systems evaluation of laser separation and economic considerations is still subject to much debate and no general agreement has been reached.

IV.   Open problems and future prospects

In addition to its potential uses in energy research, laser isotope separation must be viewed against the broader background of laser chemistry. This relationship becomes apparent from the following classification which shows the four main directions in which laser chemistry research is currently expanding (see, for instance, references [10, 12−14]).

1. Truly state-selective chemistry with laser-prepared reagents in single quantum states is obviously the ultimate goal of laser chemistry and would provide the most detailed insight into chemical reaction dynamics. This is possible, however, only for special cases and very special experimental conditions.

2. If chemical compounds cannot be made to react in single states, the next, less demanding goal is that of a bond-specific chemistry with still considerable energy localization in a molecular bond or group of bonds.

3. The next step in lowering the specificity of excitation again is connected with a compound-specific chemistry. This is the area to which isotope separation, as well as the separation of chemical and nuclear isomers, or the laser purification of materials, belongs. It basically involves the separation of one particular compound from a mixture by laser means.

4. The fourth and last area is that of a non-specific laser pyrolysis where the laser acts only as a specialized heat source.

It is fair to state that laser isotope separation is just one way of entering a broad new technology and that the development of the corresponding techniques will have many uses. This perspective provides an additional motivation and support for work in this area.

The economic magnitude of the uranium enrichment industry can be illustrated with a few figures from the US nuclear industry [10]. A typical light water reactor producing 10^9 watts requires processed fuel for an initial charge of 600 tons of U_3O_8 ore and an annual use of 200 tons. The fuel in this case contains 3 per cent U-235, as compared to a natural abundance of 0.75 per cent. The depleted ore is discarded at a U-235 content of 0.25 per cent. The enrichment is carried out in rather large gaseous diffusion plants. Investment costs for such an installation handling 20 000 tons of UF_6 annually are about \$4.5 thousand million. Similar size gas centrifuge enrichment capacity would cost perhaps 20−30 per cent less. The Becker nozzle process has entered the pilot plant stage and might in time lead to a cost reduction of a factor of two. Gaseous diffusion uses a total of 5 MeV of electrical energy for separation of each U-235 atom. The overall enrichment cost according to reference [10] is about \$5 per gram of U-235. For comparison, the energy requirement for a laser process is likely to be less than 10 eV of photons per U-235 atom. This means a few keV of electrical input to the corresponding lasers. This simple comparison should not, however, be thought to indicate that as far as the energy cost is concerned the laser would be far superior to any existing conventional process. The cost of material handling and preparation may, for instance, be higher than the running cost of the laser(s). It is also appropriate to remember that the centrifuge, employing a reversible process, in principle does not require any energy input except to overcome friction losses and yet yields the overall enrichment cost quoted above.

Capital cost requirements for a full-scale laser separation facility are not exactly known, but the comparative cost accepted in references [11b, 21] is one-twentieth of that needed for a centrifuge plant of commensurate production capability. From the investment point of view then, the laser process is likely to operate on a similar scale to that of conventional schemes.

The exact figures for the enrichment capacity needed for the next decades are at present quite uncertain. The laser isotope separation may, however, be looked upon more as a means of conserving our uranium resources by optimizing rather than competing with existing production schemes. The laser process would indeed be ideally suited for reducing the U-235 content of the tails of the diffusion/centrifuge production complexes [11b]. Such an 'advanced stripping module' could be added to a conventional plant to generate 'new feed' from the current tails stream. The amount of material available from the tails is rather large. Over 650 000 tons of U_3O_8 will have been produced by the end of 1990. If the U-235 content in this ore can be reduced from 0.25 per cent to 0.1 per cent, this could reduce the requirements for mining by about 60 000 tons of natural uranium with additional expected savings in the mining rates in the future.

To develop such laser isotope separation technologies in conjunction with conventional enrichment plants is the current policy in the US Department of Energy's Advanced Isotope Separation Program funded at the level of $54.2 million in fiscal year 1979. In addition to various laser enrichment schemes, this programme also includes the Dawson plasma separation process as an alternative.

There is justified concern about the potential for nuclear weapon proliferation through laser separation of uranium. A moratorium on this separation development has been suggested [22]. It was felt by many people that the very high cost and complex technology of the other enrichment schemes had granted some measure of control and some protection against fuel proliferation around the world. On the other hand, Krass [23] has analysed the proliferation potential and concludes that without a substantial and unexpected breakthrough, laser separation will remain among the more difficult ways of obtaining nuclear weapon-grade uranium. This is to some extent supported by the above figures showing the scale of expense to be not dissimilar from other schemes and the required technology, as far as can be established, to be on an even higher level of sophistication than that of the existing facilities. These conclusions may change, however, as further development proceeds.

References

1. Marling, J. B., Simpson, J. R., Miller, M. M. and Bauer, S. H., 'Separation of

hydrogen isotopes', ed. H. K. Rae, *ACS Symposium Series No. 68* (American Chemical Society, Washington, D.C., 1978), pp. 134 and 152.

2. Spindel, W., 'Isotopes and chemical principles', ed. P. A. Rock, *ACS Symposium Series No. 11* (American Chemical Society, Washington, D.C., 1975).

3. Benedict, M. and Pigford, T. H., *Nuclear Chemical Engineering* (McGraw Hill, New York, 1957).

4. Gunning, H. E. and Strausz, O. P. *Advances in Photochemistry*, Vol. 1, No. 209, 1973.

5. Kuhn, W. and Martin, H., *Naturwissenschaften*, Vol. 20, No. 772, 1932.

6. Physik, J., *Chemical Abstracts*, Vol. B 21, No. 93, 1933.

7. Schmidt, C. F., Reevers, R. R., and Harteck, P., *Ber. Bunsenges. Physical Chemistry*, Vol. 72, No. 129, 1968.

8. Dunn, O., Harteck, S. and Dondes, S., *Journal of Physical Chemistry*, Vol. 77, No. 878, 1973.

9. Catalano, E., Barletta, R. and Pearson, R., Paper presented at the Tenth International Quantum Electronics Conference, Atlanta, 1978.

10. Letokhov, V. S. and Moore, C. B., *Chemical and Biochemical Applications of Lasers*, Vol. III, ed. C. B. Moore (Academic Press, New York, 1977).

11. *Physics of Quantum Electronics*, Laser photochemistry, tunable lasers and other topics, Vol. 4, ed. S. F. Jacobs, M. Sargent III, M. O. Scully and C. T. Walker (Addison Wesley, 1973).
    (a) —, Aldridge, U. P., Birely, J. H., Cantrell, C. D. and Cartwright, D. C.
    (b) —, Rodes, G. W.

12. Ambartzumian, R. V. and Letokhov, V. S., Paper presented at the Tenth International Quantum Electronics Conference, Atlanta, 1978.

13. *Tunable Lasers and Applications*, Springer Series in Optical Sciences, Vol. 3, ed. A. Mooradian, T. Jaeger and P. Stokseth (Springer Verlag, Berlin, 1976).
    (a) —, Rockwood, S. D., pp. 140-49.

14. *Laser Spectroscopy III*, Springer Series in Optical Sciences, Vol. 7, ed. J. L. Hall and J. L. Carlsten (Springer Verlag, Berlin, Heidelberg, New York, 1977).

15. Brunner, F. and Proch, D., *Journal of Chemical Physics*, Vol. 68, No. 4936, 1978.

16. Fuss, W., *Annual Report of the MPG Projektgruppe für Laserforschung*, Garching, Munich, 1977.

17. Jensen, R. J., Marinuzzi, J. G., Robinson, C. P. and Rockwood, S. D., *Laser Focus*, Vol. 12(5), No. 51, 1976.

18. Svelto, H., 'High power lasers and their applications', *Proceedings of the summer-school 'Enrico Fermi'*, Varenna 1978 (North Holland Publications, 1978).

19. Wittig, C., Paper presented at the Tenth International Quantum Electronics Conference, Atlanta, 1978.

20. *Nuclear Industry*, February 1978.

21. Statement of G. W. Cunningham, DoE Program Director, before the Subcommittee on Fossil and Nuclear Research, Development and Administration, US Congress, 8 February 1978.

22. Casper, B. M., *Bulletin of Atomic Scientists*, January 1977, p. 29.

23. Krass, A. S., *Science*, Vol. 196, No. 721, 1977.

Chapter 3. Reprocessing

Paper 5. Proliferation risks associated with different back-end fuel cycles for light water reactors

K. HANNERZ and F. SEGERBERG

Square-bracketed numbers, thus [1], refer to the list of references on page 103.

I. Introduction

The question of whether or not to reprocess spent fuel from light water reactors (LWRs) has hitherto dominated the current debate on the risks of nuclear weapon proliferation. This is unfortunate because it has tended to obscure other ways and means of proliferation which represent a greater risk, but which may be unrelated to the question of nuclear fuel reprocessing or indeed to the question of whether or not to utilize nuclear power in the first place.

Those non-nuclear weapon states which aspire to manufacture nuclear weapons now, or which may later have such aspirations, will not be prevented by technical or economic difficulties which make the availability of an existing LWR fuel cycle an attractive alternative. This is so regardless of the way in which the LWR fuel cycle is implemented, that is, whether or not plutonium recycling takes place.

In assessing the problems of proliferation, there is a clear tendency among politicians to concentrate on such tangible issues as the question of reprocessing. This creates the erroneous impression that a decision on whether or not to reprocess will significantly affect the risks of proliferation as a whole.

This is not to imply that the question of reprocessing is considered to be negligible with regard to proliferation. However, to place it in its correct risk category, the following facts should be borne in mind.

1. The technological know-how needed to produce fissile material is openly available to anybody who wishes to use it.

2. From a technical standpoint, the easiest among a large number of available methods is to build an unsophisticated plutonium-producing reactor and a 'dirty' facility for separating the plutonium. This strategy can be pursued irrespective of whether or not the nation in question has a peaceful nuclear power programme.

3. An LWR programme represents an extremely unattractive means of obtaining fissile material, from the points of view of technical obstacles,

economic sacrifice, susceptibility to sanctions from other nations and avoiding detection.

The frightening truth is that even now there are no technical or resource-related obstacles to the production of fissile material. The secret is already out. Mankind will therefore have to rely on political means of eliminating the need for nuclear weapons and of simultaneously raising the political price of nuclear weapon production to a level which will deter those nations currently not possessing these weapons from acquiring them.

In this paper an assessment of various fuel cycles in relation to their proliferation risks will be made. It must be borne in mind, however, that even though one fuel cycle may represent a greater risk than another, the margin of these risks may be relatively very low. In other words and even at the risk of deterring the reader, the aim of this paper is to elucidate one of the less significant aspects of the question of horizontal proliferation of nuclear weapons.

II. The problem of defining proliferation risk levels

Even a cursory analysis of the last stage (back-end) of the LWR fuel cycle shows that the proliferation risks associated with the 'once-through' and reprocessing cycles cannot be determined in a straightforward way. An evaluation of the following factors, among others, will be crucial to the assessment: (a) the efficiency of the safeguards system; (b) the possibilities for sanctions; (c) the significance of warning time; and (d) the so-called technology factor, that is, how fast competitive technologies are being developed within the time perspective considered. Thus, for example, final disposal of non-reprocessed spent fuel may seem to be the safest alternative, provided that there is an efficient safeguards system. On the other hand, while an efficient safeguards system might be able to operate among the relatively few reprocessing facilities that will come into production before the turn of the century, if one assumes a large number of reactors and central storage facilities for spent fuel scattered all over the world, then it is likely that in the long run, safeguards-system efficiency would be inadequate. In this case, large-scale plutonium reprocessing and recycling must be accomplished as soon as possible in order to limit the size of the scattered plutonium inventory in spent fuel assemblies.

A number of technological considerations might also be in place. We emphasized above that the back-end of the LWR fuel cycle technically is even now an inconceivable means of proliferation. Less sophisticated techniques for producing nuclear weapons are already available and, as time goes on, this is likely to be even more the case. The development of enrichment technologies is especially important. Many experts regard the possibility of a large number of currently non-nuclear weapon countries having the means to make nuclear weapons based on uranium-235 by the turn of the century as entirely plausible. Such a development would eliminate the need for irradiation and thus for reactors and reprocessing. Interest in the production of plutonium would, in other words, be significantly reduced. This is especially true as regards reactor plutonium.

III. Common assumptions

Only those countries generating nuclear power in light water reactors but without domestic fuel enrichment or reprocessing plants are considered in this paper. Such nations have basically two options for the back-end of the fuel cycle. Either they can reprocess their spent fuel at a foreign reprocessing plant (in a nuclear weapon state, in another non-nuclear weapon state or, in the future, at an international fuel cycle centre) and re-use the recovered uranium and plutonium, or they can employ the once-through cycle and consider the spent fuel as waste to be disposed of. The purpose of this paper is to compare the proliferation risks associated with these two alternatives.

A nation not yet using nuclear power based on LWRs but nursing long-range plans for acquiring nuclear weapons would do better to forgo the former altogether, since these would only increase its susceptibility to international sanctions, while not significantly contributing to the procurement of weapon-grade fissionable material. For this reason, we assume that the nations under consideration started their nuclear power programmes with peaceful intentions only, and that their interest in acquiring weapons appeared later when their civilian programmes were already well under way, as a result of a change in circumstances, such as external threat or change of government. The question to be answered is: with which of the two alternatives for the organization of the back-end of the fuel cycle can the procurement of reactor plutonium for the production of (crude) weapons be made most easily, in terms of time, resource commitment, and so on.

The following general assumptions are made, pertaining to both alternatives:

1. The country in question is subject to 'comprehensive inspection' involving all facilities in which fissile material is handled. Resident inspectors can be placed where this is considered essential by the responsible body (probably the International Atomic Energy Agency (IAEA)), and their access to the respective plant is not restricted by commercial secrecy or other reasons.

2. The supply of fuel (such as that provided by enrichment services)

for the LWRs is subject to the condition that the receiving country is not engaged in nuclear weapon procurement in any way.

3. A certain amount (such as one year's supply) of spare enriched uranium fuel is on hand, that is, enough for one year's operation. This is already established practice, for example, in Switzerland, and is probably a minimum requirement from the supply security point of view, although it delays the effect of fuel supply sanctions from the outside.

The various stages of the two alternative fuel cycles will now be compared from the following points of view: (a) the probability of detection of clandestine activities; (b) the time required from a decision to go ahead to the availability of weapon-grade material for the first charge and for a small weapon programme (involving, for example, tens of charges); (c) the time required from the (probable) moment of detection to the availability of weapon-grade material for the first charge; (d) the resource allocation needed in terms of financial commitment; (e) the level of technology necessary for implementation; and (f) the consequences of the application of international sanctions.

IV.   The reprocessing route

The various steps of the fuel cycle are depicted in figure 5.1. It contains only well-known processes and plants, and needs no further description. The reprocessing route can be subdivided into three alternatives as shown in the figure. The first alternative consists of fabrication of the plutonium recycle fuel in the country owning the reactor. This would be the normal alternative, at least when the volume of business has grown to a sufficient level, and only non-proliferation arguments speak against it at that stage. This alternative obviously involves transport of unfabricated plutonium oxide. Although pure PuO_2 has mainly been considered hitherto, there seems to be little reason for not mixing it with UO_2, thereby creating at least a certain obstacle to fabrication of an explosive device. This is assumed here. In the second alternative, fabrication takes place at the reprocessing site and only the finished fuel assemblies are transported. Finally, in the third alternative, a further barrier to proliferation is created by 'spiking' the fuel assemblies with gamma activity. This could be done, for example, by exposing the fuel to a short irradiation in a special critical assembly at the reprocessing site, and serves the purpose of making the fuel virtually inaccessible to potential subnational groups of hijackers.

Turning to the reactor and reactor pool, there is general agreement that the potential for proliferation is negligible. All the facilities in a reprocessing plant with the exception of uranium purification would have to be available for extraction of the plutonium. The same applies to the transport

94

Figure 5.1.   The reprocessing route

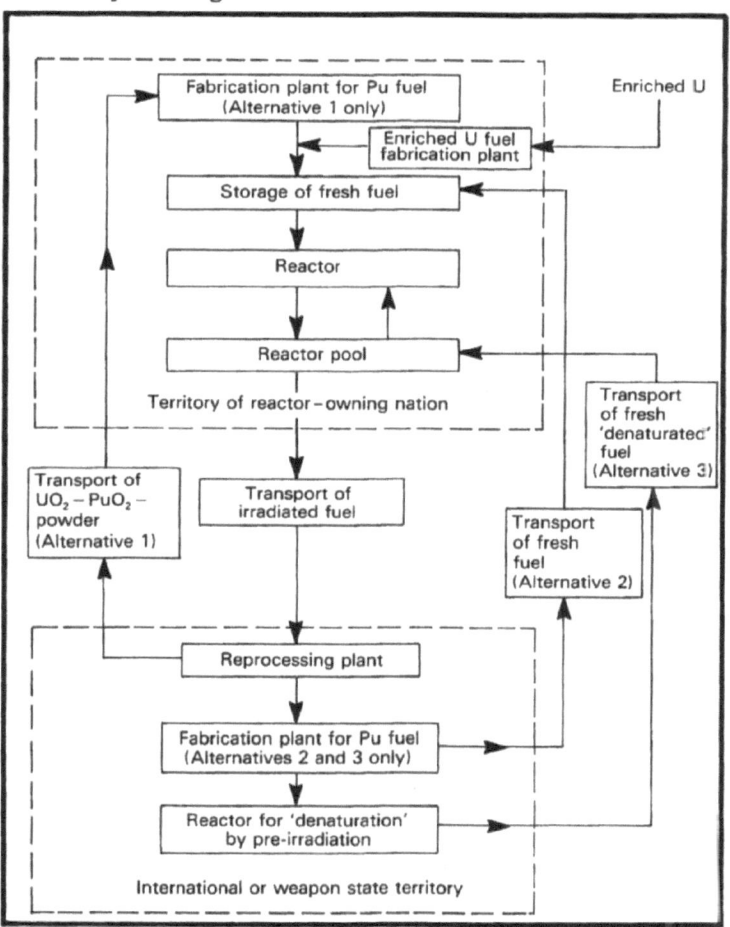

of the irradiated fuel to the reprocessing plant. In addition, reprocessing plants and other facilities on the same territory need not be considered, since we assume that they are owned by a nuclear weapon state or are otherwise under international control.

As regards the fate of the plutonium on its way from the reprocessing plant to the reactor, it is necessary to discuss the three alternatives separately.

In the international transport of the uranium–plutonium oxide mixture in alternative 1, hijacking by subnational groups has been the main consideration. However, provided there is adequate security, such dramatic exploits are highly unlikely. What matters here is mainly the cost of providing this security, which will burden the reprocessing alternative in general.

As mentioned above, it is assumed that international inspection of the fabrication plant producing plutonium enriched fuel takes place. Nothing short of a resident inspector will suffice in this case, which may obviously create problems in connection with proprietary commercial know-how. However, with such an inspector present, the chances of succeeding with a systematic diversion of plutonium (by doctoring material balances, smuggling past activity monitors, and so on) will be virtually nil, since it would be practically impossible to prevent leakage of information, unless the entire personnel consisted of trained secret service agents. (This is not to be construed as a statement that inspection procedures currently in force are adequate.)

Once the fuel is fabricated, it is presumably placed in the fresh fuel storage room of the reactor where it is to be used. At this stage, undiscovered diversion could take place through the exchange of a plutonium-containing fuel assembly with a uranium-containing one from the spare fuel inventory and the replacement of the latter with a dummy, all three assemblies being identical in outward physical appearance. However, assuming frequent checks by an inspector provided with adequate gamma-monitoring equipment, such attempts at diversion could not go undetected.

Having discussed all the stages of the fuel cycle (in alternative 1), it can be seen that the risks of undetected diversion are more or less negligible. In order to procure nuclear weapon material, overt seizure of the plutonium-containing fuel for this purpose would be necessary, which would lead to immediate discovery and implementation of international sanctions.

The plant necessary for the dissolution of the fuel in nitric acid, precipitation of plutonium as oxalate and calcination to obtain oxide is quite small and simple due to the absence of gamma-shielding arrangements. It could probably be built in a few months following a decision to proceed and would be sufficiently inconspicuous for clandestine construction and test operation to be possible. Costs would be small in the present context and the technology required does not exceed that required for handling of many toxic chemicals (except that the physicists must cater for criticality safety).

Having clandestinely installed such a plant, the time needed from seizure of the PuO_2-UO_2 fuel material to the availability of pure calcined PuO_2 is at most a few days. If the intention is to produce an explosive employing this material (presumably with a penalty in yield compared to a metal device), the first explosion could follow very shortly afterwards. (The problems and time requirements involved in producing and fabricating plutonium metal are outside the scope of this paper.) The effect of sanctions would be to limit the further generation of nuclear energy to that available from the spare fuel in storage.

Alternative 2 differs from that considered above in that the fuel arrives at the user country in a fabricated form. The risk of diversion in transport is roughly the same (negligible), while that at the fabrication plant (equally small) disappears. Otherwise there is little difference between the two alternatives, since the extra work required for cutting up and emptying the fuel-cladding tubes cannot take much time.

In alternative 3, on the other hand, handling of the fresh fuel will be roughly similar to that of spent fuel, that is, storage under water and transport in shielded casks, presumably identical to those used for spent fuel. In this case, production of clandestine weapon material would, of course, be even more difficult to effect, not only because of the more cumbersome fuel handling but also because of the much more extensive plant necessary for extraction of the plutonium.

This now becomes a reprocessing plant, admittedly somewhat simplified because of relatively low content of long-lived gamma emitters, but nevertheless extensive enough for the possibility of clandestinely constructing and test-operating it to be virtually disregarded. Hence the use of 'spiked' (that is, pre-irradiated) fuel produced at the main reprocessing plant will result in a much longer 'warning time' and is therefore to be preferred from the non-proliferation point of view. However, the costs associated with the use of 'spiked' fuel have not yet been determined and may be high enough to make this alternative unattractive. It is to be hoped that more information on this subject will accrue in the near future, for example, through the International Nuclear Fuel Cycle Evaluation (INFCE) programme.

V.   The once-through cycle

The steps involved in the once-through cycle as currently perceived are depicted in figure 5.2. The reprocessing cycle has always been the preferred alternative, until recently assumed as the logical one, and the plants and processes involved have been well described in the literature. This is not the case with the final stage of the once-through cycle, that is, the long-term pool storage needed for reduction of the decay-heat generation rate and the final disposal of the spent fuel. But before embarking on a discussion of the proliferation risks involved at the various stages, a few words may be appropriate about the way in which the practical implementation of these two stages is envisaged in Sweden.

Long-term pool storage

A review of the long-term behaviour of irradiated zircalloy-clad fuel during low-temperature pool storage has revealed that very little change takes place due to corrosion or other phenomena [1]. There is no question but that several decades of storage without significant degradation can be undertaken. In the proposal for direct disposal of spent fuel worked out by the KBS organization[1] during 1977−78, a pool-storage time of 40 years prior to

---

[1] Kärnbränslesärkerhet (KBS)—Nuclear Fuel Safety—was the *ad hoc* organization set up by the Swedish authorities in December 1976 to show that it was possible to undertake final disposal of nuclear waste without risk to the public, as stipulated in a special proposal by the newly elected government in October 1976 and adopted by the Swedish Parliament in early 1977.

Figure 5.2. The once-through cycle

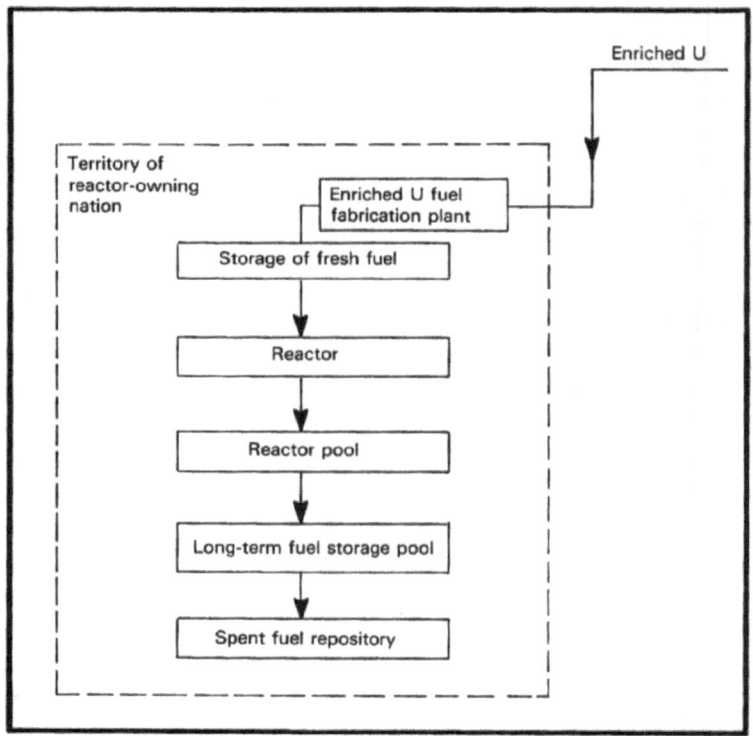

final disposal is assumed. Following initial cooling for a few years in the reactor pool, the fuel assemblies are transferred to a central fuel storage facility. This facility consists essentially of a series of storage pools with associated systems for reception of fuel, cask handling, water cooling and purification, and so on. The currently proposed facility is depicted in figure 5.3. Most of the plant, particularly the storage pools, is located underground, blasted out of the high-grade rock so common in Sweden.

A Swedish fuel storage facility is needed irrespective of whether reprocessing takes place in Sweden or not. At the moment, Sweden has no reprocessing capacity and there is therefore a need for indigenous fuel storage. Construction of a fuel storage pool is expected to start within the near future. From the surveillance point of view, the location of all fuel underground in a common pool area will make undetected diversion extremely difficult, if not impossible, if adequate inspection is in force. This is equally true in the case of reactor pools.

Final disposal

While the vitrified high-level waste emerging as the final output of a reprocessing plant is of little interest from the weapon-proliferation point of

Figure 5.3. The Swedish central spent fuel storage facility, CLAB (Centrallager for Använt Bränsle)

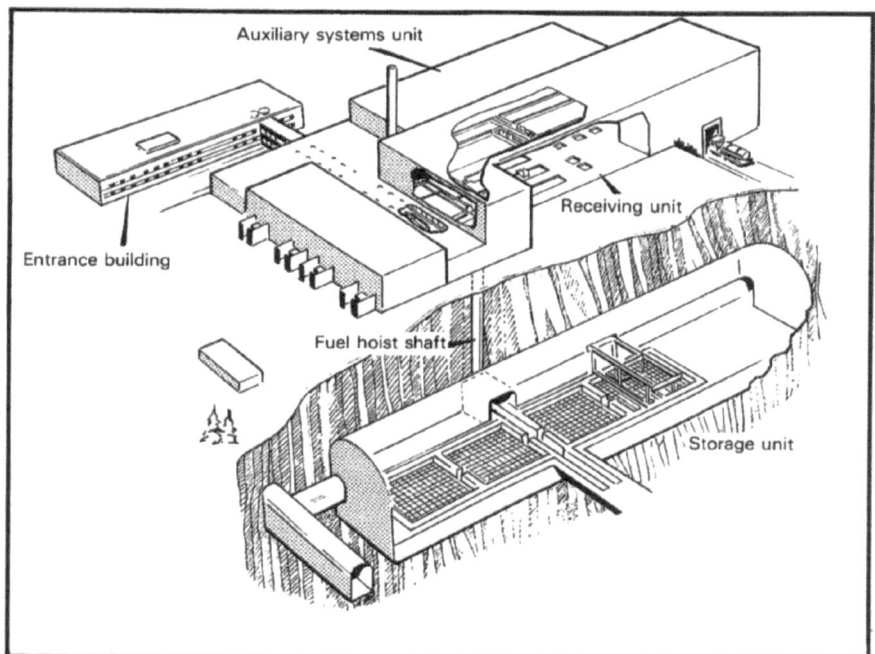

Auxiliary systems unit

Receiving unit

Entrance building

Fuel hoist shaft

Storage unit

view, the fate of the finally disposed spent fuel in the once-through cycle must of course be considered.

In the KBS proposal for final disposal [2] (which incidentally appears to be the first concrete proposal for direct disposal that has been published), the spent fuel, in the form of fuel rods removed from their assemblies but not otherwise processed, are packaged in the thick-walled copper containers. These are deposited in holes at the bottom of tunnels blasted out of high-grade rock at 500 m depth. The deposition holes are filled with a specially conditioned high-density clay, and the tunnels and vertical shafts are completely filled with a clay–sand mixture at the final shut-down of the facility.

The idea behind this scheme is based on the immunity of copper against attack by water (ground water will always surround the waste, since the repository is far below the water table). Copper corrosion can only be caused by certain impurities present in minute quantities in the ground water. By surrounding the thick-walled containers with a zone of stable clay of extremely low permeability and diffusivity, corrosion can be made so slow that a canister life of a million years can be assured. An artist's conception of the repository is shown in figure 5.4.

Proliferation risks

The proliferation risks associated with a widespread use of the once-through

Figure 5.4. Spent fuel repository

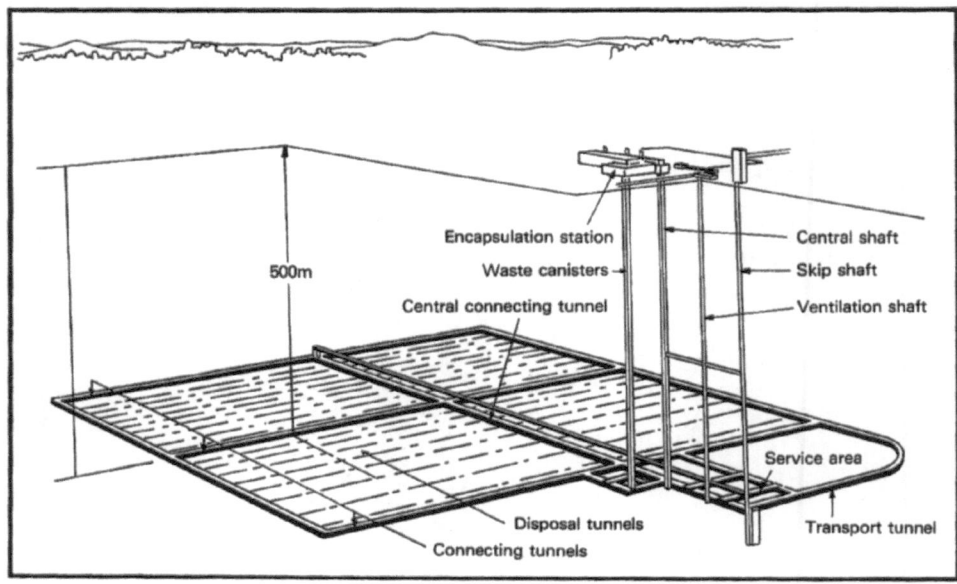

cycle have been discussed at length by Marshall [3]. His main argument is that the radionuclide decay taking place in extended pool storage gradually reduces the difficulties associated with reprocessing, so that the fuel passes from a state that Marshall terms "inaccessible" to one called "extractable".

To make such claims sustainable, really major differences must exist, in terms of resource commitment, construction time and technological level required, between a reprocessing plant (designed for plutonium recovery only) needed to handle fuel of one to two years' cooling time (as used in the reprocessing alternative) and a corresponding plant for fuel cooled for one to several decades. ASEA−ATOM has commissioned a study to determine whether this is indeed the case [4]. The answer is largely negative, that is, that there is, of course, a difference, but that it is only a matter of a few tens of per cents. This is because the amount of shielding needed does not significantly decrease due to the persistence of the main gamma emitter caesium-137. Off-gas problems are not a major factor once iodine-131 has been allowed to decay. In short, it seems to us that Marshall has over-emphasized the proliferation potential of spent fuel stored in pools during longer periods of time.

In practice, even after, say, 40 years of storing the fuel, the plant needed to extract plutonium from it is so extensive and complicated that the procurement of components and its construction could not possibly fail to attract attention, if an adequate inspection system were in force. Further-more, such a construction would take several years even if it enjoyed the highest priority. Consequently, the once-through fuel cycle would guarantee an adequate warning time for diplomatic and other manoeuvres before weapon proliferation became a fact.

As mentioned above, clandestine diversion of fuel from a central storage facility such as that envisaged for Sweden is a virtual impossibility. The practical importance of this may be limited, since again, the warning signal would have already been received from the construction of the reprocessing facility. The underground disposal facility described above, should it ever be implemented, would of course itself guarantee a long warning time so long as surveillance of mining activities is performed. The latter should be a relatively simple and inexpensive activity. In any case, the time perspective of the whole proliferation issue is such that a spent fuel repository, closed down sometime in the latter part of the 21st century, would present a most unlikely route to weapon procurement. Considering the 'operating time' of the repository, the check of transferring fuel up, down and through the main shafts is no more difficult than the corresponding activities at the long-term storage facility. Transport of fuel 'on the road' between the various facilities would have to be surveyed, as would transport from the reactor site to the reprocessing plant site in the reprocessing alternative. As mentioned previously, the first warning of intent of weapon procurement would in any case come from the construction of the reprocessing facility.

Thus it can be seen that the once-through cycle, if implemented as originally intended, presents very little proliferation risk. The problem is that such an implementation is in practice very unlikely to occur. Having decades of spent fuel readily available as an energy source will, in the long run, constitute an irresistible temptation to use it as an energy source, since all the required technology is at hand. The opposite could be true if other energy technologies, such as fusion, make such unexpected progress that fission energy becomes uneconomical. Thus a policy involving the abolition of reprocessing on the lines of the Carter Administration's proposals will cause a pent-up need for reprocessing which is likely to lead to the construction of a number of national reprocessing plants. The plant owners will have low-grade weapon material available as a normal part of their operations, and the vast amounts of plutonium available in the pools will make them virtually immune to sanctions involving a break in enriched uranium supply.

While, therefore, the once-through cycle may in the long term have some disadvantages from the proliferation point of view, it can nevertheless be a perfectly satisfactory alternative for an interim period, during which the institutional framework for a proliferation-proof reprocessing cycle is established.

VI. Summary and conclusions

The subject matter of this paper is a comparison of the proliferation risks associated with two different alternatives for the back-end of the fuel cycle for LWRs in countries not possessing their own enrichment and reprocessing facilities. One of the alternatives involves reprocessing of the spent fuel in a foreign plant (for example, in a nuclear weapon state) and re-use of the recovered uranium and plutonium in LWRs. The other alternative is the once-through fuel cycle, in which the spent fuel is regarded as waste for final geological disposal following long-term (20−40 years) pool storage for initial decay.

The principal proliferation risk associated with the former alternative involves diversion of reload fuel delivered from the reprocessing plant (or mixed oxide intended for such fuel), separation of pure PuO_2 and fabrication of a 'crude' weapon from this (with or without conversion to the metal form). Particularly if the explosive device is made from the oxide (at the expense of yield), the time between diversion, which can be assumed to be discovered immediately, and a test explosion can be quite short. However, if the fuel is 'spiked' with gamma activity before it leaves the combined reprocessing−refabrication plant, for example, by pre-irradiation, the time between discovery of intent of diversion and a test explosion can be considerably extended. Clandestine weapon-material production can be virtually excluded if there is an effective inspection system so that international sanctions involving at least a break in fuel supplied for nuclear power plants can be assumed to be effected immediately following discovery of intent of weapon procurement. The total amount of plutonium involved will be limited to that present in one, or at the most a few, reload charges for the reactors concerned, but may nevertheless suffice for tens of charges.

Using spent fuel from LWRs after a storage time of up to several decades as a source of weapon material requires an investment in sizeable reprocessing facilities. Although the long storage time does facilitate the task of reprocessing, the difference in technology and plant investment compared to processing fuel of one to two years' storage time is not really a major one. The construction of the required facilities has a negligible chance of passing unnoticed in a country subject to international inspection of its nuclear facilities. For this reason the 'warning time' available for the outer world for diplomatic and other actions is much longer.

On the other hand, if the once-through cycle is actually adopted on a wide scale and for a long period due to the unavailability of large-scale reprocessing plants for international service (resulting from decisions by the nuclear weapon states or international bodies), then rising energy prices are likely eventually to become an overwhelming argument for building national reprocessing plants on the basis of peaceful needs alone. With these in place, the amount of weapon material available to the potential new nuclear weapon states would greatly increase, and/or the effect of cutting off nuclear fuel supplies would take much longer to be felt.

On balance, the alternative involving reprocessing, refabrication and 'denaturation' by pre-irradiation (for at most a few per cent of the final burn-up) before dispatch to the power reactor appears to be the most promising from the non-proliferation point of view. It would guarantee a long 'warning time' because of the extensive long lead-time plant needed for producing the weapon material while at the same time avoiding the build-up of large stocks of spent fuel in pools with the longer-term risk of a proliferation of reprocessing plants built for peaceful purposes. The economic and political problems involved in this alternative should be further examined.

However, the once-through cycle may be a perfectly satisfactory interim solution for a couple of decades until an international agreement has been reached on how to operate a proliferation-resistant system for recovery and re-use of the fissile content of spent fuel.

In the long run, a system must be found in which all production and refabrication of fissile material potentially usable for weapons are carried out on a truly international basis. To achieve this, and at the same time preserve adequate supply, security and technological status among the participating nations may be the principal institutional problem in connection with the future peaceful utilization of nuclear energy. One can only hope that the efforts of the INFCE will be focused on these tasks and not dispersed on futile efforts to find technological 'fixes' that stand little chance of practical implementation.

That these problems will, no doubt, be subject to prolonged deliberation and much publicity in the mass media, should not be allowed to obscure the fact that civilian nuclear power makes a negligible contribution to the risks of nuclear war and, in fact, may even decrease those risks.

References

1. Vesterlund, G. and Olsson, T., *Degradation Mechanisms in Pool Storage and Handling of Spent Power Reactor Fuel*, KBS Technical Report No. 68, ASEA-ATOM, January 1978. (In Swedish.)
2. Kärnbränslesäkerhet (KBS), Stockholm, Sweden, *Final Disposal of Spent Fuel*, KBS Main Report No. 2, June 1978. (Available in English translation late 1978.)
3. Marshall, W., 'Nuclear power and the proliferation issue', *Atom*, No. 258, April 1978, pp. 78–102.
4. Eschrich, H., *Influence of the Cooling Time of Power Reactor Fuel on the Recovery of Plutonium Contained in it*, Technical Report to ASEA-ATOM, July 1978.

Chapter 4. Waste disposal

Paper 6. Reprocessing and waste management

D. ABRAHAMSON

Square-bracketed numbers, thus [1], *refer to the list of references on page 111.*

I. Introduction

The presence of the fissile materials plutonium-239 or uranium-233 produced in reactors which were either ostensibly or actually operated for the production of electricity is one factor contributing to increased risk of the proliferation of nuclear weapons. Were there no reprocessing of the spent fuel from such reactors, this special nuclear material would not be available for at least the several hundred years during which the radioactivity of the fission products effectively prevents access to the spent fuel.

Whether or not there is reprocessing of the spent fuel, a method for responsible short- and long-term management of the radioactive wastes must be developed and put into operation. The decision to reprocess is dependent on complex economic and political factors, of which the implication for waste management strategy is but one.

It is not the purpose of this paper to present the general case for or against spent fuel reprocessing. Except for some general comments on the reprocessing decision, the only questions explored are whether reprocessing is necessary or is desirable from the standpoint of waste management.

II. Is reprocessing necessary from a waste management standpoint?

It is sometimes suggested either that waste management considerations demand reprocessing, or that reprocessing would in some way simplify or improve the safety of waste management. Two allegations are frequently advanced to support the proposition that reprocessing is necessary from the standpoint of waste management.

The first is that since it has always been assumed that there would be breeder reactors, which require reprocessing, there has been no research and development on management of the spent fuel itself as the final waste form. Therefore, the reasoning goes, reprocessing must be done as only methods for dealing with the wastes from reprocessing have been developed. This argument might appear circular, and in view of recent research and proposals for long-term permanent storage of spent fuel, is now based on error.

The second argument, recently discussed by Walter Marshall and others, is that were the unreprocessed spent fuel to be placed in a repository, that repository would become, after the several hundreds of years required for fission product decay, a potential plutonium mine. Dr Marshall argues that this is an unacceptable hazard. Others have expressed concern that the proliferation hazard which would result were the plutonium isolated by reprocessing is unacceptably high.

In effect, Dr Marshall's argument is that a plutonium mine would be more dangerous than a plutonium river, and that it would be more difficult to safeguard a small number of potential plutonium mines than it would be to safeguard a very large number of access points to the plutonium river.

A rigorous examination of these safeguards implications would be helpful in making the necessary policy decisions. At first glance it seems to be at least as likely that there are safeguards problems with the river as there would be with the mine. If this turns out to be the case, and it is accepted that safeguards considerations in the commercial fuel cycle are of genuine concern, then it appears that the difficulty in safeguarding a potential plutonium mine is a convincing reason not why there should be reprocessing but rather why there should not be reactors, breeders or otherwise.

III. Reprocessing would not simplify waste management

It has been asserted that reprocessing would simplify or make safer the management of the wastes even though the purpose of reprocessing is only to recover the plutonium. On the contrary, reprocessing appears to complicate waste management and to make it less safe.

Problems posed by actinides in the waste are exacerbated by reprocessing and recycle

As plutonium-239 is one of the most toxic and long-lasting components of the spent fuel, and as reprocessing would theoretically remove 99.9 per cent of the plutonium, the wastes might be less hazardous after reprocessing. There are, however, a number of other highly active long-lasting, alpha-emitting components of the wastes so that removal of plutonium would make no qualitative difference. As the Ford/MITRE report concluded:

"Reprocessing and recycle of plutonium provide a way to reduce the long-term risks by reducing the amounts of transuranic elements in wastes. However, the magnitude of this effect is not large." [1] Rather than separating only the plutonium and uranium, all of the long-lived actinides and fission products might be isolated. Reprocessing, as currently described, does not include this detailed partitioning of the various wastes.

Partitioning is theoretically possible. A recent technical review includes:

If this second conclusion [referring to a discussion of waste storage risks] is confirmed by work during the next few years, then it can no longer be argued that actinide separation and recycling is necessary to ensure the future of nuclear power. Separation and recycling may still be preferable on economic grounds to long-term disposal of actinide-containing waste, although this seems very unlikely. Separation and recycling may also be preferable from the point of view of public acceptability. However, until a great deal more is known about disposal methods on the one hand, and actinide separation [partitioning] and recycling on the other, it is impossible to assess their relative feasibility, cost safety and acceptability. [2]

A related consideration is that, with recycle, the concentration of actinides in the spent fuel would increase by up to an order of magnitude. Therefore, the radioactive wastes with reprocessing and recycle pose a greater long-term hazard than the wastes without recycle. The technical review of the Swedish nuclear industry's 'KBS' proposal for waste management included a quantitative estimate of the possible increased radiation exposure to the public which would result from this build-up of actinides. Those studies indicated that the radiation dose rate to the public could be an order of magnitude higher from radioactive wastes resulting from reprocessing recycled fuel than from radioactive wastes resulting from reprocessing virgin uranium fuel [3].

Reprocessing would complicate waste management

Before reprocessing, all of the wastes are contained in the spent fuel in a solid, rather insoluble form. The result of reprocessing, which is a combination of several mechanical and chemical processes, would be that the radioactivity would no longer be in one form (the spent fuel) but rather would be contained in several waste streams.

After reprocessing, the highly active liquid wastes would have to be resolidified, the gaseous wastes trapped and converted into a form suitable for long-term storage, and the various solid waste streams consolidated and suitably encapsulated.

The October 1977 OECD radioactive waste report concluded: " . . . waste management would be greatly simplified and the main problem would be the safe disposal of the spent fuel [were there no reprocessing]. However, this operation would be similar to the disposal of long-lived reprocessing wastes. . . ." [4] OECD goes on, however, to

Table 6.1. Radioactive waste production in the nuclear fuel cycle

Solid or solidified waste equivalent to the generation of 1 000 MW(e)-year in a light water reactor

Origin and type	Volume (after treatment and conditioning) (cubic metres)	Activity or weight	Main disposal concepts
1. *Uranium mining and milling* Ore tailings	60 000 (40 000 if Pu recycle)	0.01 Ci/m^3	Land burial
2. *Fuel fabrication* UO$_2$ fuels UO$_2$–PuO$_2$ fuels for an annual reload of 500–700 kg of Pu)	– 10–50	Negligible 5–10 kg Pu	Geological formations Sea dumping[b]
1. *Light water reactor* Various solid wastes and conditioned resins	100–500[a]	0.1–10 Ci/m^3 beta–gamma	Sea dumping[b] Land burial at the surface or in deep formations
4. *Reprocessing* Solidified high-level waste (HLW)	3[c]	150 MCi beta–gamma[d] + actinides (2 kg Pu for UO$_2$ fuels, 5–10 kg Pu for UO$_2$–PuO$_2$ fuels)	Geological formations on land or under the ocean floor
Compacted cladding hulls	3	1.5 MCi beta–gamma[d] + actinides	Geological formations on land or under the ocean floor
Low- and medium level beta–gamma solid waste	10–100	0.01 MCi beta–gamma + alpha contamination	Land burial at the surface or in deep formations Sea dumping[b]
Solid and solidified alpha waste	1–10	1–5 kg Pu	Geological formations Sea dumping[b]

[a] Depending on reactor type and conditional process.
[b] This option is available only for those wastes having specific activities within the limits of the London Convention.
[c] From an original volume of liquid waste of about 15m.
[d] 150 days after fuel discharge from the reactor.

assume that the economic value of the plutonium will force the decision to reprocess, even though: "This alternative [reprocessing] leads to a more complicated technical solution as far as radioactive waste management is concerned since many different waste streams are generated at fuel reprocessing plants.''

The Ford/MITRE study reached the same conclusions: "Reprocessing would complicate waste management by broadening the spectrum and the potential difficulty of problems." In addition, "Without reprocessing and recycle, there would be time for a more orderly and assuredly more error-free process in both the management and disposal aspects of waste" [2a]. There are similar conclusions from other studies.

Reprocessing would increase the volume of radioactive wastes

With reprocessing, the total volume of radioactive wastes would be larger than without reprocessing. The volume of the solidified high-level liquid waste from reprocessing would be less than that of the spent fuel from which it was derived, but this is more than offset by the other radioactive waste resulting from reprocessing, from solidification of the liquid waste, from plutonium fuel fabrication, and from decommissioning of the reprocessing, solidification and fuel fabrication facilities.

A fairly detailed examination of these points, and others, is found in the 1977 OECD report, the conclusions of which are partially summarized in table 6.1, reproduced from table 1 of the report.

Occupational exposures to the waste, environmental releases of the waste, and potential for large accidental releases of the waste are likely to be increased by reprocessing

Were there no reprocessing, the spent fuel would probably be stored for an interim period, encapsulated, and then placed in the final repository. Pending development of a satisfactory method for each of these steps, it is difficult to predict with any accuracy the resulting occupational exposures, environmental releases, or potential accidents.

It is known, however, that the occupational exposures and the environmental releases which have resulted from reprocessing are considerable. No accident analysis for reprocessing comparable to, for example, the US Nuclear Regulatory Commission Reactor Safety Study [5] has, so far as I know, been made public. However, the potential for large releases, either through accident or act of malice, from a reprocessing facility is known to be large.

It is likely that the occupational exposures, the environmental releases, and the potential for large unintended releases would be considerably larger with reprocessing than with direct deposition of spent fuel.

Figure 6.1. Projected demand for electrical generating capacity 1970–2020

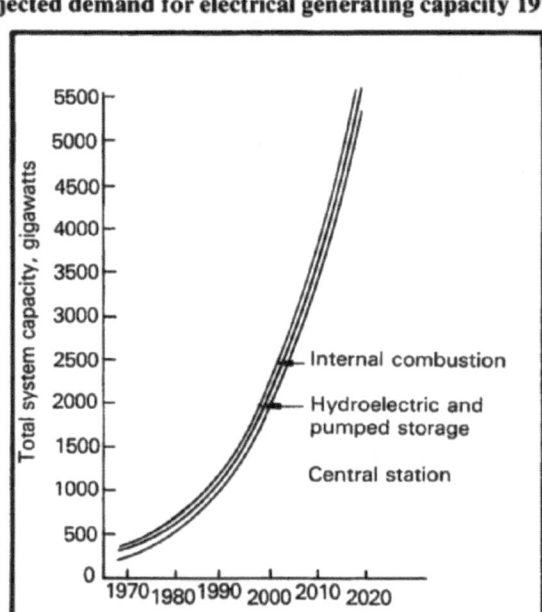

Source: AEC, *Proposed Final Environmental Impact Statement for LMFBR Program,* Vol. 4 (December, 1974), pp. 9.1–e.

IV.   Conclusions

There does not seem to be a compelling case either that reprocessing is necessary or that it is somehow desirable from the standpoint of waste management. However, this does not weigh heavily in the decision as to whether or not to reprocess spent fuel.

The fundamental argument for reprocessing is that, of known energy sources, only breeder or fusion reactors could, even theoretically, supply sufficient commercial energy to satisfy projections based on historical trends. Commercial application of controlled fusion is many decades away from realization. Thus, only the breeder remains as a viable alternative to meet these projected 'needs'.

The logic of the industrialized economies, and of those developing nations which still emulate them, rejects non-nuclear alternatives to the fossil fuels not because solar energy in its various manifestations is not a technical possibility and not because solar energy could not provide more commercial energy than is now being consumed in the world, at a competitive price, but rather because solar energy would not permit energy consumption tens of times greater than that at present. The perceived necessity derives from the dogma that per capita gross national product is a direct measure of welfare and that energy consumption is a linear function of gross national product.

Stark illustrations of this logic are easily found. For example, the president of a major international oil company in 1974 expounded: "Can anybody doubt that increased consumption makes a better way of life?" [6]

The entire justification of the United States breeder reactor program was based on the very high growth projections depicted in figure 6.1. The slopes of these curves are remarkably similar to those circulated at this symposium to support certain assertions which were made about uranium 'needs' and resources.

The logic of the industrialized economies, in both east and west, demands the plutonium economy, which in turn is possible only with reprocessing. Impediments such as possibly hazardous conditions associated with waste or plutonium management have been, are, and will continue to be, brushed aside by the 'experts' as hobgoblins of small minds. The public seems increasingly to have a different perception however. A move from reprocessing and its attendant waste management and proliferation hazards will require a change in the principle which guides our system from "the greatest per capita product for the greatest number of persons" to "a sufficient per capita product for the greatest number which is sustainable over time". It will not be an incremental change.

References

1. S. M. Kenny, Jr., *et al.*, *Nuclear Power Issues and Choices* (Ballinger Publishing Co., Cambridge, Massachusetts, 1977), (often called the "Ford/MITRE Report"), p. 249.
2. H. A. C. McKay, *Separation and Recycling of Actinides*, Report No. EUR 5801, 1977, pp. 67−68.
   (a) —, p. 267.
3. T. B. Johansson and Steen, P., *Radioactive Waste from Nuclear Power Plants: Facing the Ringhals-3 Decision*, Swedish Industry Department Report DsI 1978:36, September 1978. (Available upon request from Swedish Industry Department, Fack, S-103 10 Stockholm.)
4. OECD, Nuclear Energy Agency, *Objectives, Concepts and Strategies for the Management of Radioactive Waste Arising from Nuclear Power Programmes* (OECD, Paris, September 1977), p. 107.
5. U.S. Nuclear Regulatory Commission, *Reactor Safety Study*, Document NUREG-75/014 or WASH-1400, October 1975.
6. W. Tavoulareas, President of Mobil, National Press Club interview (Washington, D.C., 12 September 1974).

Chapter 5. Physical barriers to proliferation

Paper 7. Can plutonium be made weapon-proof?

B. T. FELD

Square-bracketed numbers, thus [1], *refer to the list of references on page 119.*

The long-term future of nuclear fission energy is still uncertain. However, whatever the objective economic and technical realities may be—even if reasonable people could agree upon them—most nations of the world are behaving today as though nuclear power is the foundation upon which their future well-being must be based. Steady economic growth remains the common goal of the developed and underdeveloped worlds; and this growth continues, in the common view, to be founded on the steady growth of energy consumption as the key to increased industrialization and the expansion of the gross national product.

Everybody recognizes that the world supply of accessible fossil fuels is bound to run out sometime in the foreseeable future—whether the prediction is sooner or later depends on some combination of one's vested practical and intellectual interests. Among the possible replacements, only nuclear fission energy is at this time generally believed to have reasonable potential for permitting the perpetuation of current life-styles and growth expectations. As a consequence, in spite of active 'ecological' movements in a number of countries of the industrialized West (here taken to include Australia and Japan), it seems to many observers as practically inevitable that nuclear fission energy will be increasingly encouraged and pursued throughout the world in the remainder of this century, and probably considerably beyond.

Furthermore, whether true or not, the preponderance of 'informed' opinion in the industrialized nations is that accessible supplies of uranium are insufficient (or, at least, its distribution insufficiently homogeneous) to avoid the need for the conversion of the non-fissionable bulk (the heavy U-238 isotope) into fissionable plutonium—that is, plutonium reprocessing for now and, eventually, plutonium breeding for the future.

The belated acknowledgement (it has long been known, but seldom discussed) of the dangers, in a widespread plutonium energy economy, of the proliferation of nuclear weapons—both to governments and to non-governmental groups—has lately led to an international re-examination of the possibilities either for alternative fuel cycles (presumably 'proliferation-proof') or of international control measures aimed at safeguarding plutonium stocks by a variety of means.

There is no doubt that appropriate reactor fuel cycles can be devised which are both technically and economically feasible (perhaps even favourable) and which could avoid, for at least an indefinite time into the future, the plutonium proliferation problem. A number of such prospects, ranging from improved once-through fuel cycles to various U-233 breeding—denaturing schemes, have already been examined closely enough to indicate their plausibility as candidates for further study. However, unless there is a drastic change in the international climate, widespread plutonium reprocessing seems to be extremely imminent—whatever the conclusions of the ongoing international studies.

The sad fact is that, with the (still uncertain) exception of reprocessing and plutonium breeding in the United States, not one of the civilian nuclear powers has in any way slowed down any current programme for plutonium reprocessing and/or breeding. Japan and the UK have both decided to go ahead with the construction of reprocessing facilities (Windscale); the French (Phénix) breeder programme is, if anything, being pursued with even greater vigour; the Brazilian–West German deal remains confirmed; India is receiving its contracted enriched uranium from the USA despite its continuing refusal to place elements of its system under international safeguards; and the socialist countries continue to advocate plutonium breeding as though it represented an ideological commitment.

In the circumstances, those who continue to be driven by the non-proliferation ideal are being forced into the last resort of the desperate 'arms-controller'—the search for some technological gimmick that, in this case, might permit plutonium proliferation for use in reactors without the danger of plutonium bombs. In the discussion below, some of these current schemes will be considered, and an evaluation given of their prospects for success.

Many of these techniques can be classified under the heading of denaturing—that is, the addition to the plutonium of some material that will render the combination incapable of being fabricated into an explosive device. The prototype example of effective denaturing is demonstrated in figure 7.1, in which is shown the critical mass of mixtures of U-235 and U-238 as a function of the percentage of U-235,[1] in the most favourable situation for a U-235 explosion. (The plutonium curve will be discussed later.)

The denaturing effect would work equally well if the U-235 were mixed with, say, lead instead of ordinary uranium (that is, U-238). However, such denaturing would buy us very little protection, since it is no great technical

---

[1] The critical mass is the minimum mass of fissionable material (U-235 in this case) capable of sustaining an explosive chain reaction. Note that the denaturing effect of U-238 on the other fissionable isotope of uranium, U-233, which can be 'bred' in nuclear reactors by neutron absorption in thorium, is very similar. Of course, concentrations of U-235 well below 20–30 per cent (even down to the natural concentration of 0.7 per cent) are capable of sustaining a chain reaction in appropriately designed nuclear reactors, but these cannot be used as explosive devices. Hence, any mixture of U-235 (U-233) and U-238 can be considered effectively denatured if the concentration of the light isotope is less than, say, 20–30 per cent (to choose an arbitrarily conservative number).

Figure 7.1. The critical masses of uranium and plutonium as functions of fissile content

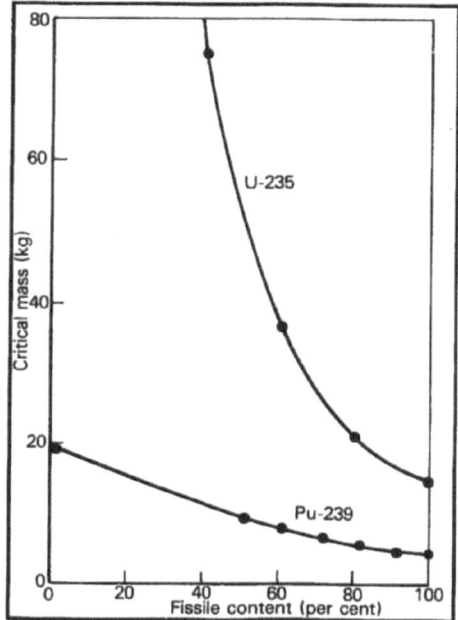

Note: The two metals are in the form of spheres enclosed in thick neutron reflectors of natural uranium. The rapid increase in its critical mass makes isotopically dilute uranium unusable as a bomb. This is not so for plutonium, making it a greater proliferation hazard.

Source: Data from references [3, 4], reprinted in reference [5].

feat (even for a group of non-specialists as might be assembled by a terrorist organization) to separate the uranium from the lead by ordinary, very well-known chemical means, and then to refabricate the U-235 into the metallic form. (If done with excessive absence of care, the group might succeed in blowing itself up, or in irradiating itself to death, but this implies a high level of technical incompetence or carelessness.)

But the advantage of using natural uranium as the denaturing material is that the chemical properties of atoms containing U-235 and U-238 as their nuclei are, for all intents and purposes, identical. It is therefore exceedingly difficult to separate the isotopes by chemical means.

It is well known, of course, that the light U-235 isotope can be and is separated from natural uranium by a number of physical techniques. But such separation has hitherto required highly sophisticated technology, very expensive to perform and requiring large expenditures of energy—none of which prerequisites are now available to any but a few of the most technologically advanced nations. One problem for the future is the probable development of new, cheap and readily accessible technologies for isotope separation (cheap and efficient ultracentrifuges or selective absorption of intense laser radiation), but this is a problem that does not at the moment need to be faced, at least not in this paper.

As far as the denaturing of plutonium by mixing with an inert isotope (that is, by isotopic dilution) is concerned, however, the problem is to find the appropriate inert isotope. Plutonium is not normally found in nature—any that was produced in the primordial processes of creation of the elements has long since decayed. The main isotope of interest, Pu-239, has a half-life of 24 400 years, which is long enough on a human time-scale, but short as measured in geological time. Other isotopes can be made in nuclear reactors—notably Pu-240 and Pu-238—and, indeed, the percentage of these isotopes (almost entirely Pu-240) contained in reactor-produced plutonium increases with the length of time that the plutonium-producing fuel elements are allowed to 'cook' in the reactor.

However, unfortunately, unlike the effect of U-238 in uranium isotopic mixtures, the plutonium-240 isotope is not an effective denaturant. Its fission properties, while not as favourable as those of the predominantly produced Pu-239, are such as only moderately to increase the critical mass of mixtures as compared to the favourably small value for pure Pu-239, as seen in figure 7.1.

Any other denaturant, even if chemically similar to plutonium, will, since it is a different element, be separable by conventional chemical processes that are widely described in the open literature.

Denaturing by isotopic dilution not being available for plutonium, it is necessary to turn to some other aspects of the fast (explosive) chain reaction for possible means of making plutonium weapon-proof. For many years, expectations were placed on another property of the heavier, Pu-240 isotope. This effect, which may be referred to as *neutron denaturing*, derives from the relatively copious spontaneous neutron emission from Pu-240 (as a consequence of its property of very occasionally decaying[2] through undergoing spontaneous fission). The consequence of this property is that the relatively large neutron background—present in any sample containing an appreciable Pu-240 content—tends to cause premature detonation during the assembly of the super-critical mass of plutonium required to produce an explosive chain reaction. It was precisely this problem that led the designers of the first nuclear weapons to devise the implosion technique for the first plutonium bombs—in contrast to the very much easier 'gun-assembly' technique that was employed for the detonation of the early U-235 bombs.

In deference to this assumed denaturing feature of Pu-240, most of the plutonium that has been produced specifically for weapon use is created in special reactors where the fuel elements, in which the Pu-239 is produced, are not permitted to remain in the reactor long enough to accumulate any appreciable Pu-240 content. Since, however, even a large Pu-240 proportion does not adversely affect the usability of plutonium in nuclear reactors, it was also considered (until relatively recently) that power reactor-

---

[2] The overwhelmingly dominant decay channel is α-particle emission, with a half-life of 6 580 years.

produced plutonium—in which the Pu-240 content was deliberately enhanced—would solve the denaturing problem by providing a weapon-proof form of plutonium that was still of reactor grade.

Alas, it is now known that Pu-240 denaturing is an illusion [1]. Although the presence of appreciable Pu-240 may render a plutonium bomb somewhat less reliable or less predictable, the resulting neutron background apparently does not make its detonation appreciably more difficult.

However, if the neutron background level associated with reactor-produced plutonium (even that in which the Pu-240 content is deliberately enhanced) is not sufficient to accomplish neutron denaturing, this does not necessarily mean that neutron denaturing is a complete *non sequitur*. The neutron background associated with any given plutonium sample can be increased by orders of magnitude through mixing in with the plutonium a substance whose nuclei emit neutrons, either spontaneously or upon the absorption of the α-particles normally and copiously emitted by the plutonium isotopes. Thus, a very large neutron background can be provided by mixing into the plutonium even microscopic quantities—parts per million or less—of the artificially created californium-252 isotope. (Cf-252, which is created in reactors by successive neutron capture, has a half-life of 2.65 years, with 3 per cent of the decays proceeding via the spontaneous fission mode.) Another possibility for neutron denaturing would be to mix the element beryllium into the plutonium, in even relatively modest proportions (by weight). This could provide such a prodigious neutron source, through alpha-particle absorption, as to render the plutonium incapable of safe handling, let alone detonation, by normal procedures.

The weakness of such denaturing, of course, is that the neutron-producing contaminant can be separated from plutonium by known chemical means. However, the presence of a fiercely high neutron background would necessitate the use of special handling techniques—involving the provision of remote control procedures carried out behind neutron-absorbing barriers—that would very substantially increase the difficulties associated with the process of bomb manufacture. However, chemical separation behind thick water barriers is still considerably easier than uranium isotope separation by available physical means.

If one is prepared to consider the 'spiking' of plutonium with a neutron-emitting source, then there are obviously other possibilities available for denaturing by *radioactivity*. As a matter of fact, the easiest and most straightforward possibility for eliminating the plutonium proliferation danger—that of eschewing the reprocessing of spent reactor fuel elements and storing the plutonium unseparated from the fiercely radioactive uranium fission products—is of course the extreme example of this mode of denaturing. However, even if it is considered desirable to reprocess the plutonium and to use it to produce power in a plutonium-fissioning reactor, it would still greatly discourage any diversion (theft) of the plutonium-containing fuel elements (say, during their transport from the fabricating plants to the reactor) and their reworking to extract the plutonium for weapon use, by making the plutonium, at an early stage after its separation, too radioactive to handle (without special facilities). One way of doing this is by 'pre-irradiation' of fresh fuel elements in a special reactor at the same location as the separation and fuel-fabrication facility. Another way that this can be accomplished is by 'spiking' the plutonium with a highly γ-radioactive denaturant, such as, for example, cobalt-60.

Again, although the handling of spiked plutonium is greatly complicated by the need for remote-handling capabilities, the chemistry of the extraction of the foreign elements is generally straightforward.

An interesting recent suggestion, which might be termed 'physical denaturing', would involve making the plutonium literally too hot to handle, by greatly increasing the content of the short-lived Pu-238 isotope (an α-particle emitter with a half-life of 86 years). The effect of only a few per cent of this isotope, normally present in reactor-grade plutonium, is already felt in an increased ambient temperature of metallic plutonium fabricated from it. An increase of the Pu-238 content to the order of 15−20 per cent, it is claimed [2], could lead to the melting of the metallic plutonium core of a 'conventional' plutonium bomb. Such an increase in Pu-238 content can be achieved by adding a sufficient quantity of neptunium-237 (which becomes Pu-238 through neutron capture) to the original uranium fuel elements in which the plutonium is to be bred.

However, it can be argued that such a physical constraint (that is, excessive heat production) could be overcome by appropriate physical alterations in the bomb design (heat-dissipating fins, and so on). Nevertheless, a large intrinsic heat source, not separable by chemical means, would undoubtedly place serious constraints on the usability of such plutonium for weapons by relatively unsophisticated groups. However, this same impediment would also complicate its handling as a reactor fuel, and this could render such plutonium unacceptable in normal reactor commerce.

Aside from the various spiking possibilities described above, there are other forms of non-isotopic denaturing conceivable. For example, if it were possible to find an element that tended strongly to absorb neutrons in the (high) energy region relevant to the fast (explosive) chain reaction, but which behaved relatively benignly with respect to the low-energy neutrons that dominate the operation of power reactors, such a substance might provide a useful form of denaturing by neutron poison. However, generally speaking, there are very few effective neutron absorbers in the high-energy range which are not also very strong slow-neutron absorbers; in fact, no likely candidate for this form of denaturing is known to this author.

Finally, it might be noted that in the absence of any possibility of simple denaturing by dilution with an appropriate plutonium isotope, one might seek a substitute (that is, the equivalent of the U-238 denaturant that works so well with U-235 or U-233) in some element whose chemistry is sufficiently similar to that of plutonium as to make the chemical extraction laborious and complex. In fact, in any of the modes of denaturing considered above, the ideal denaturant of choice is one whose chemical properties are as close to those of plutonium as can be managed.

Generally speaking, while none of the available denaturing methods can approach in effectiveness a true isotopic denaturant, if there is to be plutonium reprocessing for use in nuclear reactors, the objective should be to render as difficult as is conceivable the diversion, extraction and refabrication into weapons of any plutonium found outside a nuclear reactor. The real problem, of course, will be to provide sufficient pressures and/or incentives on all the nations with plutonium handling capabilities to adhere to a régime which insists on the application of discomforting and complicating handling procedures when life would be infinitely simpler (albeit more dangerous) without them. The temptation to cut corners will be overwhelming.

Once plutonium is in wide circulation in world-wide national and international traffic, the problem of preventing the further proliferation of nuclear weapons will have been rendered extremely difficult, and the chances of successful solution of the problem cannot be made satisfactorily large by any known means, certainly not by any simple means. But by a combination of the various available partial measures of denaturing, plus continuing stress on eternal physical vigilance over the plutonium stocks, it may perhaps be possible to continue to postpone the day when it is an easy task for anyone to obtain a crude nuclear bomb.

References

1. Mark, J. C., 'Nuclear weapons technology', *Impact of New Technologies on the Arms Race*, ed. B. T. Feld, G. W. Rathjens and S. Weinberg, Pugwash Monograph (MIT Press, Cambridge, Mass., 1971), pp. 137–38.
2. 'Allied-General Nuclear Services proposes proliferation-resistant fuel cycle', *Nuclear Industry*, September 1978, p. 24.
3. Taylor, T., *Annual Review of Nuclear Science*, Vol. 25, No. 407, 1975.
4. Seldon, R. Unpublished communications.
5. Moniz, E. J. and Neff, T. L., 'Nuclear power and nuclear-weapons proliferation. Breeder reactors easy fuel supplies but at increased risk of diversion', *Physics Today*, Vol. 31, No. 4, April 1978, pp. 42–51.

Part II

Introduction

B. JASANI

Square-bracketed numbers, thus [1], refer to the list of references on pages 125−6.

A commitment to generating energy from nuclear fission has been made by several nations. Their decisions are largely based on the fact that the average amount of energy released in various fission reactions is about 200 million electron-volts (MeV) compared with about 2−4 electron-volts (eV) released in a chemical reaction. Even though this serious commitment to fission energy has been made, concern is still expressed in the industrialized countries about the safety and the environmental effects of the use of such a form of energy. However, as shown in Part I, international concern is to a large extent centred on the hazards of the proliferation of nuclear weapons associated with the use of the largely man-made element plutonium as a nuclear fuel. Such concerns became even more intense with respect to the fast breeder and hybrid reactors. In this Part, attention is focused on the issues arising from the use of these types of reactor.

I. Breeder reactors

In the thermal reactors currently in operation, natural uranium (~ 0.711 per cent U-235) or slightly enriched uranium (1−3 per cent U-235) is used as the fuel. However, this is a relatively inefficient process compared with fast breeder reactors, in which fast neutrons are used so that during fission nearly three neutrons are released for every neutron absorbed. Thus, the excess fast neutrons available would convert the U-238 in the blanket into fresh fuel, that is, into Pu-239.

This ability of fast reactors to breed more fuel is the basic argument used for their development and deployment, since it is argued that the supply of uranium is limited and in some countries even non-existent (see Paper 8). In a second argument put forward in favour of fast reactors, it is reasoned that such reactors alone have the capability of burning the recovered plutonium, thus reducing the inventory of that material. Moreover, it is argued that the fast reactor can be used either to produce slightly more or slightly less plutonium than it burns so that the growth of plutonium can be checked. However, Paper 9 argues that there is no need, at present, to deploy breeder reactors but that development should continue until the cost of electricity produced from such reactors becomes comparable to and eventually cheaper than that produced from thermal reactors. Moreover, in the Paper it is pointed out that it is useful to deploy breeder reactors using moderately enriched uranium (some 20 per cent U-235). The technology of such a system is available now.

Although some of the above arguments are essentially valid, there are difficulties in a total commitment to energy produced by fast breeder reactors. For example, the costs—capital, the running costs and the political costs—are not fully worked out. The question of safety has also not been resolved. Current estimates of the capital costs indicate that the ratio between the breeder reactor and the light water reactor (LWR) is about 2:1. Secondly, a breeder reactor fuel cycle requires reprocessing facilities to extract the plutonium fuel from the irradiated fuel and the blanket. Plutonium fabrication plants are also required. This means that large quantities of plutonium may be available in relatively easily accessible form for diversion by a governmental or a non-governmental organization. This is clear from table II.1, which shows that whereas in a thermal reactor about 330 kg of plutonium (with 57 per cent Pu-239) are produced per year, 2 580 kg of plutonium are available from a fast reactor when the spent fuel has been reprocessed. Moreover, at least about 400 kg of almost pure Pu-239 are available from the reactor blanket when plutonium is separated from U-238, so that a net of some 3 000 kg of plutonium is produced in a fast breeder reactor. There are technical measures available to make it difficult but not impossible to extract this plutonium, as discussed in Part 1.

While breeder reactors have implications for the horizontal proliferation of nuclear weapons, the developments in laser applications to fusion reactors and hybrid reactors affect the vertical proliferation of nuclear weapons. Developments of these reactors are considered below.

II. Fusion and hybrid reactors

The possibility of fusion energy released as a controlled power source has been under intensive study since the early 1950s, mainly in the United States and the Soviet Union. It is only now that we are beginning to see the feasibility of this technology. This is indicative of the complexities of fusion power technology.

The need for energy and the fact that the present sources of energy are limited gave considerable impetus to the development of fusion energy, as has been the case with the breeder reactor developments. However, an

Table II.1. Production of plutonium in a pressurized water reactor (PWR) and a fast breeder reactor (FBR) for each GW(e) year

Reactor	Reactor core		Reactor blanket		Total plutonium production
	Fuel input	Net plutonium produced	Fuel input	Net plutonium produced	
PWR	1 110 kg/year (fissile)	330 kg/year Pu-219 57% Pu-240 24% Pu-241 14% Pu-242 5% (150 kg/year fissile equivalent)	–	–	330 kg/year
FBR	9 500 kg/year (U-235 0.4%) 2 800 kg/year Pu-239 57% Pu-240 24% Pu-241 14% Pu-242 5%	2 580 kg/year	16 900 kg/year (U-235 0.4%)	409 kg/year (Pu-239 ~97%)	2 989 kg/year

Source: See reference [1].

additional argument made in favour of fusion reactor development is that in a pure fusion reactor, horizontal proliferation problems do not occur. However, this depends on how the reactor is used.

In a fusion reaction, atoms of light element join or fuse together, releasing a considerable amount of energy. The least technically demanding reaction is the one in which heavy hydrogen atoms, deuterium (^{2}H, or D) and tritium (^{3}H, or T), fuse (see equation 1).

$$D + T \rightarrow {}^4He + n + 17.6\,MeV, \qquad (1)$$

where He = helium, n = neutron. The conditions needed to achieve net power output are less demanding than, for example, the following reactions:

$$D + D \rightarrow {}^3He + n + 3.2\,MeV \qquad (2)$$
$$D + D \rightarrow T + p + 4.0\,MeV \qquad (3)$$
$$D + {}^3He \rightarrow {}^4He + p + 18.3\,MeV, \qquad (4)$$

where p = proton. Deuterium occurs in nature in abundance and tritium can be produced in a reactor.

The reactions (2) and (3) occur with equal probability and the products of these reactions form the fuel for the reactions (1) and (4). The net reaction produces two helium nuclei, two protons, two neutrons and a net energy release of 45.1 MeV. This is represented as follows:

$$6D \rightarrow 2{}^4He + 2p + 2n + 43.1\,MeV. \qquad (5)$$

In practice, lithium deuteride is used, so that tritium is produced according to

$$n + {}^6Li \rightarrow {}^4He + T + 4.8\,MeV, \qquad (6)$$

provided energetic neutrons are slowed down.

Production of useful power from such fusion reactions requires that small amounts of fuel be heated to temperatures of at least 40 million kelvin for a long enough time to obtain reaction of a substantial fraction of the fuel. At these temperatures the fuel cannot be confined by material walls but, since the fuel nuclei are electrically charged, a magnetic field can be used to contain such hot plasma. In a magnetic confinement system, the density of plasma typically has to be of the order of 10^{14} ions/cm^3 for a few seconds [2].

A much higher and a more useful plasma density (10^{24} ions/cm^3) for a much shorter time (10^{-10}s) is achieved in an inertial confinement system [2]. However, this technology has raised some concern from the point of view of the proliferation of nuclear weapons. The inertial confinement fusion (ICF) requires beams of radiation with an instantaneous power of the order of 10^{14} W to be deposited on a target a few millimetres in diameter and a few milligrams in weight. A high-power pulse of energy is directed at a fuel pellet of deuterium. The energy is absorbed by the pellet, thus heating it and causing explosion. The inward component of the force of the explosion compresses the deuterium to about one-thousandth of its original size and raises its temperature 5−10 million degrees kelvin [4].

There are three main types of beam under consideration for igniting a thermonuclear reaction. These are laser, electron beams, and light-ion or heavy-ion beams. Lasers suitable for ICF could use CO_2 gas as the lasing medium [3]. However, they produce relatively long-wavelength radiation (10 mm), so that coupling to the target plasma may be difficult [2]. Other types of gas are being investigated. The electron beam ICF system is identical in concept to the laser system. The electron energies are about 1 MV [2].

In the ion beam ICF, ions can be accelerated to high energies and focused on a small target using conventional accelerator technology. The latter has reached a considerably advanced state for high-energy physics research. Several techniques are being developed. In one, a pulse of ions is injected into a magnetic mirror so that the ions are trapped in the form of a ring. The ring is then compressed magnetically, thus increasing the energy of the ions and reducing the dimensions of the ring. Such a ring is then accelerated axially to impact a deuterium−tritium pellet [5]. A problem with such a scheme is that, after every pulse of ions, a lithium guide tube which directs the ion ring to the pellet has to be replaced.

It is useful at this stage to compare the costs of fusion reactors to those of other types of reactors. For example, it is estimated that the construction cost of fusion reactors would be some \$4 450 per kW compared with \$780 per kW for fossil power plant, \$975 per kW for thermal fission reactors and the estimated \$1 820 per kW for a fast breeder reactor [7].

Proliferation aspects of fusion and hybrid reactors

From equations (1) and (2) it can be seen that, as a result of the fusion of deuterium and tritium or deuterium and deuterium, neutrons are produced; neutrons in equation (1) have energies of 14 MeV each. If the core of a fusion reactor is surrounded by a blanket in which neutrons are allowed to react with fertile or fissile ions, then one of the following three processes could be made to occur: (1) neutrons could be used to cause fission reactions in an otherwise subcritical system so that the energy output from the fusion system could be increased; (2) neutrons could produce fissile material from a fertile blanket for the purpose of fuelling pure fission reactors; or (3) neutrons could be used to transmute the long-lived radioactive wastes produced in fission reactors to shorter-lived radionuclides. Systems (1) and (2) are the principles of the hybrid reactors described in Paper 11.

With such hybrid reactors, it is easy to see the existing danger of the horizontal proliferation of nuclear weapons. A blanket containing either U-238 or Th-232 round a fusion core could be used to produce Pu-239 or U-233. The Paper also points out, quite correctly, that a further link exists between fusion (particularly laser fusion) and hybrid reactors and fusion weapons. However, with the present highly sophisticated state of fusion technology, it is not very likely that non-nuclear weapon states will use the fusion route to proliferate nuclear weapons. However, the possibility of vertical proliferation through the use of ICF using high energy beams is more real.

The implication of this technology for the vertical proliferation of nuclear weapons lies in the fact that the ICF device would reproduce on a laboratory scale much of the basic physics of nuclear weapons. Moreover, if the system is built on a sufficiently large scale, many of the radiation effects of nuclear weapons could be studied in laboratories. Experiments involving the interactions between high energy beams and target materials, for example, would provide data required for the development of new types of nuclear weapon. In fact, the early high-energy particle beam systems were developed for studying the effects of nuclear weapons rather than for fusion power reactors [8]. Certain measurements can be made only in a laboratory using the ICF technology. Data obtained from an underground nuclear test is limited since the measuring instruments can be destroyed in the explosion. In the event of a comprehensive test ban treaty, ICF technology presents the greatest hazard, since it would allow nuclear weapon designers to continue designing new weapons and get much of the data needed about the weapon without conducting underground tests [3].

References

1. Marshall, W., 'Nuclear power and the proliferation issue', *Atom*, No. 258, April 1978, pp. 100–101.

2. Arnold, R. C., 'Heavy-ion beam thermal-confinement fusion', *Nature*, Vol. 276, No. 5083, 2 November 1978, p. 19.
3. '10% efficiency is deemed "possible" with CO_2 fusion laser at Los Alamos', *Laser Focus*, Vol. 14, No. 11, November 1978, pp. 28–32.
4. Wetmore, W. C., 'U.S. electron beam test trigger fusion', *Aviation Week and Space Technology*, Vol. 107, No. 1, 4 July 1977, p. 22.
5. 'New fusion scheme', *Laser Focus*, Vol. 14, No. 11, November 1978, p. 79.
6. 'Progress in fusion reactors', *Nuclear Engineering International*, Vol. 23, No. 271, May 1978, p. 60.
7. Holdren, J. P., 'Fusion energy in context: its fitness for the long term', *Science*, Vol. 200, No. 4338, 14 April 1978, p. 173.
8. Stickley, C. M., 'Inertial confinement fusion program', statement before the US Senate Armed Services Committee, 25 March 1977 (Energy Research & Development Administration, Washington, D.C., 1977).

Chapter 6. Breeders

Paper 8. The proliferation aspects of breeder deployment

B. BARRÉ

I. Background

Current concern about nuclear proliferation contains a number of paradoxes. For example, most people seem to worry more about the possibility of an individual country acquiring a handful of kiloton-range bombs in the future than about the thousands of megaton weapons which are already accumulating in the arsenals of the major nuclear weapon states. In addition, while acknowledging that the spread of nuclear weapons during the period 1945–73 seemed to have no destabilizing effect on world security and in fact may even have been a stabilizing factor, perhaps helping to prevent a major conflagration, nevertheless the majority of people, including myself, now agree that any new proliferation should be prevented.

Furthermore, it is a well-known fact that, in every nuclear weapon state, military establishments go to great lengths in terms of manpower, time and expense to produce military-grade plutonium and to design sophisticated equipment, while most people take for granted that half a dozen terrorists could easily and rapidly manufacture weapons with the kind of 'dirty' plutonium that is produced by reprocessing light water reactor fuel.

This paper will not try to solve these paradoxes. Thus, for our purposes, no distinction is made between different types of plutonium, and it is argued for the sake of simplicity that horizontal proliferation is more crucial than vertical proliferation.

But there is one thing that should be stressed, which is that war is the ultimate evil, not proliferation. The danger with proliferation is that it could make war uglier and deadlier, but if, in attempting to reduce the risks of proliferation, states took steps which were ultimately to increase the risk of war, the results would be devastating to mankind.

Historically, no war has been caused by one single factor, but in most cases competition has played a major role, whether for food, status, land, wealth, or other resources. Any competition almost inevitably stems from scarcity, which is why one of the main causes of present-day rivalry, although by no means the only one, is the competition for energy. There has been a temporary oil glut but every expert agrees, and most governments are convinced, that oil shortages are not far away; it is a fact that the rate of production has exceeded the rate of discovery since 1972 (see figure 8.1). Although prospects for the more distant future are optimistic, the only alternative now available is nuclear energy, whose technology is reasonably mastered, and whose economy compares favourably in most countries to that of fossil energy.

There is no doubt that there is a link between the civilian and military applications of nuclear energy, stemming from similarities in knowledge, material and technological background, but this link is weak. The Hiroshima bomb was exploded about 10 years before the Calder Hall power station became operational, and until now, civilian nuclear power has never been the basis of military development. Under these circumstances, renouncing nuclear power would not prevent further proliferation but by increasing competition for energy, would lead to disruption of the fragile international balance, and might cause the nuclear war it was intended to prohibit. By promoting the development of nuclear power we will have to live with a certain level of proliferation risk, but this is definitely the lesser evil.

II. Why the breeder?

Our present position can be summarized thus: (a) a certain level of proliferation risk exists already irrespective of whether or not the development of nuclear power continues; (b) development of nuclear power for civilian purposes to some extent increases this risk; and (c) the benefits of nuclear power by far outweigh the dangers of increasing this proliferation risk.

Point (a) is obvious, since methods of producing weapons or explosive devices through uranium enrichment, production reactors using natural uranium and even in one case a research reactor, are already well known and have actually been the basis of all weapon development hitherto.

Point (b) is certainly also indisputable. Here it should be emphasized that the increase in the level of risk largely depends on the type of reactor chosen and on the associated technologies; reactors which use natural uranium and on-load refuelling are much more proliferation-prone than reactors using low-enriched uranium and off-load refuelling. Some current efforts to reduce the refuelling time of light water reactors and to design easily dismantled fuel assemblies would reduce this difference, and their potentials should be carefully evaluated. On the other hand, enrichment techniques have quite different characteristics, and it may be unwise to develop such 'ambiguous' technologies as centrifuge and laser enrichment, which enable one to obtain highly enriched uranium from a civilian plant without major modification to its layout.

Figure 8.1. Rate of discovery of world crude oil reserves, excluding socialist countries

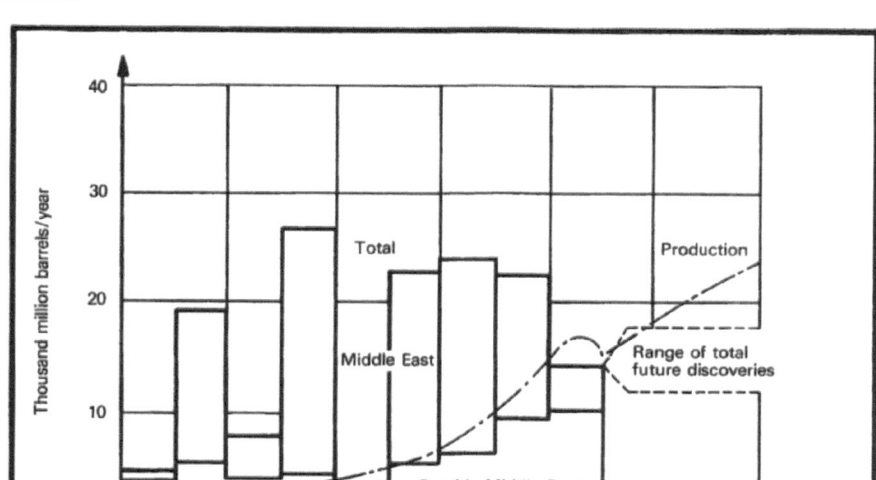

Source: *World Energy Outlook*, Exxon Background Series, 1978.

For anybody not agreeing with point (*c*), there is no point in discussing breeder deployment at all. International deployment of light water reactors associated with non-proliferating enrichment plants—preferably multinational—is thus considered as an acceptable incremental proliferation risk. It is against this background that the problems associated with breeder deployment will be considered.

Plutonium breeders have been studied and developed for some years in a number of countries, as can be seen in figure 8.2. Of all the countries listed, only the United States seems to have recently reduced its efforts in this field, and even now it leads the world in terms of R&D expenditure. What lies behind so vast an effort and why are so few countries content with deploying current commercial reactors? The answer is uranium economy.

Light water reactors are very uneconomical uranium users, and if we were to rely upon them alone uranium resources would not last much longer than petroleum and gas. There is room for improvement in that respect, but not to such an extent as to change the scale of the phenomenon. The only solution to the problem is breeders (see figure 8.3) which are urgently needed to curb world uranium consumption, as illustrated by figure 8.4 (from the last World Energy Conference in Istanbul). By bringing down the consumption of uranium to a minimum, and by allowing the possibility of burning the current stockpiles of depleted uranium, breeder reactors would free nuclear energy from any concern about uranium reserves in the immediate future, and would leave plenty of time for future generations to develop post-nuclear energies. It would also free uranium-poor nations

Figure 8.2. World fast neutron reactors

Reactors in operation or closed down

Name	Location	Country	Power capacity	1950 1955 1960 1965 1970 1975 1980
EBR-1	Idaho	USA	0.2MW(e)	
BR-5	Obninsk	USSR	10MW(th)	
DFR	Dounreay	UK	15MW(e)	
EBR-2	Idaho	USA	20MW(e)	
Enrico Fermi	Michigan	USA	66MW(e)	
Rapsodie	Cadarache	France	40MW(th)	
BOR-60	Melekess	USSR	12MW(e)	
Sefor	Arkansas	USA	20MW(th)	
KNK-II	Karlsruhe	FRG	20MW(e)	
Phénix	Marcoule	France	250MW(e)	
BN-350	Shevchenko	USSR	150MW(e) + desalination	
PFR	Dounreay	UK	250MW(e)	
Joyo	Oarai	Japan	100MW(th)	

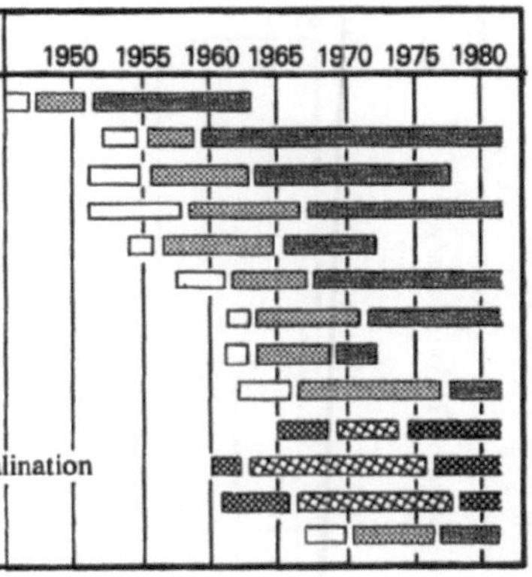

Reactors under construction or projected

Name	Location	Country	Power capacity	
FFTF	Hanford	USA	400MW(th)	
BN-600	Beloyarsk	USSR	600MW(e)	
PEC	Brasimone	Italy	116MW(th)	
FBTR	Kalpakkam	India	15MW(e)	
SNR-300	Kalkar	FRG	300MW(e)	
Super-Phénix	Creys-Malville	France	1 200MW(e)	
CFR	..	UK	1 300MW(e)	
Monju	..	Japan	300MW(e)	
SNR-2	..	FRG	1 300MW(e)	
CRBR	Tennessee	USA	300MW(e)	
PLBR		USA	1 200MW(e)	?

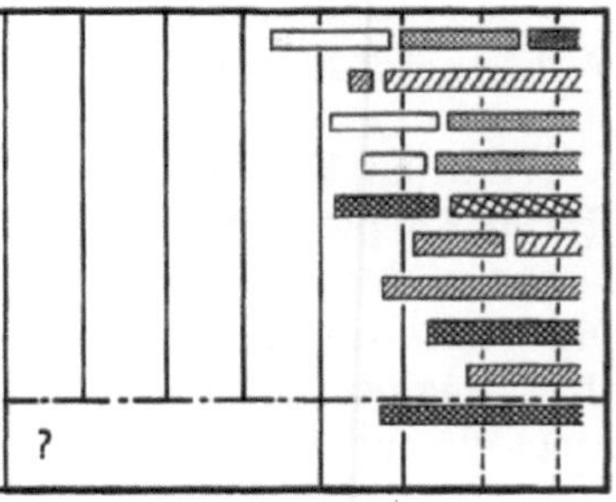

Type	Project	Construction and start-up	Production
Experimental			
Demonstration			
Commercial			

130

Figure 8.3. Comparison of uranium consumption in LWRs and FBRs

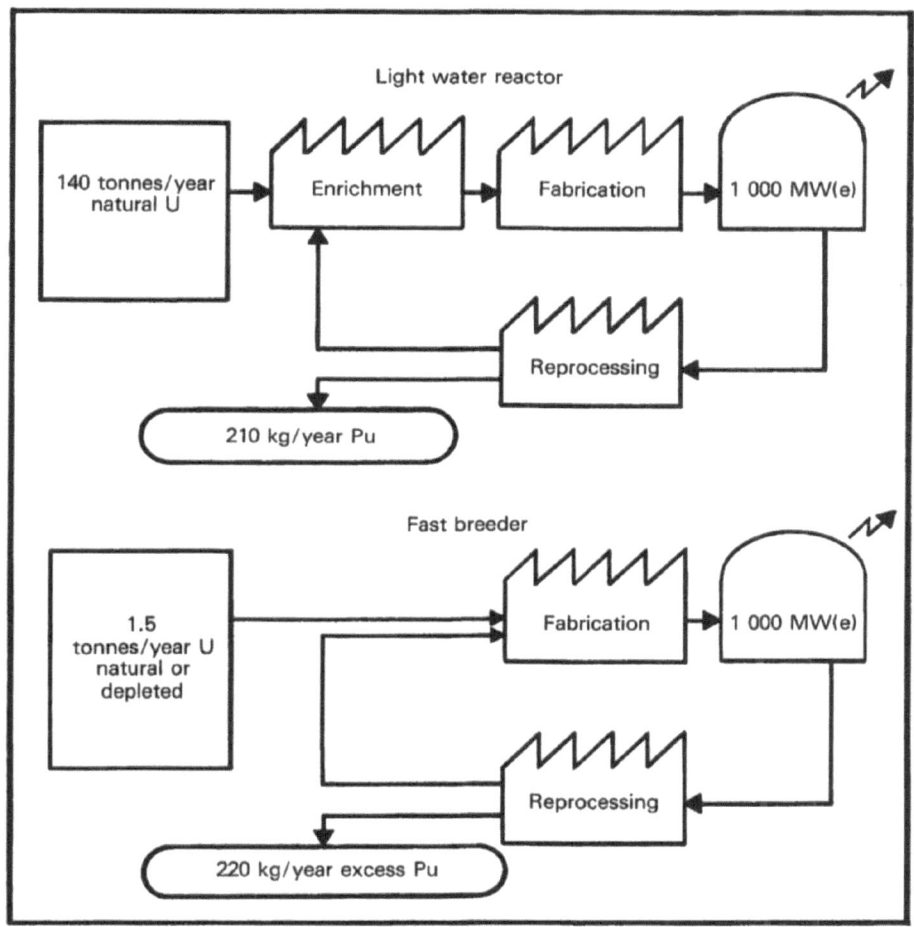

from any fear of being deprived of a vital energy source. Of course, the need for breeder deployment largely depends on the relative size of the nuclear programme and the indigenous uranium resources of each country, which explains why the feeling of urgency is stronger, for instance, in France than in the United States. However, irrespective of these differences, it is essential that breeder technology is shown to be a viable alternative. This can only be done through practical operation of a large-scale prototype power station.

By itself, breeder development has no bearing on proliferation, but breeder deployment is another question, and has to be assessed against the 'acceptable' light water reactor.

Figure 8.4. Effect of breeders on annual uranium demand

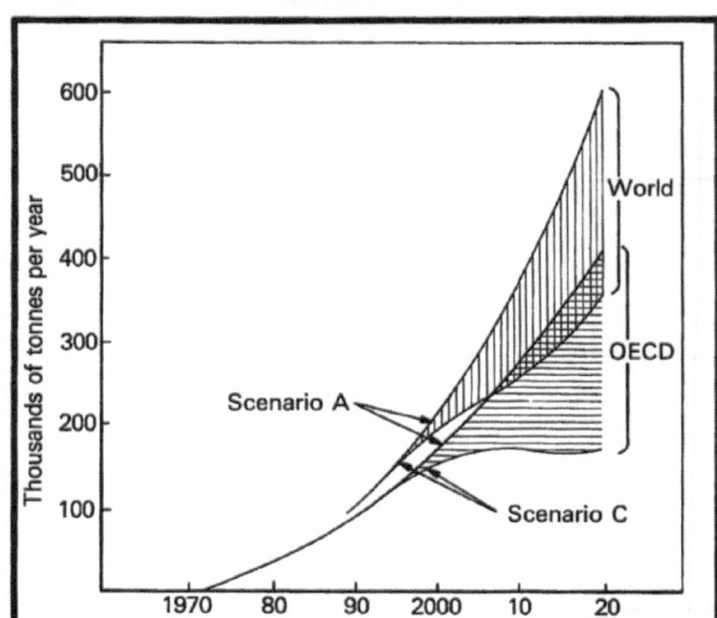

A. The 'no reprocessing' scenario—in this case only thermal converter reactors (TCR) are installed in all regions. Light water reactor characteristics are used for these converters, since this is the most prevalent commercial reactor-type in use today.

C. Base case—commercial breeders with a fuel doubling time of 24 years (at 100 per cent factor) are installed in 1993 in North America, in 1987 in Western Europe, in 2000 in Japan, and in 1995 in the USSR.

Source: Conservation Commission Report on Nuclear Resources 1985–2020, World Energy Conference, Istanbul, 1977.

III.   The uranium cycle versus the plutonium cycle

Figure 8.5 shows what is known as the nuclear fuel cycle. Considering the cycle only as far as the light water reactor (LWR), and assuming that the enrichment plant uses gaseous diffusion or, better still, the French chemical process, then the only part of the cycle that might involve proliferation is the enriched uranium between the enrichment plant and the reactor.

Low-enriched uranium may constitute a short cut to highly enriched uranium, as it has already required two-thirds of the final separative work. This is not a negligible consideration, but is probably of secondary importance.

The important point lies elsewhere, that is, in the cooling pool of the reactor. A 1 000-MW(e) light water reactor rejects annually about 30 tons of a mixture of uranium, fission products and plutonium. The yearly yield of plutonium stands at around 250 kg. According to a recent International Atomic Energy Agency (IAEA) paper, approximately 90–100 tons of

Figure 8.5. The nuclear fuel cycle

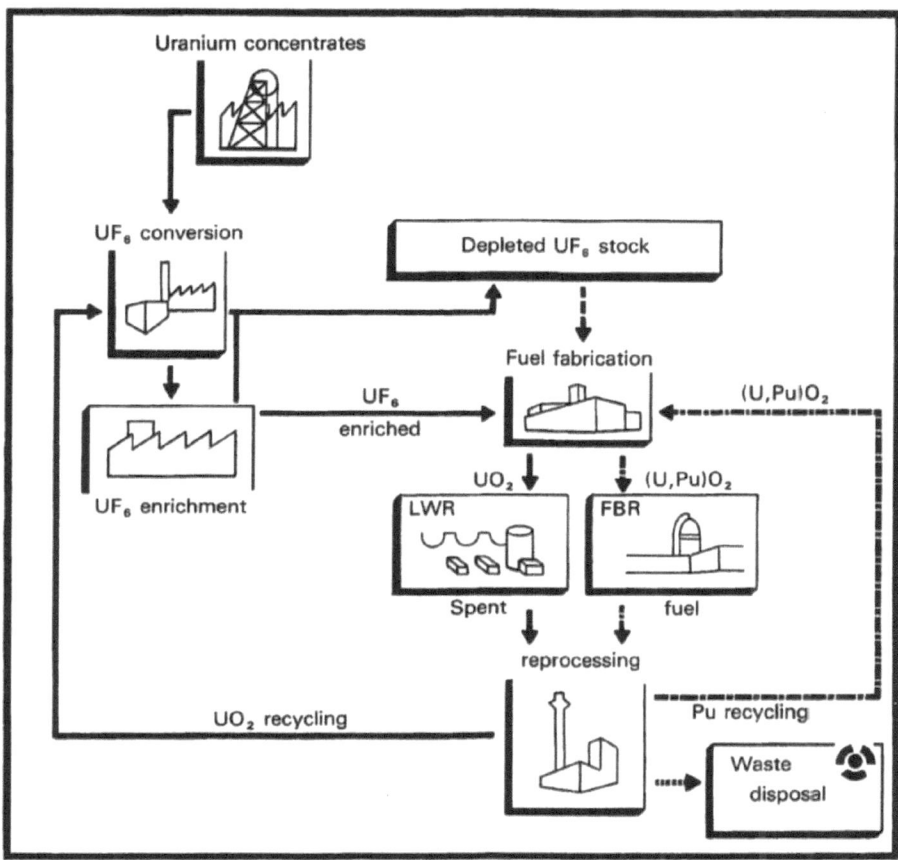

plutonium have already been produced throughout the world through 'civilian' power production, both in the East and West and a possible 20–25 tons have already been extracted.

Extracting plutonium from spent fuel is not a difficult operation from a purely chemical point of view, nor does it call for exotic or sophisticated reactants. What makes commercial reprocessing difficult is the requirement of long-term, reliable and economical processing of vast amounts of spent fuel, with minimum risk to production workers, and subsequent contamination-free waste disposal.

However, anyone unconcerned with the economics of the process and with the plutonium contamination of the waste who wants to produce only a limited quantity of plutonium can easily design and construct a 'disposable' plutonium-extraction laboratory, able to process fuel assemblies that have been cooled down for a dozen years or so. By that time, radioactivity will have decayed to one-tenth of its value after one year of cooling time, and even the Pu-241 content will have dropped by a factor of two.

Thus the cooling pools of the reactors constitute a proliferation hazard, which greatly increases with the passing of time, the decay in radioactivity and the spread of reactor sites throughout the world.

On the other hand, an examination of the closed plutonium cycle illustrated in figure 8.5 shows that instead of stopping at the cooling pool of the reactors, the fuel assemblies will be shipped when their radioactivity is still highly self-protective, and gathered in the cooling pools of the reprocessing plant. For economic reasons, a commercial reprocessing plant must have a large capacity, somewhere between 500 and 2 000 tons of heavy metal per year. Each plant will collect and process fuel elements from 20−80 reactors, which means that the number of pools to be tightly safeguarded will be reduced by a factor of 20−80. In the future, we should endeavour to build large rather than small reprocessing plants, with multinational rather than national status.

Once the plutonium is separated and breeders have been deployed to use it, it becomes a valuable fuel instead of a cumbersome waste product, and the economic incentive to recycle it in reactors will be great. Once back in reactors, the plutonium again becomes inaccessible and self-safeguarded. This is a most important point, since it emphasizes that the only place where plutonium is diversion-proof is inside an operating reactor, or shortly afterwards.

The weak points in this scheme are the back-end of the reprocessing plant, plutonium storage, the plutonium-bearing fuel fabrication plant and, to a lesser extent, the transport of fuel assemblies back to the reactor. But before embarking upon a discussion of these weak points, the terms of the alternative should be clearly delineated.

If thermal reactors are operated in open cycle, hundreds of tons of plutonium will be accumulated in many disseminated reactor sites, and the accessibility of this plutonium for diversion will increase as time passes.

If, on the other hand, thermal reactors are operated together with plutonium-burning breeders, the world inventory of out-of-reactor plutonium will be minimized. Eventually the total plutonium inventory can be reduced by ceasing to breed when appropriate. Figure 8.3 illustrates this. The bottom of the figure indicates the plutonium balance for the whole reactor—the core plus the blanket. As a matter of fact, the 220 kg per year of plutonium (expressed in equivalent Pu-239) are the balance between the 400 kg bred in the blanket, and the 180 kg burned in the core, in round numbers. If the U-238 blanket is suppressed, plutonium can be burned more efficiently than in any thermal reactor, where Pu-242 would accumulate and poison the reactions.

(Incidentally, one could, if future economic conditions were to provide the incentive, replace the depleted uranium by thorium in the blanket. Technically such a switch would pose no problem, but so far as non-proliferation is concerned it would make no difference: uranium-233 is as good—and as bad—as plutonium.)

A fundamental feature of the fast neutron reactor is that it can be operated at will both as a plutonium producer and as a plutonium

Figure 8.6.   The share of breeder reactors in French nuclear power

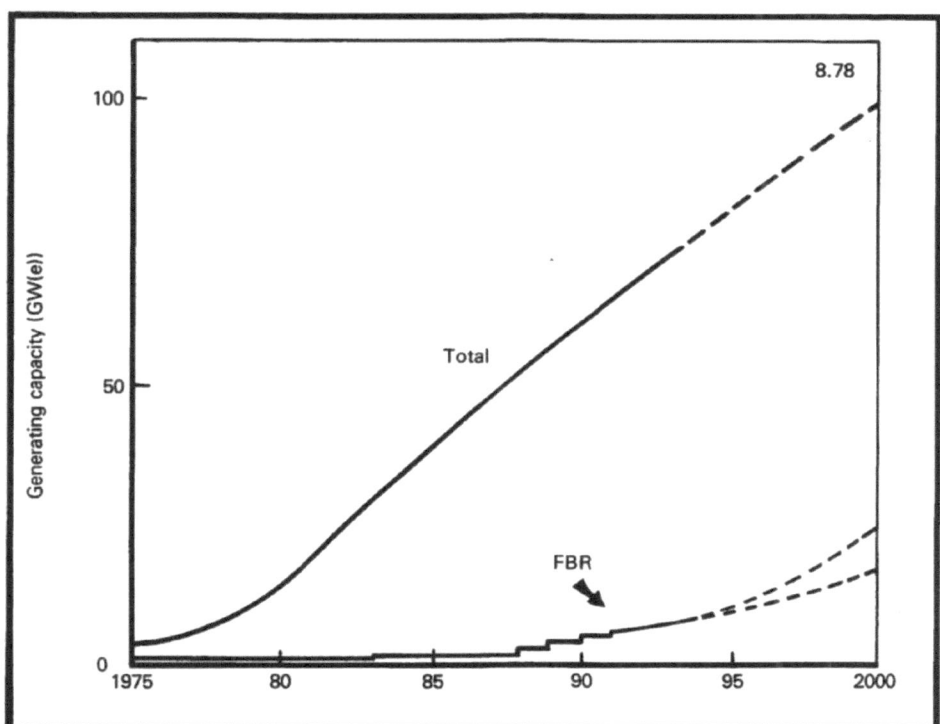

destroyer. It is against this clear advantage that the only real drawback of fast breeder reactors should be juxtaposed, which is the high isotopic quality of any plutonium that may be bred in the blanket. Although it was initially assumed, for the sake of argument, that all types of plutonium are alike, it is acknowledged that this creates a special problem, although not an insoluble one: for the sake of non-proliferation, attention must be focused on the prevention, through strict controls, of any segregation and separate reprocessing of blanket assemblies.

IV.   How to strengthen the 'weak points' in the plutonium cycle

Having clearly advocated the case for deploying breeders, even on purely non-proliferation grounds, the weak points associated with such a deployment will now be discussed. These should not be overemphasized, however. Any breeder development will be gradual, and will match the earlier deployment of thermal reactors (see figure 8.6).

The reprocessing plant

In the August 1978 issue of *Nuclear Engineering International*, 32 reprocessing techniques which have been carried to different stages of development are listed, but the only one which is being developed on an industrial scale is the aqueous liquid—liquid extraction process known as Purex. This process, originally developed for military purposes and involving metallic uranium fuels, is the one used at the La Hague plant in France for civilian reprocessing of LWR oxide fuels. The process itself includes five stages: (*a*) mechanical chopping of the fuel assembly; (*b*) dissolution of the oxides in nitric acid; (*c*) removal of fission products and high actinides by solvent extraction; (*d*) U/Pu partition' by solvent extraction; and (*e*) final decontamination of both U and Pu.

During stages (*d*) and (*e*), plutonium no longer contains any intrinsic protection. (In fact, in the La Hague plant, control and physical protection eliminate all diversion risks, even at these stages.)

In February 1978, the joint communications of Dr Marshall of the United Kingdom Atomic Energy Authority (UKAEA) and Dr Starr of the Edison Power Research Institute (EPRI) presented a new conceptual reprocessing method called Civex which may in the future overcome all the weak points of the plutonium cycle in a breeder-dominated nuclear system. Subsequently at the July meeting of the Uranium Institute, Dr Marshall himself emphasized that this promising concept was still a 'paper-plant' and that it was neither necessary nor useful for the transition period during which the first breeders would be launched, and some limited plutonium recycling possibly done in thermal reactors. The reprocessing of breeder fuel in specific plants is a concern for the 1990s, but today we are building plant to reprocess fuel from LWRs which will constitute the plutonium inventory on which future breeder deployment will depend.

To improve the proliferation resistance of those plants which are now at the drawing-board stage, extensive studies are being carried out, the idea being to use only proven technologies, since industrial development of even promising laboratory-scale processes would entail excessive delays. Any realistic proposal should then be based on the Purex method, more or less modified or adapted to reduce its drawbacks.

Thus, for example, if an appropriate plant design could ensure the physical continuity of the contaminant, from only one inlet—for fuel assemblies and reactants—to only one outlet for the purified streams, we would overcome any risk of 'on-line' diversion.

In this way, we could also allay current fears about material unaccounted for (MUF), which arise because the sheer precision of the measurements will never allow absolute certainty merely by checking the input/output balances. This is no easy task, and the problems associated with unavoidable on-line sampling for analysis are great, but our specialists at the CEA and COGEMA are optimistic of achieving such a design, which, among themselves they have christened 'Pipex' (such a plant would, in effect, constitute a leakproof pipe).

Plutonium storage

Recycling plutonium in thermal reactors requires fabrication of mixed oxides $(U/Pu)O_2$ with a variety of Pu/U ratios around 3 per cent. Mixed oxide fuels for fast breeders require higher plutonium content, but large-scale commercial reactors should not need ratios above 25 per cent.

Thus, for civilian uses, there is no need for pure plutonium: a (U/Pu) mixture containing only 30 to 35 per cent Pu could be adjusted to any useful value by adding uranium oxide during the fuel-fabrication process.

Whether this 30 per cent mixture is produced by co-precipitation from an aqueous nitric solution, or whether it is obtained by mixing UO_2 and PuO_2 powders after two separate precipitations, only mixed oxides should flow out of the contained part of the Pipex plant.

Storing mixed oxides instead of pure plutonium oxide would increase the storage capacity needed, and thus storage costs, by a factor of around three, but it is probably worth the added protection against diversion. This would in no way complicate the fabrication process as Pu dilution would have to take place anyway.

Among the customers for the large reprocessing plant, some would have no immediate use for the recovered fissile materials. To disseminate mixed oxide would definitely create proliferation hazards; it would thus seem logical to call for some kind of international storage, preferably on the reprocessing site itself to avoid useless transportation.

Plutonium-bearing fuel fabrication

Mixed oxide gives a slightly better level of protection than does pure plutonium oxide: to make weapons-grade material from it, chemical separation is still needed which, due to a certain amount of hazard in handling plutonium, would increase the delay between diversion and bomb-making. But it would be still better to transport only completed fuel elements, which are discrete entities and therefore easier to control. Furthermore, removal of the stainless steel cladding and hex tube[1] would still increase the delay if there were concern about subnational 'private' diversion.

For these reasons it appears preferable to locate the plutonium fuel-fabrication plant on the same site as the reprocessing plant. In the process, the construction of a separate chemical extraction unit for recycling the fabrication wastes could be avoided, which could otherwise create a moderate proliferation hazard.

In the future breeder-dominated system, refabrication will probably be integrated with reprocessing. This would vastly reduce the problems associated with 'spiking' the mixed oxide with some fission product so that the protection of the refabricated fuel would for some weeks or months be

---

[1] This is a hexagonal stainless steel sheath around the fuel pin bundle.

Figure 8.7. Improving the proliferation resistance of the plutonium cycle

Pu cycle steps	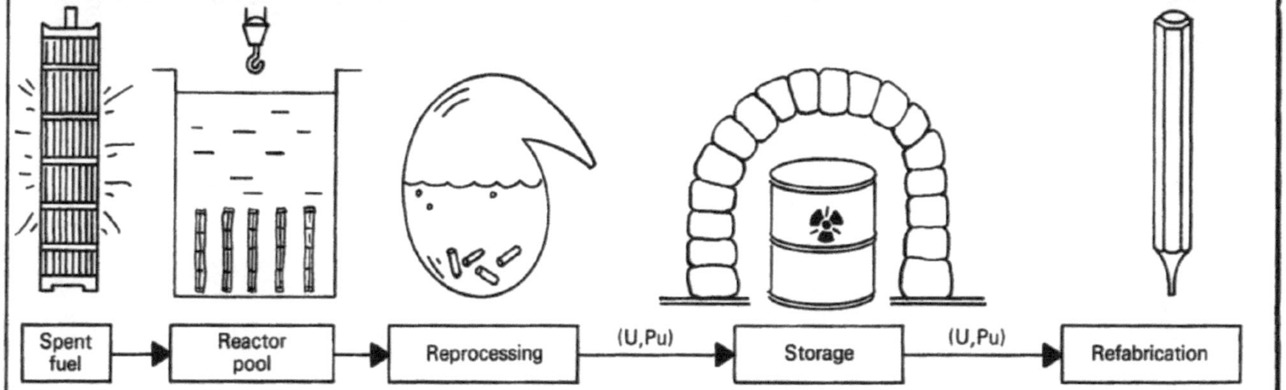 Spent fuel	Reactor pool	Reprocessing		Storage		Refabrication
				(U,Pu)		(U,Pu)	
Proliferation hazards		Pu extraction increasingly easy Pu accounting increasingly difficult	'On-line' Pu diversion	Diversion during shipment	Diversion Covert / Forcible	Diversion during shipment	Diversion in the plant or during shipment
Protective measures	Self-protected for a few years	Limited cooling time and ship to reprocessing	Pipex and mixed oxides	Physical protection Co-location with reprocessing	International control / Physical protection	Physical protection Co-location with storage	Pu return only when needed Limited stocks Physical protection (spent fuel casks?)

equivalent to that of the initial spent fuel. Nowadays, spiking proposals do not seem to be very realistic, but the picture may change if and when it proves feasible to 'break up' long-lived actinides by recycling them in fast breeder reactors. More attainable today would seem to be the short irradiation of freshly fabricated fuel assemblies, but this might entail building a number of yet-to-be-designed irradiation reactors at the fabrication plants, and, so far as is known, detailed assessment of this possibility has not yet been made.

Fuel shipping

Having temporarily discarded any form of spiking, to which the power plant operators would object as strongly as would the fuel fabricators, we are left with plutonium-bearing fuel elements with very little intrinsic protection. The transport from the fabrication plant to the reactor will then have to be submitted to strong physical protection. This will remain a rather weak point, but let us consider the eventualities:

1. The possibility of forcible seizure by a terrorist group. In this case, the delays created by chemical mixing and stainless steel canning should be enough for national or international police forces to intervene. Super Phénix-type fuel assemblies cannot be transported unobtrusively by the 'little man with the big case', nor easily dismantled on the site of the hold-up. But their protection can be improved: a realistic proposal could perhaps be to ship the refabricated fuel elements together with a strong γ-emitting source inside the spent fuel casks, which will have to return to the reactor anyway. From the point of view of both fabrication and reactor operation, use of a γ-source has none of the problems associated with spiking, but protection during transport is almost equivalent.

2. The possibility of diversion planned by the national government itself for the purpose of proliferation. However, if a state has reached the stage of nuclear knowledge and general technology at which it can operate a breeder reactor safely, such a state is very unlikely to choose to divert discrete and controlled fuel elements in order to proliferate, and thus overtly renounce its international commitments, when so many other better and easier routes are open.

In both cases, stocks should be kept very low, both at the fabrication plant and at the reactor site.

V.   Conclusion

Without repeating the reasons for advocating breeder deployment, it may be useful to summarize the measures which could render such a deployment more proliferation-resistant. Let us now refer to figure 8.7.

Initially the spent fuel is self-protected during the first few years. To leave it too long in the cooling ponds of the reactor site would increase the proliferation hazard, so it is proposed to ship it to a reprocessing plant as soon as possible. Owing to the present world-wide lack of reprocessing capacity, we must realize that throughout the world by 1990 there will be vast quantities of fuel already cooled for 10 years or more, and this is a real cause for concern.

Reprocessing should proceed only in a few large plants with special attention given to reduce any risk of on-line diversion. This has been summarized in the figure by the word Pipex which refers to the studies mentioned above.

Any stock of non-radioactive plutonium should be kept as low as possible and as a mixture of (U/Pu) oxide. Some international storage should be arranged to protect the purified plutonium waiting for re-use in reactors.

Co-location of the refabrication plant on the storage site would definitely be preferable. This would, of course, imply difficult industrial problems, since few countries would be ready to concede to the reprocessor, under the guise of non-proliferation, an actual monopoly on fabrication. International industrial agreements will have to be designed to solve this problem and to prevent unnecessary transportation of bulk fissile materials. And at the end, it will be essential to put the fuel safely back in the reactor, where its proliferation hazard returns to zero.

Such a set of measures may be pragmatic, and may lack philosophical 'grandeur', but could, I believe, be internationally acceptable; and without some kind of international consensus, any protective measure is bound to fail.

Paper 9.   The role of the breeder reactor

R. GARWIN

Square-bracketed numbers, thus [1], *refer to the list of references on page 153.*

I.   Background

There is near-universal consensus that world security would not be benefited by nuclear weapons in the hands of terrorists or other non-national groups. There is also wide agreement (reflected in the Non-Proliferation Treaty (NPT) and other efforts to control the spread of nuclear weapons) that world security (and especially regional and local security) is likely to be imperilled if nuclear weapons spread to many more nations.

II.   Types of nuclear weapon proliferation

We have already distinguished between (*a*) national, and (*b*) non-national possession of nuclear weapons. In addition we should also distinguish between (*a*) one weapon versus many, and (*b*) optimized nuclear weapons versus terror weapons. Furthermore, one can have (*a*) overt possession of nuclear weapons; (*b*) covert possession of nuclear weapons; (*c*) possession of separated plutonium or other fissile material in metallic form, under international safeguards which would warn of diversion; (*d*) possession of fissile material in less immediately usable form; (*e*) possession of spent fuel rods, together with fuel reprocessing facilities; and (*f*) possession of fresh breeder fuel without reprocessing capability.

Even this familiar but incomplete list reminds us of the complexity of the proliferation problem. Other considerations are the sizeable effort required to design, build and test the non-nuclear portions of the weapon, and the political incentives or disincentives to proliferation.

141

III.   Breeder fundamentals

The production of nuclear power in the light water reactor (LWR) is possible only because the typical fission event produces more than one neutron. On average, precisely one neutron from each fission goes on to cause another fission.

The fission rate and hence the reactor power (200 million electron volts (MeV) of energy from each fission event) is maintained constant by the very slow motion of 'control rods' changing the parasitic absorption of neutrons, or by thermal expansion or other means for changing neutron leakage, or in other ways; slow motions suffice because 1 per cent of the neutrons from fission are delayed one second or more and thus allow plenty of time for control. But fission not only produces more than the one neutron required to continue the chain reaction; fast fission in any of the fissile isotopes (Pu-239, U-233, U-235) gives considerably more than two neutrons per neutron absorbed. Thus, from the earliest days of fission, the possibility of a breeder reactor, in which one neutron per fission would continue the fission chain reaction in the reactor and another neutron per fission would be captured in U-238 eventually to give another fissile atom of Pu-239 has been recognized. Neutron energies above about 0.4 MeV will yield more than two neutrons per neutron absorbed in U-235; neutron energies above about 0.04 MeV will do the same in Pu-239. In addition, Pu-239 has substantially higher neutron excess (over 2.00) per incident neutron above 1 MeV than does either U-235 or U-233 (but U-233 gives somewhat more than two neutrons over the entire energy range, extending to the slow neutrons used in water-moderated reactors). Neither Pu-239 nor U-235 yields enough neutrons per neutron absorbed in the thermal and intermediate energy range to allow a 'thermal breeder'—that is, one in which the fission neutrons are slowed before being captured.

Breeder proliferation hazards in perspective

It is clear that a nation desiring a few nuclear weapons early would not undertake the construction or purchase of a breeder reactor, but would instead build or buy a research reactor to produce plutonium at a rate of about 1 g/megawatt-day of operation. An alternative would be the construction of a centrifuge plant to produce high-enrichment U-235. On the other hand, a nation engaged in long-range planning, especially one with internal pressure groups both in favour of and against getting 'closer to a nuclear weapon capability', could well opt for breeder reactors prematurely simply because of their proliferation potential. Without exhaustive discussion of all cases, we note that while for the LWR, fuel reprocessing and recycling are optional (and at present probably uneconomical), they are essential for the breeder. Furthermore, recycled fuel for the breeder reactor contains large amounts of chemically separable fissile material, protected by a relatively small amount of penetrating radiation. Thus while one can argue whether a given breeder produces more or less net plutonium annually than a given LWR, the breeder Pu stock (as well as its net production) must be purified and recycled if the breeder is to do its job. From the points of view both of scale of operations required and of protection against penetrating radiation, it would be more convenient to extract the 15 per cent Pu content from breeder fuel than the 0.6 per cent Pu from LWR spent fuel.

Nuclear power benefits

My own judgement, expressed in *Nuclear Power Issues and Choices* [1], is that electrical energy from LWRs is competitive with electrical energy from fossil fuel (coal) in large countries with a strong electricity grid. Because of the smaller size of an economical unit of fossil electrical capacity, electrical energy from the combustion of coal is much more economical than nuclear power in small countries, even if the coal has to be imported and stockpiled. Thus, although I believe the proliferation hazards of LWRs can be managed (primarily by delaying reprocessing and recycling of LWR fuel until it is clearly and demonstrably profitable), there would be insignificant impairment of the world's economic well-being if, for some reason, nuclear power did not exist for the next 20 or 30 years. However, by the year 2100, the known reserves of high-quality coal might near exhaustion, and the known reserves of fissile uranium would be an insignificant supplement. Furthermore, it may be important to be able to reduce substantially the input of carbon dioxide into the atmosphere, in order to avoid injurious effects on the world's climate. Two energy resources can serve from the mid-to-long run—solar energy and the breeder reactor. Of the two, paradoxically, the breeder is far more certain to be able to provide electrical energy at near-current costs.

Two advantages have been claimed for the breeder reactor—reduced cost and less national dependence on external 'sources of energy'. Recent prototype large breeder reactor (PLBR) studies in the United States estimated the capital cost of a breeder reactor to be at least $600 million more than that of an LWR. Present uranium costs of some $30/lb would have to rise to some $130/lb before such a breeder could compete with the LWR, even if no economies were achieved in LWR operation. One economy in the present fuel cycle that is likely within the next 10 years is a reduction in the cost of isotope enrichment from the present $80−100/kg SWU to around $20/kg SWU.

But perhaps a nation should buy one or more breeder reactors in order to be independent of a continued supply of low-enriched uranium (LEU)— about 3 per cent U-235 in U-238? In my opinion, a nation would serve its citizens better by deploying LWRs and buying ahead a stockpile of either fuel rods or LEU material from which to fabricate reactor cores for the next 10 years or so. The small size and inert nature of either the LEU material or the fuel rods means negligible physical space and cost required for such stockpiling. The relatively low fraction of the nuclear energy cycle represented by LEU investment means that it is economical to buy fuel five or ten years in advance. Further economies could be realized by stockpiling only natural uranium, depending upon one or another competitive suppliers of enrichment services to enrich and to fabricate fuel a few years in advance.

On the other side of the balance, possession of an operating LMFBR or other breeder reactor in no way guarantees energy independence. For example, the majority of countries operating LMFBRs will probably not have reprocessing and refabricating plants on their territories. Breeder reactor economics are seriously impaired by delay in reprocessing (or, alternatively, by a security requirement to maintain several years' core load in order to cope with interruption of supply of reprocessing/refabrication services).

In the case of LWR fuel, buying an eight-year stockpile of LEU (at an 8 per cent annual interest rate) would only double the LEU-associated portions of the energy costs (raising the busbar cost of electricity by 20 per cent). An LMFBR, on the other hand, is estimated to have a fuel inventory out-of-core ranging between 150 per cent and 50 per cent of the in-core inventory. For a 3 000-kg Pu LMFBR core, this would correspond under normal reprocessing assumptions to a total inventory between 4 500-kg and 7 500-kg Pu. If one required a similar independence of eight years' replacement of half the core annually, the Pu inventory would be increased by something like 12 000 kg.

The most significant effect of the increased inventory would not be on the cost of electricity but on the feasibility of deployment of breeder reactors. Figure 9.1 shows for the United States the plutonium stockpile as a function of time for different assumed Pu inventories for an LMFBR deployed at a very nominal rate of 15 GW(e) per year net increase in nuclear power capacity, beginning in the year 2000. This corresponds to about 3.5 per cent per year net increase in the nuclear component, without major replacement of fossil fuel-generating capacity by nuclear fuel, and without significant transition from direct heat to electricity. Even under these very modest assumptions, only those LMFBR designs significantly more advanced than those now available will avoid a plutonium inventory limitation on their deployment rate. Furthermore, although the plutonium may exist in spent LWR fuel, there may also be a shortage of reprocessing capacity, since there will be no market for the LWR plutonium until the breeder deployment begins.

Clearly, there is an economic incentive for a nation operating breeder reactors to have indigenous reprocessing capacity, but this would forgo economies of scale and would further increase the cost of the breeder reactor and delay the date at which it could compete with the LWR.

Figure 9.1.   US civilian plutonium stockpile versus year of FBR introduction

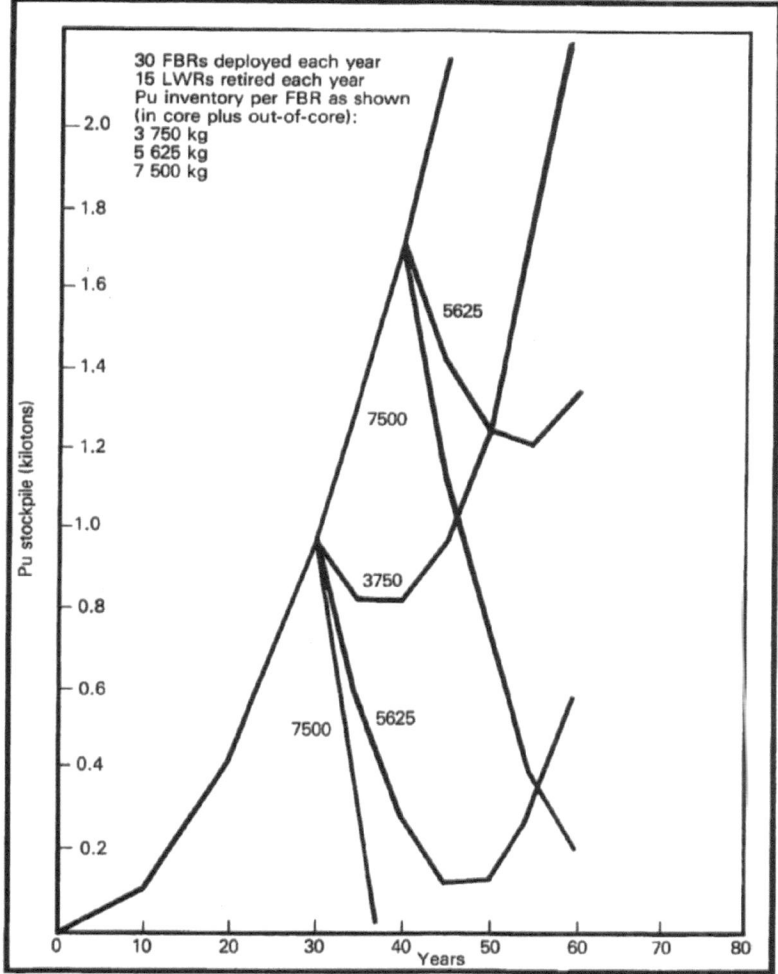

Note: The year of introduction of the FBR is taken as 2000 or 2010, with three assumptions about the required Pu inventory of each FBR. This figure was given to the author by the Director of Energy Research (DoE), 20 June 1978. Note that while NASAP assumes "a 3 750-kg FBR", a memorandum of 7 July 1978 to the Director of Energy Research (DoE) from the acting Director, Office of Fuel Cycle Evaluation (DoE) notes only that "it is reasonable to assume that lower inventory FBRs, such as the 5 625 kg/GW(e) design shown in your curves, could be available if required".

IV.   The proper role of the breeder reactor

In my opinion, the LWR as a producer of commercial electricity has a relatively minor but useful role to play for a short time. The breeder reactor, on the other hand, is an important insurance policy—against a carbon dioxide catastrophe, or against having to pay possibly relatively high costs for solar heat and electricity. It may be important to be able to deploy breeder reactors rapidly, and to provide a considerable portion of the world's energy needs for several centuries.

Introduction

Limited uranium resources

The technical achievement of producing commercial electrical power from nuclear fission may have limited economic and social benefits in view of the limited resources of high-grade uranium ore. The light water reactor which produces most nuclear power uses uranium enriched to about 3 per cent U-235 from its normal isotopic abundance of 0.7 per cent. With an enrichment plant leaving 0.25 per cent concentration of U-235 in the 'tails', a conventional pressurized water reactor (PWR) of 1 million kW peak electrical output (1 000 MW(e)) and operating at full capacity 80 per cent of the time, consumes in its nominal 30-year life 6 970 tons of uranium ore.[1] Thus only about 502 1 000-MW(e) reactors could be fuelled for their 30-year lives by the presently considered uranium resources of the United States.

In view of the present low rate of deployment of nuclear power reactors, there is no possibility that the low-cost uranium will be exhausted by the year 2000, but there is also no considerable opinion that the currently contemplated reserves will last to the year 2100. Therefore, nuclear power in the United States would be a brief and minority contributor to electrical power supply unless some way could be found to (a) find and mine considerably more uranium at an economically and environmentally acceptable cost; (b) develop and deploy nuclear power plants with a substantially smaller appetite for uranium; or (c) provide an alternative source of fissile material to the simple extraction and enrichment route.

Start-up on LWR plutonium

A substantial fraction of neutrons even in an LWR are captured in U-238 to produce additional fissile Pu—typically about 60 per cent—some of which is later fissioned. In fact, 44 years of operation (of a 1 000-MW(e) LWR at 80 per cent capacity factor, assuming that the LWR operates to recycle its uranium) together with 8 536 short tons of U_3O_8 are required to produce the net 7 500 kg of fissile plutonium needed to start one 1-MW(e) LMFBR. Had the Pu also been recycled in the LWR, 1 530 tons of U_3O_8 would have

---

[1] A 'ton of uranium ore', as used in this paper, means a short ton of yellowcake (U_3O_8) of which domestic US resources (known, probable, possible, and speculative) may be 3.5 million tons at a cost of production below \$30/lb. A ton of yellowcake contains 1 700 pounds (769 kg) of uranium metal. Except as noted, all data dealing with alternative nuclear fuel cycles are taken from reference [2].

been saved, together with 1 020 MgSWU,[2] but, of course, no Pu would then have been available for investment in a breeder.

The two extremes for fuelling first generation breeders from LWR plutonium are equally improbable—the first being to accumulate a stock of plutonium from an individual LWR and also from its successor (44 years of operation) and then use it to fuel one LMFBR. The product plutonium would be idle for an average of 22 years (alternatively, the discounted present value of the energy produced by the LMFBR would be much reduced by the delay). The other extreme is to collect all Pu produced by the entire population of LWRs and 'immediately' to invest it in new LMFBRs. This assumes no delay in reprocessing, in fuel fabrication, and in scheduling the start of new LMFBRs. The APS study assumed process times allowing for about two years between the discharge of fuel from an LWR and the incorporation of that Pu-bearing fuel in an LMFBR. Thus one might imagine a new LMFBR to be spawned every two years by each set of 22 operating LWRs, and one every two years by each set of 16 operating LMFBRs, but with a two-year delay in both cases.

Under these assumptions, the growth rate for breeder reactor power is clearly inadequate for nuclear electric generation to take over rapidly from fossil fuels, much less to replace non-electric uses of fossil fuel.

V. What to do?

The Report of the Nuclear Energy Policy Study Group, *Nuclear Power Issues and Choices*, discussed, among other things, the role of nuclear power in the US economy. It notes:

One feature of our analysis, which is particularly important from the standpoint of policy, is the assumption that an 'advanced technology' will be available around 2020. This advanced technology could be an advanced breeder or it could be solar or fusion, advanced coal production and use, or some combination. The prospects of these advanced technologies and the nature and timing of the U.S. breeder program are discussed in some detail in Chapters 4 and 12. For the perspective of this broad economic analysis, the precise costs of these advanced technologies are not important. It is important, however, that one or more of these technologies be available within the next fifty to seventy years to provide assured energy supplies and keep energy prices from increasing rapidly. [1a]

---

[2] A megagram SWU is a 'separative work unit' which, for many kinds of isotope separation (including gaseous diffusion and centrifuge), is directly related to cost. The APS study [2] assumes a cost of $75 000 per Mg SWU—approximately that charged now by the Department of Energy. Note, however, the existence of substantial private and government efforts in the United States and elsewhere to reduce the cost of isotope separation by the use of lasers (laser isotope separation (LIS)), which might result in reduction in cost by a factor of four and a substantially greater reduction in energy requirements.

The main point of this note is to put some flesh on the 'delayed breeder or alternative' and to suggest a useful way to think about the importance of the breeder reactor. Such an application also has important implications for the US breeder research and development programme.

Proposal

The conservative proposal is simply to fuel 'normal' first-generation LMFBRs with 11 250 kg (4 500 kg in-core and 6 750 kg out-of-core—that is, available for the replacement cores while the first core is being reprocessed) of U-235 as 20 per cent in U-238. The LMFBR would be operated at the same power level as usual, and the fuel would be reprocessed and recycled in the normal manner, feeding the LMFBR annually with 1 200 kg of depleted or natural U. We assume that the breeder core portion of the fuel rods will, for the most part, be reprocessed into the next core, to avoid putting the enriched U-235 into the blanket, where it would be less valuable. This would be a minor requirement to incorporate into the reprocessing plant.

At 0.25 per cent tail concentration of U-235, the assumed amount and concentration of U-235 would be obtained from 3 110 short tons of U_3O_8. The uranium mined and stripped of U-235 to fuel the LMFBR would then suffice to sustain it (or its successors) for more than 2 000 years of operation at 1 000 MW(e)) and 80 per cent capacity factor even if there were no excess plutonium production in the mature LMFBR. Thus, instead of fuelling 500 LWRs for thirty years, an assumed uranium resource of 3.5 million short tons of U_3O_8 would fuel 1 077 LMFBRs for more than 2 000 years even if their breeding performance were far worse than has already been demonstrated.

There is nothing new about starting LMFBRs with enriched U-235; most LMFBRs in the world have in fact been started in this way because of the lower cost of fabricating fuel with enriched uranium than with plutonium, and because of the limited availability of LWR plutonium. This possibility gives greater flexibility for deployment of additional, enduring nuclear power. In particular, one does not need to recycle Pu now (or to avoid recycling Pu now) in LWRs with a view to the LMFBR future; one need not deploy first-generation LMFBRs now, in order to (very slowly!) breed the LMFBR population to an appreciable level—which from the above, would take a long time. Scenarios exist which show a transition from LWRs to LMFBRs within the limitation of available plutonium; these show a 30-year period of low growth of the nuclear power sector for precisely the reasons I have given. I consider it unlikely that such a long transition (with simultaneous building of LMFBR and new LWR capacity) would be economically desirable. It should be cheering to observe that success in designing an economically superior breeder could result in total displacement of new LWR construction by LMFBRs, by virtue of the flexibility of U-235 investment, as needed, of LMFBRs.

Table 9.1. Economic penalty of starting 1 000 MW fast breeders with enriched U-235

	Water reactor plutonium	20 per cent U-235 in uranium (separate core reprocessing and recycle)
Fissile amount required from external source for start-up and replacement loadings (kg)	7 500	11 250
Value of fissile material ($/kg fissile)[a]	19 900	31 000
Total cost of fissile material ($mn)	149	349
Loss of breeding-gain fissile production		
(kg fissile Pu)	0	1 700
($mn)	0	34
Contribution to fuel cycle cost levelized over 30-year breeder plant life[b]		
Purchase of fissile material for start-up (mill/kWh)	2.2	5.3
Loss of breeding-gain fissile production (mill/kWh)	0	0.3
Relative total (mill/kWh)[c]	2.2	5.6
Levelized fuel cycle cost (mill/kWh)[c]	2.0	5.4

[a] Plutonium value is calculated for alternative use as a water reactor fuel.
[b] Calculated from time schedule of fissile purchases and sale, using utility discount factor of 0.0755/year.
[c] The relative total *not* the total fuel cycle cost. Later credits from breeding gain fissile production and cost of fabrication and reprocessing result in an estimated LMFBR levelized total fuel cycle cost of about 2.0 mill/kWh [2a].

Cost of LMFBR start-up on enriched uranium

The APS Study Group has compared the cost of starting a 1 000-MW(e) LMFBR with enriched U-235 with the cost for fuelling it with water-reactor plutonium. Such a comparison has validity only if the desired deployment rate of LMFBRs can be accommodated by the stock of LWR plutonium—otherwise the choice is between having electrical power from breeders invested with U-235 or not having it from breeders at all.

Table 9.1 assumes that water-reactor plutonium is bought at the price which would be paid for it for recycle into water reactors. (Actually, there is no guarantee that it could be produced for this price, in which case it might not be available at all or could be more costly. According to the assumptions of table 9.1 there would be approximately a $200 million penalty for starting an LMFBR with U-235. Taking the estimate at face value, there is an expected penalty of about 3.4 mills/kWh in starting an LMFBR with enriched uranium instead of LWR plutonium when the latter is available. (It should be noted that reduction by a factor of four in cost of isotope separation should reduce this penalty to about $75 million—some 1.1 mill/kWh.)

Other alternatives

I have already noted that LWR plutonium will probably not be available in sufficient quantity when one wants to build LMFBRs. I further note (as commented by the APS Study Group) that an LMFBR core could be reopti-mized to require less U-235 (the calculations were done for U-235 loading of a core optimized for steady-state plutonium operation); if the cost incurred per reactor in changing core configuration after two years or thereabouts is considerably less than the value of the U-235 saved, then one clearly would want to change the configuration (or use a more complicated interactive system in which some specialized LMFBRs processed enriched U-235 into a smaller amount of Pu for conventional equilibrium LMFBRs). But substan-tial cost reductions are possible simply from relaxing the requirement for a breeder with a conversion ratio exceeding 1.00. In the conventional LMFBR with as high a breeding ratio as possible, considerable constraints and costs are incurred in order to obtain the net breeding excess of 8.4 per cent (net annual fissile plutonium production divided by fissile *core* inventory) which results in the 3.3 per cent maximum LMFBR population growth rate when one considers out-of-core Pu as well. If one relaxes the requirement that the LMFBR produce more Pu than it consumes (because one plans to invest each LMFBR in the future with its own core of U-235), one can use wider coolant channels, thicker fuel cladding, and in general do things which will reduce the capital cost of the LMFBR (which is four or five times the $200 million 'penalty' from start-up of U-235). Part of this new-found flexibility can be traded off in increased (or more assurance of) safety.

In fact, no magic attaches to a conversion ratio of 1.00. A conversion ratio of 0.99 is almost as good, and even significantly lower conversion ratios such as might be obtained with modified Canadian heavy water (CANDU) reactors would be of benefit, if only to give more time for the achievement of conversion ratios of 1.00 or better.

Benefits

In all, the potential for starting LMFBRs with U-235 allows: (*a*) an arbitrary deployment rate (independent of LWR or breeder history); (*b*) lower-cost, greater safety potential, possible higher efficiency because of the absence of a constraint of high conversion ratio. A conversion ratio of 1.00 is fine, although one would not reject a breeder with a higher conversion ratio; (*c*) earlier availability of a useful breeder; and (*d*) potentially increased benefits from lower-cost advanced isotope separation techniques and from lower inventory breeder/converters. The costs for these benefits cannot exceed and may be less than the 3.4 mill/kWh which is computed for LMFBR deployment conditions ideally suited to the Pu-based LWR–LMFBR transition.

Non-proliferation characteristics of the proposal

Investing LMFBR or other breeders with U-235 instead of Pu has some modest anti-proliferation effect in comparison with the normal view of the fast breeder reactors.

1. Initial fabricated cores would require isotope enrichment to yield weapon-usable fissile material, as is the case with LWR fuel or natural uranium.

2. Pu recovery from irradiated LWR fuel could and should be deferred until a large number of U-235 invested fast breeder reactors were in operation, thus reducing the early availability of large amounts of Pu. Spent LWR fuel should be kept in interim or recoverable storage until the breeder era.

3. Such a system is amenable to so-called 'Civex-type' fuel reprocessing (if the Civex concept could be developed as a practical, economical process), which, by reducing the reprocessing delay, could also reduce the out-of-core fissile requirement and thus the U-235 cost.

Implications for the US breeder R&D programme

Already mentioned is the possible benefit of redesign of the LMFBR core to minimize U-235 inventory and thus reduce costs. Additional analysis and some experimentation should be done on this possibility. The likely scale of these benefits is not known (or if known, is not published).

New interest might focus on the molten-salt breeder (MSBR) with continuous reprocessing and *no* out-of-core inventory. Aside from the corrosion problems of the MSBR (which have not been solved to the degree necessary for a viable commercial technology), the system may be ideally suited to the non-breeding role envisaged for U-235 start-up. This benefit would arise from the low in-core U-235 inventory of about 2 500 kg (and the absence of out-of-core fissile material), which would allow the previously assumed 3.5 million short tons of U_3O_8 to fuel 5 000 MSBRs for more than 500 years (even if during all that time no one had a better idea as to how to make a breeder with a conversion ratio significantly exceeding 1).

It should be noted that on such multi-century time scales, even a growth rate of 1 per cent per year (of breeder reactors) would provide a substantial increase in electrical power availability (if population growth can be held to zero). In any case, humanity can afford to pay a far higher cost for the small amount of uranium needed to continue operation of these breeders than it can for the large amounts of U_3O_8 necessary to fuel LWRs or to start breeder reactors. Thus, sustaining breeder operation on uranium costing $5 000/lb would contribute about as much to the cost of breeder electricity as the present $30/lb uranium contributes to the cost of electricity from LWRs.

VI. Conclusion

It is most important to look at the breeder reactor as a means of using most of the fuel value in our uranium resources, but we should not ignore the existence of U-235 in this uranium which makes possible deployment of an arbitrarily expanding breeder population. It is *not* important that the conversion ratio exceed 1, and certainly not that it be much bigger. Abandoning the requirement to start breeders from LWR plutonium or to produce Pu from breeders to fuel other breeders allows one to modify the design of the LMFBR in order to reduce the cost of breeder electricity, to improve safety, and to further reduce the uranium investment required to fuel a new LMFBR.

Furthermore, this approach will encourage the broadening of the present breeder R&D programme to emphasize rapid acquisition of knowledge about breeders and near-breeders of low capital cost and low fissile inventory. It is possible that such a reactor will turn out to be cheaper than LWRs for the production of electrical power; it is not important, however, to have a cheaper source of electricity or to hasten the deployment of a more expensive source of electricity which uses less of our uranium resources. What is important is to have the knowledge so that we can plan for the eventual transition from inefficient consumption of U-235 to the more efficient consumption of U-238 as the cost of producing uranium rises.

It seems to me that this view of breeders (and even an eventual reduction by a factor of four in enrichment costs by the use of laser isotope separation) is much more plausible and thus worth much more emphasis than other means of expanding fuel supply such as electro-nuclear 'breeding', the 'fusion hybrid', or the obtaining of energy from pure fusion. This does not imply that we should retard the acquisition of knowledge about the potential feasibility of these alternatives—simply that we should not prematurely go beyond the test-bed stage into much more costly and uneconomic subsidized prototypes.

Deployment of breeder reactors

Were there no potential contribution of the breeder reactor to the proliferation of nuclear weapons, there would still be important questions regarding the introduction of the breeder. Enthusiasts for a new technology like the commercial supersonic transport (or, for that matter, those opposing the introduction of a new technology) may propose courses of action which imply the commitment of society to expenditures per marginal unit of product many times larger than the alternative, thereby making society poorer as a whole.

Premature introduction of an early breeder will have just this effect, since its electricity will be more costly than that from an LWR. It is difficult to see a highly competitive supply of breeder reactors, so one might expect only a slow improvement in the technology and slow reduction in cost of the breeder. It would be far better, in my opinion, to support work on breeder concepts, experimental work on breeder fuels, research into economical reprocessing and refabrication, and to delay the deployment of the breeder reactor until private suppliers (or in the case of non-market economies, generally accepted and not-too-distorted economic analyses) showed a substantial benefit of the breeder over LWR or fossil plants in the short term.

Also tending in several ways to increase the cost of the breeder is the general argument that breeders must be deployed early in order to gain experience and in order that their (not very great) breeding rate provide the plutonium stock to support exponential growth of the breeder economy. The preferable alternative, as described above, is to deploy breeder reactors of low total inventory, with initial cores of medium-enriched uranium (MEU). We have a technology to do this now; we could do it now and we should do it when the resource cost of uranium supply to an LWR becomes high enough to outweigh the greater capital costs of the breeder.

Note that additional flexibility is available in this way. For instance, if an economical module of breeder fuel reprocessing plant could handle the continuing needs of 50 breeders, approximately two such plants would have to be built every three years to handle the 30 LMFBRs deployed per year under the assumptions of figure 9.1. But the construction and actual operation of these reprocessing plants might economically be delayed several years in view of the fact that the early breeders could be supplied for several years with fresh MEU cores, building a stock of spent breeder fuel which could be economically processed by the plants coming into being about the time of breeder introduction. Spending delayed is almost as good as spending avoided.

References

1. Keeny, S. M. *et al.*, ed., *Nuclear Power Issues and Choices* (Ballinger Press, Cambridge, Mass., 1977).
   (a) —, 'Energy and the economic future', chapter 1, p. 68.
2. Hebel, L. C., 'Report to the APS by the Study Group on Nuclear Cycles and Waste Management', *Reviews of Modern Physics*, Vol. 50, No. 1, Part II, January 1978.
   (a) —, p. 7.

Chapter 7. Hybrid reactors

Paper 10. Fusion — fission hybrid reactors

V. KULESHOV

I. Nuclear fuel supply problems

Many countries have been studying controlled thermonuclear reactors (CTR) on the basis of either magnetic or inertial plasma confinement. Progress made in this field suggests that the first industrial-scale fusion reactors should be created in this century. Not only can new fuels be used in the fuel-power balance by the utilization of fusion energy, but also this creates a qualitatively new potential for satisfying the fuel balance requirements. For this purpose, hybrid reactors incorporate many advantages. Much more efficient conversion of nuclear raw material into Pu or U-233 fuel is possible in hybrid reactors. The reduction of radioactive wastes by orders of magnitude is inherent for 'pure' fusion reactors. Fusion reactors could produce very high-temperature heat, both for direct utilization of heat and direct conversion into electricity, and as a result considerably increase thermal efficiencies and reduce thermal pollution of the environment. They also open the possibility of producing some valuable isotopes, and so on.

Long-term nuclear power, whether fission or fusion, will represent an inexhaustible energy source and will be able to maintain the progress of mankind without any restriction of fuel supply in the long run.

In this respect, nuclear energy presents a most convincing case for assuming that the so-called 'limits of growth' associated with the finite reserves of our planet can be overcome by scientific achievements. Realization of the advantages of nuclear power depends, however, on overcoming the difficulties characteristic of present-day developments. It is known that in many countries, the rate and scope of nuclear plant construction have turned out to be lower in the present decade than was planned at the end of the 1960s or the beginning of the 1970s.

Although the building of nuclear power plants has not decelerated due to insufficient reserves of nuclear material, the fuel supply will have a decisive effect on the rate, scope and structure of nuclear power engineering and application later on.

Nevertheless, the problem of satisfying the future nuclear fuel balance requirements cannot be considered as solved either in practice or in principle.

The principal way to solve the nuclear fuel problem is to increase the natural uranium utilization factor from the present 1 per cent to 60—70 per cent by converting U-238 (and later on Th-232) into the man-made nuclear fuels—plutonium and U-233. Taking into account the advantages in using the increasing amounts of depleted uranium from production of enriched uranium and the development of the fuel base directed towards the uranium —plutonium cycle (uranium production, radiochemical reprocessing and fuel fabrication), it may be expected that in a few decades uranium will remain the basic fuel of nuclear power plants.

Improved LWRs could make it possible to increase the natural uranium utilization factor. The potential for reactors with the heavy water moderator is still higher. However, only the most optimistic estimates would suggest that optimization of the designs and operating conditions could increase the utilization factor more than 2—3 times that of present-day thermal reactors.

In spite of different opinions concerning the solution of the fuel problem, the dominant one concerns the role of the fast breeder reactors (FBR). Such reactors operating on the uranium—plutonium cycle could reliably ensure a conversion factor in excess of unity, thus permitting maximum use of uranium resources. Progress made in developing the sodium-cooled FBRs (BN-350 in the USSR, Phénix in France, PFR in the UK) suggests that such reactors should be mastered on an industrial scale in the next one or two five-year plans. But the recently dominant concept of concentrating all nuclear power engineering on the FBR now appears to be far from optimum. FBRs have the unquestionable advantage over thermal reactors of high conversion efficiency. However, FBRs can lose this advantage in other engineering and economic respects if they are used for peak load generation in electricity-producing systems, or for high-temperature heat supply, or for marine power supply on large ships. Therefore, thermal reactors of various types may be expected to be a considerable or even the main constituent of future nuclear power engineering. In this case, additional amounts of plutonium (or U-233) produced in FBRs will be required to provide the necessary fuel for thermal reactors. The plutonium doubling time is estimated as 10 to 20 years in typical FBR designs, taking into consideration the present-day fuel cycle potentials. With such long doubling times, the FBRs are unable to cope either with the required rates of development (their contribution to the nuclear generating capacities remains small both now and for the next century) or, especially, with a transition of thermal reactors to burn the plutonium fuel.

If the present situation does not change, either we have to reduce the growth rate and scope of nuclear power generation, confining its role to particular problems and retaining fossil fuel as the primary source of energy, or we must continue the development of nuclear power plants on U-235, increasing the production rates of natural uranium and developing new technologies to use poor ores.

The application of advanced fuels, for example, carbides or nitrides instead of oxides, in principle makes it possible for FBRs to reduce the fuel doubling time to approximately 5−7 years. However, this would require the simultaneous reduction of time for external fuel cycle operation. New transport means and improved technology for chemical processing of highly radioactive fuels after a short (3- to 4-month) cooling period would be required. To solve the nuclear fuel problem completely requires additional advances in thermal reactors to improve their fuel balance, the application of heavy water moderators, increased burn-up, the utilization of new fuel and structural materials, the realization of the Th−U-233 fuel cycle, and so on. In other words, the solution of the nuclear fuel problem requires the modernization of practically all constituents of the nuclear fuel cycle and its engineering. Many of these measures will certainly cause an increase in the cost of electricity.

Taking into account the above considerations, a search for other ways to satisfy nuclear power plants with fuel is quite justified.

II. Hybrid reactors

There are two principal methods of converting nuclear raw material into nuclear fuel, without involving a reactor. These are electronuclear and fusion methods.

The main problem of the electronuclear method is the creation of a 1-GeV accelerator capable of a beam current of about one ampere and a conversion for electric energy into proton beam energy of $\approx$ 0.5 kA. Research in some countries has shown that such an accelerator is feasible.

When the electronuclear method is used, a high-energy proton produces in a uranium target neutrons which are absorbed in U-238 to form Pu-239. If 40−50 neutrons are produced per proton absorbed, the method provides a positive power balance, taking into account the subsequent plutonium burning in fission reactors.

In fact, if the heat−electricity conversion efficiency is 0.3, E_p = 1 000 MeV is the proton energy, E_f = 190 MeV is the fission energy, the fission probability in 'burning out' the Pu-239 nucleus is 0.7, CR $\approx$ 0.6 —the Pu conversion ratio for LWRs, and E_n = 50 MeV—the average energy released in the uranium target per neutron, then the total energy multiplication factor of the system, including an accelerator plus thermal LWR, is k = 2.5.

This value k is not very large if additional indirect consumption of energy as well as high prices for the accelerator and reactors are taken into account. Such an energy balance will result in a high cost of power production, and this fact is the main disadvantage of the electronuclear method.

Table 10.1. Characteristics of a hybrid reactor, obtained from estimates based on the Tokamak system and metal uranium gas-cooled blanket

Characteristic	Unit
Fusion power	690 MW
Total thermal power under steady-state conditions	5 100 MW
Maximum current of 14 MeV neutrons through first wall	4.6×10^{13} n/cm^2 s (100 W/cm^2)
First wall surface area	410 m^2
Magnetic field on plasma axis	50 kG
Chamber dimensions	$R = 6.4$ m; $a = 1.6$ m
Tritium breeding ratio, K_T	1.2
Mean power release density in blanket	90 W/cm^3
Maximum plutonium accumulation in uranium fuel	1.1 per cent
Plutonium delay time in blanket and in external cycle	3 years
Plutonium production rate per year	4.1 t
Specific production rate Pu, r_{po}	0.8 t/GW(th) year or 2.7 t/GW(e) year with thermal efficiency of 30 per cent

In a hybrid reactor, the 14 MeV neutrons from a D−T reaction penetrate the first wall to the uranium blanket where, due to fission reactions, for the first collision they can produce about 3.1 neutrons per D−T reaction. After one neutron is absorbed in lithium to reproduce the 'burned out' tritium nucleus, about one neutron is captured by U-238 to form Pu-239, which can then be used in fission reactors. Simultaneously, the energy E of about 130 MeV is released in the reactor. If E' is the energy consumed for the ignition and maintaining the fusion reaction, per D−T reaction, then the electric power multiplication factor of the system, including a hybrid reactor plus fission reactor, is

$$k = K \cdot K_i \left[E - (n_n - 1)E_f / (1 + \alpha)(1 - CR) \right] / E'.$$

If the Lawson criterion is satisfied (to ignite a reaction), E' can be made very small and $k \rightarrow \infty$. However, k can be rather high due to the large energy release in uranium and plutonium production, even without achieving the self-sustaining fusion reaction conditions if the required plasma temperatures are maintained, for instance, due to the injection of accelerated deuterons. (For example, for $E = 10$ MeV and injector efficiency $K_i \approx 0.5$, $k \approx 13$. Note that the second k turned out to be a few times larger than the first one.) It follows that the main role of hybrid reactors in the total power balance of a nuclear power system is not to generate the energy but rather to produce plutonium for fission reactors to operate. The plasma parameters necessary for a fusion reactor to operate represent a rather long theoretical extrapolation of the experimental values obtained in the magnetic confinement systems. Experimental studies of inertial plasma confinement are still at the earlier stages of research. The creation of fusion reactors requires the solution of a number of problems. They include the development of magnetic systems based on the use of superconductors with high magnitudes of the critical magnetic field and acceptable mechanical properties. Also, the first containment wall should be able to sustain strong radiation, both internal and at the surface. Plasma heating and fuel injection methods, as well as the removal of the reaction products and impurities from the first wall, should be developed.

The favourable power balance of the fusion reactors and the new prospects they open for power generation indicate, however, that the idea of a fusion—fission hybrid reactor is highly attractive. Therefore, the efforts to create such systems are completely justified.

The hybrid reactor has a better neutron balance compared with the FBR. It is therefore able to provide appreciably higher breeding rates for artificially made nuclear fuel.

More detailed studies have shown that a hybrid reactor burning, like an FBR, either depleted or natural uranium will produce 6—7 times more 'marketable' plutonium per unit of heat output than an FBR. Unlike FBRs, the construction of hybrid reactors would not be limited by the plutonium doubling times. As for tritium, doubling times in hybrid reactors can be made sufficiently short, due to a large excess of neutrons, to supply all that is needed.

It may be expected that the specific capital cost could be reduced for the hybrid reactors compared with 'pure' fusion reactors because the power output in the hybrid reactors is many times higher in relation to the similar expensive 'thermonuclear' equipment.

The principal factor determining the economic viability of hybrid reactor construction, however, is their high rate of plutonium production. Due to this fact, a relatively small number of such installations are capable of having a strong influence on the fuel balance, and thus on the entire structure of the power generation system.

It should be mentioned that if the present relatively low prices for uranium and plutonium are taken as a basis, an elementary economic estimate suggests that it is not economically profitable to produce plutonium in expensive plants. The first-generation hybrid reactors are likely to be such plants. Such estimates, however, cannot be extrapolated to the future since the price for uranium and plutonium is likely to change considerably. This change will reflect not only the costs for a given type of fuel but also the relationship between them and increasing prices for the fuel they supplant (oil, gas, coal or natural uranium).

To estimate the measures to be taken to satisfy the fuel supply requirements of future nuclear power generation, it is necessary to take into account the costs required for the entire fuel cycle and power production system as well as the variations of these costs when one or another measure is realized.

For example, if a 1-GW hybrid reactor produces plutonium in an amount sufficient to fuel four fission reactors with a total power output of 4 GW, but it costs twice as much, this would cause a 20 per cent increase in the cost of the power production. Other ways of meeting the requirements in fuel necessary for generation of the same power (utilization of coal or uranium obtained from poor ores, or reactor and fuel cycle modernization) will in many cases lead to still higher costs.

Hybrid reactors are capable of accumulating sufficient quantities of fissionable material in fabricated fuel elements containing raw materials so that these fuel elements can be used directly in fission reactors. This possibility should be studied in detail because it would allow, for example, U-233 produced in a thorium fuel element to be burned directly in a high-temperature graphite reactor, without special reprocessing and refabrication. In addition, it would in principle be possible to achieve 10−15 per cent burn-up by using depleted or natural uranium in the hybrid reactor blanket. Such a possibility, if realized, could solve the nuclear fuel problem almost completely without any chemical processing for recycling of fissionable material. In this case the fuel fabrication is considerably simplified.

The development of hybrid reactors is simpler from the point of view of technology than is the development of 'pure' fusion reactors. It is not necessary for the plasma parameters of the hybrid reactor to satisfy the Lawson criterion. Therefore, the fusion power and plasma dimensions can be reduced for moderate magnetic fields. On the other hand, the use of the uranium blanket may reduce the neutron loading on the reactor first wall, and help to solve this major fusion reactor problem. For the same reason, it is possible to expect that specific capital costs will be decreased. At the same time, the production of plutonium in quantity should cause an increase in the level of these costs allowable from the standpoint of economic conditions. Taking the above factors into account, it is evident that the hybrid fusion−fission reactors designed primarily to produce plutonium may become the first application of a fusion reactor in practice. It is also a step towards the future development of a pure fusion reactor.

Paper 11. Laser fusion and fusion hybrid breeders: Proliferation implications

D. WESTERVELT and R. POLLOCK

Square-bracketed numbers, thus [1], *refer to the list of references on page 172.*

I. Introduction: energy in the very long term

Inevitably, world needs for energy will increase and, equally inevitably, sustainable sources will have to be found to replace the energy capital that is now being dissipated. Such sources are limited to solar, geothermal, fuel-efficient fission, and fusion. Ultimate choices must lead to environmental and societal conditions that are acceptable, controllable and stable. At one extreme in energy-policy advocacy are those who propose abandonment of 'hard technologies' entirely [1] in favour of 'soft technologies'. This approach would render the subject of this paper irrelevant. By focusing on a single facet of the hard technology régime, the authors do not intend to discount societal arguments, since these are highly germane to broad considerations involving nuclear weapon proliferation.

We shall nevertheless focus on a substantial world-wide effort in a particular high-technology area that carries with it strong technical implications for proliferation: that of laser fusion and of fissile-material production in fusion−fission 'hybrid' reactors. Until an optimum mix of sustainable sources becomes clear, it is likely that all promising avenues will be followed; each has advantages and handicaps. The costs and societal (but probably not environmental) impacts of fusion are the least clear among the sources listed, while fusion is the most difficult technically and the implications for proliferation cannot be ignored. Proliferation concerns are the primary subject of this paper.

II. Fusion as a process

Thermonuclear fusion can occur with a number of the lighter isotopes; table 11.1 lists a few of the more interesting reactions. In practice, the fundamental problem is to reach a high enough temperature to drive the reaction,

161

Table 11.1. Possible fusion reactions

Reaction	Characteristic temperature (keV)	Confinement parameter nt (sec/cm^3)
$D + T \rightarrow He^4 + n + 17.6\,MeV$	25	2×10^{14}
$D + D \rightarrow T + p + 4.0\,MeV$	120	2×10^{16}
$D + D \rightarrow He^3 + n + 3.2\,MeV$	120	2×10^{16}
$D + He^3 \rightarrow He^4 + p + 18.3\,MeV$	100	6×10^{14}
$p + Li^6 \rightarrow He^4 + He^3 + 4.0\,MeV$	450	4×10^{16}
$p + B^{11} \rightarrow 3\,He^4 + 8.7\,MeV$	$\simeq 300$	$\simeq 5 \times 10^{15}$

and simultaneously to confine the reactants at high enough density long enough for appreciable burn-up. Temperatures required are extremely high by ordinary standards: several tens of millions of degrees Celsius. The characteristic temperatures shown in table 11.1 are typical. At the temperatures listed, the product of density and confinement time needed for energy break-even (reaction yield equal to initial material energy) is a minimum, shown in the third column. An operating power reactor must of course exceed these parameters.

Research on controlled fusion began in earnest in the 1950s. Contrary to the expectations of that era, the task has proved extremely difficult. There is still no proof of scientific feasibility, although fusion specialists now feel that success cannot be far away.

This slow progress is due not to lack of efforts but simply to the extraordinary difficulty of creating the conditions for fusion—which in nature only occur in the centre of stars. A measure of this difficulty is given by the characteristic temperatures and confinement parameters listed in table 11.1. Clearly the DT reaction is the most accessible, and this reaction forms the prime focus of current fusion research. Unfortunately, the product neutron carries 80 per cent of the energy yield, so use of this reaction for practical power requires conversion of neutron energy to thermal energy. And, of course, tritium must be bred, and a tritium inventory managed.

Reactions involving more highly charged nuclei, such as lithium or boron, will be much harder to drive simply because the higher collision velocities needed to overcome the coulomb forces demand even higher temperatures. With only charged reaction products (no appreciable neutrons), conversion to useful energy would be simplified and the worrisome problems of neutron activation and structural damage could be avoided. However, it seems clear that, for the indefinite future, practical fusion applications must be based on the DT reaction, or perhaps eventually the DD reaction. In either case, the power economics of fusion, as with fission, will depend very strongly on the exploitation of neutrons.

Any energy technology addressed to the long-term future must be based on a fuel or energy supply that is inexhaustible (in a practical sense) and widely available. Table 11.2 illustrates that fusion and its applications fulfil these conditions very well indeed. World energy use in 1976 was 8.5

Table 11.2.   Potential nuclear fuel resources

Resource	Basis	Fuel (10^{12} g)	Energy yield (TW-year th)
Uranium—US	Ores > 25 p.p.m. U	12	18 000
Uranium—World terrestrial	(Reasonably assured)	170	250 000
Deuterium— seawater	33 g/m^3, 50% recovery	23 000 000	250 000 000 000
Lithium—US	To three times 1975 price	6	7 000
Lithium—World terrestrial	(Reasonably assured)	84	100 000
Boron-11—US	At 1971 prices	30	70 000
Boron-11—seawater	4 g/m^3—50% recovery	2 700 000	6 500 000 000

Source: Reference [2].

TW-years (10^{12} watt-years); clearly for a nominal 10 TW-year world, all of the materials (including fissionable but largely non-fissile uranium) listed are in effect inexhaustible. One of the principal applications of fusion to be considered, however, is conversion of non-fissile uranium to fissile plutonium, and the proliferation implications of that conversion will not easily be ignored.

At present there are two principal approaches to the fusion problem, magnetic confinement fusion (MCF) and inertial confinement fusion (ICF), each of which includes a variety of specific alternatives. Both approaches are discussed briefly, and their relation to nuclear weapon proliferation is examined.

Magnetic confinement

This approach relies on strong magnetic fields to hold a hot, fully ionized fuel plasma long enough for a substantial fraction of the fuel to react. The leading contender here is the class of toroidal machines called Tokamaks, with back-up efforts in mirror machines and pulsed high-energy pinch devices. Typically, a Tokamak reactor would operate with a plasma ion density of approximately 10^{14} cm^{-3}, and a ratio of plasma pressure to magnetic pressure of 5 per cent. At this low particle density the burn must last for several seconds to produce net power. Ideally, continuous operation would be desirable. It remains to be seen just how closely this ideal may be approached.

Although recent results from the Princeton Large Torus appear encouraging, proof of scientific feasibility is not yet available. As recently as March 1978, in a report concentrating on Tokamak confinement problems, the Electric Power Research Institute (EPRI) concluded:

The scientific problem then appears to be the establishment of the physics principles necessary to create, contain, heat, shape, and keep clean, a high temperature plasma in a fusion environment. To proceed towards this goal from present-day experiments, the plasma physics issues that must be addressed are:

1. demonstration of satisfactory equilibria and their gross stability.

2. determination of plasma transport scaling laws and their limitations.

3. development of high power supplementary heating methods along with the knowledge necessary for extrapolating their results.

4. determination of proper descriptions and methods of controlling the interaction of the plasma with its containment environment, including neutrals, particles, impurities from the walls or limiter, and radiation [3].

Nevertheless, there is great confidence that these scientific issues will be laid to rest by the mid-1980s.

Scientific feasibility is of course not the whole story. Beyond this lies technological feasibility—building a device that produces net output in usable form and incorporates practical fuel handling and energy conversion equipment. And beyond this lies commercial feasibility—building a reliable machine capable of economical operation in competition with alternative sources of power.

Most of this remains for the future, but at least the general characteristics of a Tokamak reactor may be visualized. Table 11.3 shows the major characteristics of a few conceptual reactor designs [3]. To a large extent these conceptual studies have identified technological problems—such as nuclear damage to the reactor structure—and established their impact on reactor characteristics and potential costs. It is clear that Tokamak reactors would be large and costly.

Table 11.3. Conceptual reactor designs

Conceptual reactor designs	R (metres)	B_{tor} (teslas)	I (megamperes)
UWMAK-I	13	8.7	21
UWMAK-III	8.1	8.8	16
CULHAM II	7.4	8	12
CTRD	16.7	8	12
JAERI	10	12	8
High density	5	7.7	. 8.4
EPR (GA)	. 4.5	7.86	11.4
DPR (GA)	7	7.5	19

Source: Reference [3].

It is interesting to contrast these conceptual designs with similar characteristics of operating Tokamak experiments (table 11.4) and of the next generation experiments (table 11.5). Tables 11.4 and 11.5 show the breadth of interest in magnetically confined fusion; the effort is world-wide and intense. Several of the proposed experiments in table 11.5 would cost in the neighbourhood of $200 million.

164

Table 11.4. Some of the world's Tokamak experiments

Operating experiments	R (m)	B_{tor} (teslas)	I (megamperes)	Country
LT 3	0.4	1.4	0.033	Australia
TFR	0.98	6	0.4	France
Petula	0.72	2.5	0.1	France
Pulsator	0.70	2.8	0.3	FR Germany
Tokamak	0.83	10	1.0	Italy
JFT-L	0.90	1	0.07	Japan
CLEO-TOK	0.90	2	0.120	UK
DITE	1.13/1.17	2.8/2.7	0.216/0.218	UK
ATC	0.90	15/35	0.060/0.140	USA
ST	1.09	4.5	0.120	USA
PLT	1.32	(5)	1.6	USA
ORMAK	0.80	2.5	0.40	USA
ISX	0.92	1.8	0.20	USA
Microtor	0.30	6/25	0.020/0.10	USA
Macrotor	0.90	2/7	0.040/0.25	USA
DOUBLET IIA	0.66	0.8	0.140/0.30	USA
ALCATOR	0.54	8	0.640	USA
FT-1	0.625	1	0.030	USSR
TM-3	0.40	4	0.050	USSR
T-4	0.90	4.5	0.200	USSR
TO-1	0.60	0.7	0.020	USSR
T-6	0.70	1.5	0.060	USSR
T-10	1.50	1.0	1.6	USSR

Table 11.5. Proposed MCF experiments

Proposed experiments	R (m)	B_{tor} (teslas)	I (megamperes)	Country
DOUBLET III	1.4	2.6	5	USA
JET	2.96	3.4	4.8	Euratom
TFTR	2.48	5.6	2.5/1.0	USA
T-20	5	3.5	6	USSR
JT-60	3	5	3.3	Japan
ORMAK Upgrade	1.5	3.5	1	USA
PDX	1.45	2.4	0.5	USA

It is clear that there still is a long way to go before magnetic confinement fusion becomes a potentially attractive energy source *per se*; it is included in the present discussion for another reason: its potential as a source of fissile material. It is reported, for example, that the Soviet Union intends to turn its largest Tokamak (T-20) into a hybrid [2]. Thus while MCF has little implication for development of weapon technology, its connection to the proliferation problem is evident.

Inertial confinement

Inertial confinement fusion is the alternative approach. Its relation to nuclear weapon work poses a potentially serious proliferation problem. The confinement is described as 'inertial' because its duration is determined primarily by the mass of fusion fuel or its immediate surroundings. Burn occurs during the small, finite interval of time required to accelerate and move material through a distance comparable to its initial dimensions. At the temperatures and pressures necessary for fusion, this time is very short—of the order of nanoseconds (10^{-9} s) or less.

It is easy to show that the mass of material—deuterium and tritium, for example—required for fusion under inertial confinement varies inversely with the square of the fuel density. Considering that one gram of DT burned to completion yields 3.5×10^{11} joules (about 80 tons of explosive yield) and so poses macroscopic containment problems, it is apparent that for applications other than weaponry, small masses, and therefore high compressions, are desirable. Not until the arrival of the laser with its intrinsic ability to concentrate energy in space and time was inertial confinement fusion regarded as a serious candidate. The concept of the ablatively driven implosion brought the additional interest to ICF [4].

Today the field of possible 'drivers' for ICF includes pulsed lasers in the wavelength range of 0.25 to 10 μm, and pulsed beams of $1-1$ MeV electrons, $5-10$ MeV protons, or heavy ions in the tens of GeV energy range. All of these function by heating and ablating the surface of a pellet containing DT fuel, blowing off surface material and consequently compressing and heating the fuel.

Fusion implosion experiments have to date been carried out only with 1 μm and 10 μm lasers, and with electron beams. Because of beam power limitations, pellet designs tested so far are intrinsically incapable of reaching ignition or producing significant fuel burn-up. Nonetheless, thermonuclear neutrons have been observed and the results of experiments are in accord with theoretical expectations. A more detailed review of ICF is available which states that, in addition to the large US programme, major programmes in inertial confinement fusion are under way in the USSR, France and Japan, while the UK, FR Germany, Israel and Canada have significant but smaller programmes [4]. With new high-power laser and electron beam drivers now coming into operation, it is reasonable to expect dramatic progress toward significant fusion yields in the next few years.

There are still major problems to be addressed and overcome before scientific feasibility, let alone technical or commercial feasibility, can be demonstrated. Heading the list, and of greatest significance with regard to proliferation, are the problems of fuel-pellet design. For successful ignition and efficient fusion at manageable total yields, compressions to densities several thousand times that of normal liquid DT are needed. This will demand precise control of the driver power history, full knowledge of all modes of energy re-distribution occurring in the target, and the techniques for avoiding catastrophic growth of instabilities. Much has been written about this in the context of simple spherical targets [5]. However, it appears that the expected solution to these problems lies in a body of knowledge about classified targets, of which little has been released other than an anticipated gain ratio [6]. This aspect of the ICF–proliferation problem is discussed later in this paper.

Power reactors using inertial confinement technology will differ strikingly from magnetic confinement designs. Perhaps their major advantage will lie in physical separation of the driver system from the reaction chamber or nuclear 'island'. This feature should materially ease the problems of engineering neutron capture blankets and thermal recovery systems, without concern about magnetic coils.

A number of conceptual studies have been carried out to date, with emphasis on laser fusion. The studies have tended to converge toward similar designs, and therefore highlight similar problems.

A good example is the SOLASE study carried out at the University of Wisconsin [7]. Characteristics of SOLASE are:

(a) 1 000 MW(e) fusion power;
(b) net efficiency of 30 per cent;
(c) 1 MJ/pulse, 20 Hz laser, 6.7 per cent efficient; and
(d) 6 metre reactor cavity, graphite wall, Li_2O blanket.

Although SOLASE differs in many specifics from other conceptual designs, some general conclusions about a laser-driven fusion power plant can be drawn. The SOLASE authors see the following as major issues (in rough order of priority):

(a) fission feasibility: $G = 150$ at 1 MJ;
(b) laser efficiency;
(c) pulsed power technology with 10^9 cycle lifetime;
(d) first-wall protection;
(e) pellet manufacture; and
(f) tritium inventory.

III. Fusion/fission hybrid: fissile material production

It has been pointed out that "fusion is under fire" [2]. Early claims that fusion power would be cheap, clean and absolutely safe have not survived the course of 25 years of hard labour. Why then should there be increasing interest in fusion–fission hybrid reactors which, at first glance, would seem to combine the worst features of both?

The answer depends on one's point of view. Fundamentally, the attractions of the hybrid stem from the tenfold energy advantage of fission over fusion; 180 MeV per plutonium or uranium atom fissioned versus 18 MeV per DT reaction. The 14 MeV neutron from the DT reaction can be used to induce a fission for prompt energy, or it can be multiplied and captured in a fertile material to produce fissile atoms for later use in a fission reactor. At

167

least from an energy standpoint, the hybrid turns the neutron yield of DT fusion into an asset rather than a liability.

The authors are not aware of any serious interest in the simple fusion—fission reactor with enough fissionable material to greatly enhance the reactor's power. This would combine the complexity of fusion with the radio-activity of fission, and offer little advantage. However, the fusion/fissile material breeder, sometimes called the fusion fuel factory [8], is another story. This system will be referred to in the discussion about the hybrid reactor. It can, as already seen, involve either MCF or ICF.

For the fusion proponent, the hybrid offers the potential for earlier entry into economically feasible programmes; the tenfold energy multiplier could make practical a fusion reactor which would not otherwise be economical. And, even better, this could be achieved without any significant diversion from the main path of development towards an ultimate pure fusion power plant.

For the electric utility, the hybrid gives fusion higher credibility. Utilities on the whole have rather little confidence in fusion, particularly as a direct power source. It is at the outer reach of their present view—something that will be handled by the next generation, or perhaps the one after that. One concern follows from possible problems of coupling a pulsed system into an electrical grid. Another has to do with the projected size of fusion power plants. Focusing on the economical 1 000-megawatt reactor eliminates most of the customers. No utility will commit more than 10 per cent of its capacity to any one installation, or at least any one generating system within an installation. No utility will commit more than 5 per cent of its capacity to a relatively untried technology. Examining the size distribution of all US utility companies, one finds that approximately 90 per cent of the entire base is removed if one considers only 1 000-megawatt or larger systems [9].

The hybrid, however, presents quite a different picture. Off-line breeding with no direct connection to the power grid could be quite attractive. The EPRI has forecast a substantial shortage of fuel—either nuclear or non-nuclear—for electricity generation to meet US demand by the year 2010 [10]. Fifty hybrids, each supporting 5–10 light water reactors, could relieve this shortage.

Finally, for those concerned about nuclear weapon proliferation but willing to accept the notion that fission must play an ever-increasing role in energy supply, the hybrid offers the possibility of breeding reactor fuel in only a few carefully guarded nuclear parks [4, 8, 11, 12].

The general features of the hybrid fusion fuel factory have been discussed, giving particular consideration to the breeding of U-233 in thorium, taking advantage of the small fission cross-section of thorium to hold down fission heating in the fusion reactor [8]. Copious (n, 2n) reactions on Th-232 will lead to contamination of the U-233 by strongly gamma-active U-232, providing an intrinsic safeguard against diversion; the concentration of U-233 will be higher than in a thermal thorium cycle reactor.

Taking into account the conversion factor of the fission reactor (employing reprocessing), it is concluded that one hybrid with a fusion power of 1 GW can supply make-up fuel for 12–25 light water reactors of similar power, or for 50 or more advanced converters like the CANDU. Cost estimates are necessarily quite imprecise, but a price of $120 to $450 per gram of U-233 is arrived at, corresponding roughly to U_3O_8 at $180 to $700 per pound. This is a fairly high price; according to the figures quoted, net fuel cost in a light water converter reactor would be 0.6 to 2.2 cents per kWh(e), in current dollars [8]. To achieve these costs requires that the 60 per cent conversion to fissile material in the LWR be exploited by reprocessing. Without reprocessing, the costs would run to 1.5 to 5.5 cents per kWh(e) and a given FFF would be able to support only 5 to 10 LWRs.

Fissile plutonium production

The preference of one author [8] for the thorium–U-233 cycle is due to its resistance to proliferation, although some of the U-238/Pu-239 schemes described by other authors have similar characteristics. Other schemes depend directly on production of fissile Pu-239 in depleted U-238, which is available in great quantities, followed by reprocessing to separate the Pu-239, the very sequence thought most proliferation-prone. This scheme would, for each hybrid breeder, provide fuel for six light water reactors, and it would increase the supply of U-based fuel up to 100-fold. Some version of it may well be the scheme of choice for hybrid utilization in the long term by reason of its comparative economics. The problem will be to make it proliferation-proof [13].

The economic factor is strong. According to a recent report, the busbar cost of nuclear-derived electricity to the Commonwealth Edison Company (CECo) in 1977 was 13.3 mills/kWh, compared with 24.1 mills/kWh for coal-fired systems [14]. Nuclear fuel cost was 3.5 mills/kWh versus 12.1 mills for coal. CECo's estimated fuel costs per kWh for the late 1980s are 7 mills (nuclear) and about twice that for coal. Because of the capital-intensive nature of the electric power industry, carrying charges will become an increasingly large fraction of total power costs in the next decade, but this is inescapable while casual acceptance of substantially increased fuel cost— beyond that due to inflation—is not. The equivalent U-235 cost quoted by the author referred to above [8], based on the U-233 scheme, would raise the predicted CECo nuclear cost from 7 mills to 20–65 mills/kWh and the total cost of electricity by 40–170 per cent. On the other hand, fuel bred in the U-238/Pu-239 cycle is estimated by at least one source to raise the cost of electricity by only 20 to at most 40 per cent depending on the cost of the hybrid facility [13].

It is not possible, within the scope of this study, definitively or even authoritatively to sort out the many conflicting estimates of future costs for all the hybrid fuel schemes under consideration. It is enough to note that a one-mill busbar cost difference in a 10 TW-year society translates to a nearly

$90 thousand million swing in the world cost for power, and we are talking of many millions and perhaps trillions of dollars annually. While the United States remains uncertain about reprocessing and a plutonium economy, the rest of the world seems to suffer no such uncertainty [15, 16]. The US Department of Energy is funding hybrid studies at the level of $1.8 million in 1978 fiscal year; one objective is the identification of fuel cycles addressing the goals of the Non-proliferation Alternative Assessment Programme [12].

Without attempting to predict the outcome of US policy, it is believed that aggressive pursuit of technical alternatives to the present open-ended fuel cycles, capable of minimizing proliferation risks, can succeed, and that the gigantic scale of world-wide energy costs will provide sufficient incentives, if not imperatives, for success. If this estimate is correct, Pu-239 production in hybrid reactors can be expected to play a role determined mainly by economic factors, without substantial implications for nuclear weapon proliferation in nations where proliferation is not the national intent and safeguard procedures are accepted, or even where they are not. Further, it is suggested that fusion hybrids are unlikely to be available outside countries that already have a fission-weapon capability in less than 30 years, by which time the proliferation problem will have been resolved one way or the other [2].

Thus any danger is not immediate. However, we foresee that the ICF hybrid could some day prove to be a fairly direct path not only to fission, but also to thermonuclear weapons. This possibility is discussed in the next section.

IV. The question of knowledge and technology

It has been observed that "inertially confined fusion . . . provides a capability for studying the physics of nuclear weapons and simulating their effects on a laboratory scale", and that conditions for fusion "are attained in a nuclear weapon by use of the energy release from a fission reaction, which drives a fuel mass to fusion conditions" [4]. In a discussion of the link between fusion activities (ICF in particular) and fusion weapons there are concluded to be two issues: tritium and knowledge. The first is dismissed, perhaps too casually, as of little importance. The second is considered more serious:

Whereas lack of access to fissile material has been a significant technical barrier to the spread of fission weapons, and getting the fissile material is a significant part of the task of getting a fission bomb, having fusion fuels (of which tritium is only one of the possibilities) is by comparison a very minor part of the much more difficult task of getting a fusion bomb. The spread of knowledge is another matter . . . To the extent that lack of certain insights and degrees of technical sophistication have limited the spread of fusion bombs until now . . . the spread of inertial confinement fusion research may be spreading a limiting ingredient for fusion weaponry [2].

In noting the tight classification shroud covering certain aspects of ICF work, it is inferred that, until proven otherwise, the linkage between ICF technology and that of fusion weapons must be considered a major liability of ICF [2]. A third issue is added to the two identified in the passage above [2]. The importance of the 'fission trigger' for thermonuclear weapon operation [4] is well known. All nuclear weapon states now in possession of thermonuclear weapons started with fission weapons.[1] The time-lapse between fission-weapon capability and thermonuclear-weapon capability has tended to diminish since the original development of thermonuclear weapons by the United States. These facts suggest that denial of fissile material to a non-nuclear weapon state is equivalent to denial of a thermonuclear-weapon capability as well as a fission-weapon capability. Thus there are two 'limiting ingredients' for weaponry. ICF hybrids uniquely threaten to supply both ingredients.

V. Conclusion

Most of the concern expressed about proliferation has focused on fission weapons (hence reference to the "juxtaposition of thousands of existing megatons on the one hand and a few hypothetical kilotons on the other" [16]). The prospect of kiloton proliferation is bad enough; a Hiroshima-sized bomb on the centre of the District of Colombia would severely damage an area bounded by M Street on the north, the Jefferson Memorial on the south, the Kennedy Center on the east and the National Gallery on the West. But most of the District would escape with minor damage (the Pentagon and Embassy Row would be relatively unscathed). On the other hand, a weapon of just over a megaton would destroy almost the entire District, or the city of San Francisco, or most of Stockholm or Damascus (all remarkably similar targets). In our view, the link between fission-weapon and fusion-weapon capability, and the vastly greater danger inherent in the latter have been subjects not adequately exposed by the arms control community.

To the extent that ICF may contribute to facilitating that link, or even possibly provide an effective (and legitimate) means to short-circuit the historic linkage, its technology deserves to be controlled. Success in ICF may

---

[1] In a press conference in September 1954 explaining the role and conduct of Los Alamos in connection with the US thermonuclear weapons programme, Norris E. Bradbury, Director of the Los Alamos Scientific Laboratory, stated that "Technically, the development of fusion weapons is so inextricably allied with and dependent on the development of fission weapons, that great success in the former had to follow success in the latter."

establish a substantial technical basis useful for thermonuclear weapon capability. It is possible that success in ICF will not occur until long after the more direct historic route to thermonuclear weapons has been followed by all who make the choice to do so. Nevertheless, the ICF hybrid, should it exist, might uniquely supply—either in its design and development or in its product— essential ingredients for a full fission and thermonuclear weapon capability. Unsafeguarded ICF hybrid activities should therefore be regarded with grave suspicion.

References

1. Lovins, A. B., 'Energy strategy: the road not taken?' *Foreign Affairs*, Vol. 55, No. 1, October 1976, pp. 65–96.
2. Holdren, J. P., 'Fusion energy in context: its fitness for the long term', *Science*, Vol. 200, No. 4338, pp. 168–80, 14 April 1978.
3. *Assessment of the Tokamak Confinement Data Base,* Report No. ER-714, Electric Power Research Institute, March 1978.
4. Stickley, C. M., 'Laser fusion', *Physics Today*, Vol. 31, No. 5, May 1978.
5. *Laser Programs Annual Report—1976*, UCRL 50021-76, Lawrence Livermore Laboratory, 1976.
6. Maniscalco, J. *et al.*, 'A laser fusion power plant based on a fluid wall reactor concept', Paper presented at the Annual A.N.S. Topical Conference on Controlled Nuclear Fusion, Santa Fe, New Mexico, 9–11 May 1978.
7. Conn, R. W. *et al.*, *SOLASE, A Laser Fusion Reactor Study*, Report No. UWFDM-220, University of Wisconsin, December 1977.
8. Bethe, H.A., 'The fusion hybrid', *Nuclear News*, May 1978, p. 41.
9. *Utility Requirements for Fusion Power*, Report No. EPRI ER-714, Electric Power Research Institute, March 1977.
10. *Research and Development Program for 1978–1982, An Overview*, Electric Power Research Institute, 1 September 1977.
11. Booth, L. A. and Frank, T. G., *Commercial Applications of Inertial Confinement Fusion*, Report No. LA-6838-MS, Los Alamos Scientific Laboratory, Los Alamos, New Mexico, May 1977.
12. Kintner, E. E. and Stickley, C. M. 'Fusion energy applications in the world's energy future', Paper presented at the Annual A.N.S. Topical Conference on Controlled Nuclear Fusion, Santa Fe, New Mexico, 9–11 May 1978.
13. Maniscalco, J., 'A conceptual design study for a laser fusion hybrid', Paper presented at the 3rd Topical Meeting on the Technology of Controlled Nuclear Fusion', September 1976.
14. Rossin, A.D. and Rieck, T. A., 'Economics of nuclear power', *Science*, Vol. 201, p. 582, 18 August 1978.
15. *World Armaments and Disarmament, SIPRI Yearbook 1978* (Taylor & Francis, London, 1978, Stockholm International Peace Research Institute).
16. Rose, D. J. and Lester, R. J., 'Nuclear power, nuclear weapons and international stability', *Scientific American*, Vol. 238, p. 45, April 1978.

Part III

Introduction

J. GOLDBLAT

Under the Treaty on the Non-Proliferation of Nuclear Weapons (NPT), which was signed in 1968 and entered into force in 1970, the nuclear weapon states are committed not to transfer nuclear weapons or other nuclear explosive devices, while the non-nuclear weapon states are under obligation not to receive, manufacture or otherwise acquire such weapons or devices or control over them. In addition, the nuclear weapon states are not allowed to assist, encourage or induce any non-nuclear weapon states to manufacture or acquire the devices in question. Cooperation among nuclear weapon powers in the field of nuclear arms, with the exception of transfer of the weapons themselves, has not been prohibited.

More than any other arms control agreement in existence, the NPT depends for its effectiveness mainly upon universal acceptance. This goal has not been achieved. By December 1978, only three nuclear weapon powers—the UK, the USA and the USSR—out of five had joined the NPT, while a dozen non-nuclear weapon countries with significant programmes for peaceful uses of nuclear energy had kept their nuclear weapon option open, and most of them are unlikely to join the treaty in the foreseeable future. The non-parties include India, which has exploded a nuclear device termed 'peaceful', and Israel, which has been reported to possess several untried nuclear bombs, as well as South Africa, which was rumoured to have been on the brink of testing a nuclear weapon.

In assessing the durability of the NPT, account must be taken not only of the fact that almost one-third of all the nations are still outside the treaty, but also of the constant danger that the behaviour of non-parties may influence the behaviour of the parties. The NPT is not of indefinite duration: 25 years after its entry into force (of which 10 will have passed in 1980, when the second NPT Review Conference is convened), the parties will decide whether or not its validity is to be extended. An erosion of the treaty may begin even sooner, since each party has the right to withdraw from it at any time, invoking its "supreme interests". Thus, the international legal barrier against further nuclear weapon dissemination is far from being impermeable.

The temptation to 'go nuclear' may grow in parallel with the spread of nuclear technology for peaceful purposes, especially with the increasing availability of fissionable materials. To guard against possible diversion of these materials to the production of nuclear explosives, the NPT requires that non-nuclear weapon states accept international safeguards on all their peaceful nuclear activities. Agreements providing for such safeguards are in force between the International Atomic Energy Agency (IAEA) and over 60 individual countries. The IAEA safeguards techniques, which include on-site inspection, have developed considerably over the years, reaching a high degree of sophistication in surveillance and containment, as described in Papers 12 and 13. Further improvements could be achieved through changes in the design, construction and operation of nuclear facilities. The authors of the Papers envisage the possibility of safeguarding even large uranium enrichment, reprocessing and plutonium fuel fabrication facilities.

It should be kept in mind that only parties to the NPT are obliged to accept safeguards on their entire peaceful nuclear effort. Non-parties are subject to safeguards of a different type, which apply solely to IAEA projects or to projects carried out under bilateral agreements concerning individual plants; their indigenous nuclear activities may not be covered by international supervision.

But even the effectiveness of full-scope safeguards is difficult to evaluate. Some participants in the SIPRI symposium pointed out that the margin of error in material accountancy, especially in large civilian nuclear programmes, may create a loophole facilitating diversion of 'significant' quantities of fissionable material, and that possible military uses of nuclear energy for non-explosive purposes, for example, for the propulsion of war-ships or submarines, are not to be covered by IAEA controls performed in accordance with the NPT. Others considered it inappropriate for one organization, the IAEA, to carry out both promotional and regulatory acti-vities. Still others raised the question of the fallibility of IAEA inspectors.

The main problem, however, is that the function of IAEA safeguards is merely to detect diversion quickly enough for some kind of response. And in the case of plutonium or highly enriched uranium, it may require no more than weeks, if not days, for diverted nuclear material to be transformed into an explosive. Indeed, it would be prudent to assume that certain non-nuclear weapon states have already designed nuclear weapons, and perhaps even developed their non-nuclear components, since there is nothing in the NPT or in the existing agreements on nuclear transfers to prevent such activities. All that will be needed for such countries to produce a nuclear weapon, if they ever make a political decision to do so, is to get hold of the necessary amount of weapon-grade fissionable material.

But supposing that a diversion of the fissionable material has been detected well in advance of the time at which a state could exercise its nuclear weapon option, what kind of response could stop the defaulter from producing the weapon? The IAEA Statute provisions, applicable in such a case, are as follows:

The inspectors report the non-compliance to the Director-General who transmits the report to the Board of Governors; the Board may call upon the recipient state to remedy the non-compliance "which it finds to have occurred", and report to IAEA members as well as to the UN Security Council and General Assembly. In the event a corrective action is not taken by the state in question "within a reasonable time", the Board may direct curtailment or suspension of assistance provided by the IAEA or a member state, and call for the return of materials and equipment made available to the recipient state. The IAEA may also suspend a non-complying member from the exercise of the rights of membership.

The procedures described above are exceedingly cumbersome. *First*, it may be difficult for the inspectors to acquire evidence which would be convincing enough for the IAEA Director-General and the Board to set the sanction machinery in motion. *Second*, in view of its politically delicate nature, the action envisaged is bound to be slow, leaving a violator the time needed to carry out his design. And *finally*, if ever applied, the penalty is rather weak as compared to the gravity of the violation. However, in the present state of the world, more effective international measures to force a nation to desist are hardly feasible, while a collective punitive action against a transgressor is unthinkable. This reasoning brought one of the participants in the symposium to express the view that undue significance attached to international safeguards may create a false sense of security. (However, no one disputed the value of national safeguards against theft of fissile material by subnational groups.)

Even those who consider it important strictly to control the flow of fissile materials for deterrence purposes agree that clandestine diversion is not the only problem, and that an open abrogation of safeguards, under any pretext, by a non-nuclear weapon state possessing the wherewithal to manufacture nuclear weapons is also a possibility. Therefore, more far-reaching constraints than safeguards are needed to meet the danger of proliferation or at least to maximize the time needed to produce a nuclear weapon. In this connection, the possibilities of limiting or even denying transfer of enrichment and reprocessing facilities or technology, as well as weapon-usable materials, were discussed. Such restrictions can be introduced through a concerted international action of supplier states or through unilateral action of such states, or both.

An international action taken by a group of supplier states, the so-called London Club, has already led to the adoption of *Guidelines for Nuclear Transfers*, tightening the terms for transfer of nuclear items and technology, and reducing the advantages that non-parties may derive from remaining outside the NPT. But the guidelines are still insufficient to ensure that no further proliferation will occur as a result of transfers, because they do not require safeguards on all nuclear activities in recipient states as an absolute condition for nuclear supplies, and because they do not definitely preclude exports of highly sensitive facilities.

As regards unilateral action, the US Nuclear Non-Proliferation Act of 1978 is the most recent example of limitations imposed by a supplier state upon recipient countries. In particular, the act requires full-scope safeguards; a guarantee that no nuclear material transferred to a country or derived from US exports will be re-transferred or reprocessed, enriched or otherwise altered without prior approval of the USA; as well as a guarantee that no plutonium or highly enriched uranium transferred or derived from US exports will be stored in any facility that has not been approved in advance by the USA. As an incentive for the recipient states, the act envisages the development of international approaches for meeting world-wide nuclear fuel needs. (For an overview of the act, see Paper 14.) Certain other supplier countries have decided to follow a similar restrictive export policy without adopting special legislative acts.

It is common knowledge that many recipient states resent any con-straints that go beyond the usual safeguards requirements. They consider them discriminatory and therefore unacceptable. For some, permanent dependence on nuclear material supplies in general, and on uncertain, changeable export policies of the supplier states in particular, may prove intolerable—economically and politically—whatever the incentive. Export controls, based on the present monopolistic position of certain countries, may buy time but in the longer term they may become counter-productive in stimulating efforts by individual states or groups of states towards nuclear self-sufficiency, and in thereby weakening the non-proliferation régime.

It would stand to reason that, to be fully effective, non-proliferation policies must be acceptable to the supplier as well as the recipient states. The International Nuclear Fuel Cycle Evaluation (INFCE), which was set up in 1977 to study ways of making nuclear energy for peaceful purposes widely available, while minimizing the danger of nuclear weapon proliferation, has provided a framework within which relevant problems, equally important to both categories of states, can be explored. However, INFCE has not generated radically new ideas or solutions, and the chances of achieving broad agreement on alternative, so-called proliferation-resistant fuel cycles, which would reduce the probability of nuclear power programmes being utilized for military ends, are rated low. (See Paper 17.)

New possible anti-proliferation institutional arrangements were also discussed at the symposium. One concept, that of multinational nuclear fuel cycle centres, where technological and economic advantages could be combined with easier means of supervision, was considered attractive, but only if such centres were established soon enough to prevent further signifi-cant spread of enrichment and reprocessing facilities, if they were widely recognized as the sole legitimate means of meeting nuclear fuel cycle needs, and if appropriate controls on the stockpiling of nuclear materials were instituted. However, all these conditions are not likely to be met and, even if they were, a possible competition among centres could lower the standards of control, and give rise to conflicts of loyalties, as was pointed out in Paper 18.

Internationalization seems to be a better way to reduce the incentives to develop the sensitive parts of the fuel cycle. In this connection, one partici-pant in the symposium suggested that the possibility provided in the IAEA

Statute to make, under the aegis of that organization, arrangements for the international management and storage of plutonium and spent fuel, should be seriously considered and wherever possible realized (see Paper 15). Another participant went further by proposing that the operation of enrichment, fuel fabrication, reprocessing and waste disposal plants be carried out only by, or under a licence of, an International Nuclear Fuel Agency (INFA). According to this proposal, measures would have to be taken to ensure that no weapon-grade materials from civilian programmes were in existence anywhere outside the control of the international authority. And since the undertaking to submit to INFA control would apply to all nations, the nuclear weapon states would have to separate their military programmes from the civilian use of nuclear energy (see Paper 18).

An alternative non-proliferation arrangement, which would avoid the emotional issue of surrender of sovereign rights inherent in other proposals, was also considered. The idea is to establish on a voluntary basis one, world-wide cooperative open to all states, parties and non-parties to the NPT, with a view to ensuring the availability of adequate and reliable supplies of uranium and enrichment and reprocessing services. The members of the cooperative would undertake not to export nuclear materials, services, facilities or technology to non-members. Enrichment and reprocessing would remain under national control, and member states would be free to construct new capacity. But proliferation of nuclear weapons would be restrained through an undertaking by the members of the cooperative to deny themselves access to the services of enrichment and reprocessing facilities beyond that required to meet the immediate needs of their power programmes, and to accept safeguards developed to verify observance of their undertaking. This would preclude the stockpiling of enriched uranium and plutonium under the control of member states and would remove the danger of stockpiles being diverted to nuclear weapon production. (See Paper 16.) The concept of a nuclear fuel supply cooperative was found useful, provided that the cooperative was joined from the beginning by a group of states at least as large and influential as that constituting the London Club.

The participants in the symposium were unanimous in recognizing that the problem of non-proliferation is basically political—that progress in disarmament and the resolution of the most acute regional conflicts would be needed to diminish the stimulus and pressure for proliferation. Some participants, however, argued that it was impossible under any circumstances to sever the link between civilian nuclear power and the spread of nuclear weapons. It was pointed out that the power reactor cycle may not be the easiest route, technically or economically, towards a nuclear weapon capability, but it certainly is a convenient one in the sense that a political decision to produce a bomb is not required at the time a civilian nuclear programme is embarked upon. The decision may be just deferred.

Many experts felt that excessive emphasis is placed on nuclear energy in general, and in relation to the needs of underdeveloped countries in particular, without a comparably large effort being invested in developing alternative sources of energy. To remedy this situation, one participant proposed setting up a World Energy Organization, as a UN specialized agency, to promote and encourage research on various forms of energy other than nuclear and on their utilization (see Paper 18).

Still others expressed the view that, from the technical point of view, nuclear weapon proliferation is difficult, if not impossible to stop—among other reasons, because the spread of nuclear know-how cannot be stopped. But even they conceded that stricter control could delay the process and that this, in itself, was a worthwhile endeavour.

Chapter 8. Safeguards technology

Paper 12. IAEA safeguards technology

A. VON BAECKMANN

Square-bracketed numbers, thus [1], *refer to the list of references on page 185.*

I. Introduction

More than 20 years ago, during the discussions leading to the founding of the International Atomic Energy Agency (IAEA) and the adoption of its statute in the United Nations, serious concern was expressed that the promotional activities of the IAEA might also lead to an undesirable spread of nuclear weapon capabilities. It was agreed that the IAEA should be authorized to complement its promotional activities by adopting appropriate control measures. The provision of these control measures—normally called international safeguards or IAEA safeguards—was embodied in the statute of the IAEA and reflected in all its project agreements. It was also expected that, as more international cooperation in the nuclear field was channelled through the Agency, IAEA safeguards would extend even further so that the Agency would become a powerful bastion against the proliferation of nuclear weapons.

The application of IAEA safeguards, however, was not limited to its own projects. Under its statute the Agency can also assume safeguards responsibilities whenever this is requested by a member state (unilateral submission) or whenever this is foreseen in bilateral agreements for cooperation between states in the nuclear field (safeguards transfer agreements). In order to standardize IAEA safeguards, to make them universally applicable and to improve their effectiveness, guidelines for their implementation were developed and finally approved by the IAEA Board of Governors in September 1965 (the 'INFCIRC/66/Rev. 2' System for Agency Safeguards [1]). Due to the fact that the system was developed to enable the Agency to apply safeguards to its own projects or to projects promoted under bilateral agreements for cooperation in the nuclear field, which cover as a rule an individual facility or a limited number of facilities, the system was primarily facility-oriented.

II. The NPT and safeguards

A new situation was created, however, when in March 1970 the Non-Proliferation Treaty (NPT) came into force. By 1978 this treaty had 101 non-nuclear weapon state parties, in addition to the three nuclear weapon states—the USSR, the UK and the USA. It has proven since to be a most important international instrument against the proliferation of nuclear weapons. The treaty specifies, among other things, that each non-nuclear weapon state party to the treaty undertakes (*a*) not to receive from any transferor whatsoever either nuclear weapons or other nuclear explosive devices, or control over such weapons or explosive devices either directly or indirectly; (*b*) not to manufacture or otherwise acquire nuclear weapons or other nuclear explosive devices; (*c*) not to seek or receive any assistance in the manufacture of nuclear weapons or other nuclear explosive devices; and (*d*) to conclude an agreement with the IAEA submitting all nuclear material in all peaceful nuclear activities to IAEA safeguards. The structure and content of these agreements between the Agency and states, required in connection with the NPT, were discussed in 1970 and 1971 by the Safeguards Committee, which advised the Board of Governors on safeguards-related questions. The report of the committee (INFCIRC/153) was adopted by the Board as the basis for negotiating agreements required by Article III of the NPT. (A SIPRI book published in 1975 [1] describes in considerable detail this history of IAEA safeguards and related international agreements.)

Under the NPT-type safeguards agreements, the Agency is responsible for independent verification of compliance with the provisions of the safeguards agreements. The wide international acceptance of these obligations has caused a major shift in the Agency's safeguards activities from a facility-by-facility approach to a state-wide, full-scope, nuclear material-oriented approach.

III. The implementation of safeguards

INFCIRC/153 specifies that IAEA safeguards should be implemented for the exclusive purpose of verifying that the safeguarded material is not diverted to nuclear weapons or other nuclear explosive devices. The technical objectives of the IAEA safeguards system are (*a*) the timely detection of the diversion of significant quantities of nuclear material from peaceful nuclear activities to the manufacture of nuclear weapons or of other nuclear explosive devices or for purposes unknown; and (*b*) the deterrence of such diversion by the risk of early detection. Material accountancy, complemented by containment and surveillance, is used to achieve these objectives.

The implementation of IAEA safeguards requires the establishment of a report and record system based on measured nuclear material flow and inventory data and on a material balance area (MBA) concept with one or several MBAs for each nuclear facility. The responsibility for this system of nuclear material accountancy and control lies with the national authorities. Comprehensive recommendations for establishing and operating such a system are under preparation in the IAEA. Based on verified design information, safeguards procedures for each individual facility are developed and agreed upon by the IAEA and the national authorities. The safeguards measures are then laid down in a specific facility attachment which becomes part of the subsidiary arrangements to the safeguards agreement with that state.

The national authorities submit periodic reports on inventory changes, inventory taking and material balances for each material balance area. The IAEA Safeguards Department evaluates these reports for completeness and correctness and sends its inspectors to the nuclear facilities or other locations where nuclear material is reported to be present, to verify independently the correctness of the reports and compliance with agreements.

At the facility, the inspector compares the reports with relevant internal records and source data to verify that the reports are not falsified in order to conceal diversion, and independently verifies the presence of the reported quantities of nuclear material. After return to headquarters, the inspection reports are evaluated and the conclusions of the Agency's verification activity are transmitted to the state. These technical conclusions are statements, in respect of each material balance area, of the amount of material unaccounted for (MUF) over a specific period, giving the limits of accuracy of the amounts stated. Specific information relating to the implementation of IAEA safeguards is given annually to the Board of Governors.

Material accountancy requires measurement of material inventory and the flow of material in and out of material balance areas. INFCIRC/153 specifies accounting with respect to amounts of uranium and plutonium elements as well as U-235 and U-233. These measurements are performed by the plant operator, who reports the results through the national authorities to the IAEA for independent verification. Methods used for verification measurements aim primarily at determining the quantities of uranium and plutonium as elements and of the specific isotopes U-233 and U-235. Other parameters such as isotope composition of plutonium, impurities, and so on, may be determined if required for the determination of primary data. For example, if calorimetry is used for the determination of plutonium, its isotope composition and the date of last separation from Am-241 must be known and are verified.

Verification techniques should, as far as possible, be simple, tamper-resistant, accurate and reliable. The simultaneous and sufficient accomplishment of these four objectives is not always feasible.

Verification techniques

For verification purposes, two kinds of measurement technique are normally employed: non-destructive assay techniques (NDA) and chemical analysis (DA) of representative samples, combined with a determination of batch volume or batch weight. Although non-destructive assay techniques do not always allow for clear-cut conclusions and although their results are often not as accurate as results from chemical analyses, NDA techniques are normally preferred because of the lower cost, the fast availability of results, and the ease with which these techniques can be used during inspections.

NDA techniques for determination of nuclear materials [2] are as a rule based on measurement of characteristic decay features of the material (passive techniques) or on measurement of induced radioactivity (active techniques). In addition, absorption or reflection of nuclear radiation or X-rays is also used for analysis. It is sometimes difficult to employ active techniques in the field because of difficulties with regard to transport of the radioactive sources and the required administrative effort (environmental protection).

The most important NDA instrument used by IAEA inspectors is the 3 AM-II (stabilized assay meter)—a relatively simple two-channel γ-spectrometer used, as a rule, together with a sodium iodide detector. By proper setting of the two channels, uranium, uranium-235 or plutonium can be identified and semi-quantitatively determined with this system. A more sophisticated instrument, based on the same principle but with improved electronic and detector capability (high resolution γ-spectroscopy), will be available for in-field testing in the near future. For more complex determinations, for example, burn-up measurement, Pu-isotopic analysis or the measurement of UF_6 in transport cylinders, high-resolution γ-spectrometry is employed. Intrinsic germanium detectors in combination with 1 000- to 2 000-channel analysers are available for these measurements. Because of the complex data reduction required for this type of measurement, the information is stored on magnetic tape cassettes and processed in a central computer at IAEA headquarters in Vienna.

All these instruments are portable so that the inspectors may carry them from one facility to another. Although special transport containers have been developed for safe shipment of equipment—mostly as air-freight —damage during transport sometimes creates considerable difficulties, particularly with respect to the intrinsic germanium detectors.

Other NDA techniques used for verification measurements are less universally applied than γ-spectrometry and generally less well implemented. Occasionally calorimetry and neutron-coincidence counting are used for the determination of plutonium, and β-reflectometry is used for the analysis of UO_2 pellets. Other NDA safeguards techniques under development are based on auto-radiography, γ-absorptiometry, thermal imaging and active neutron interrogation.

The most simple method for verifying the presence of at least a minimum quantity of special fissionable material in a small research reactor is to ascertain that the reactor is in operation. In many research reactors this can be done by observation of the Čerenkov glow. The total nuclear material inventory in such installations is often not significantly larger than that required to maintain criticality, so that the observation of reactor operation is in itself sufficient to support a conclusion of non-diversion.

Further development with respect to the use of NDA techniques for safeguards purposes is directed towards: (a) improvement of those features which facilitate in-field application (for example, portability, robustness, ease of operation, maintenance and repair, and so on); (b) simplification of calibration procedures and minimization of physical standards requirements; (c) investigation of scope and limits of application; (d) development of procedures for proper interpretation of results; (e) provision of adequate training of inspectors in the use of NDA instruments; and (f) investigation of the possibility to utilize installed instrumentation for international safeguards purposes.

In the open part of the fuel cycle, that is, in those facilities in which nuclear material is handled in bulk form rather than in sealed item form, destructive analysis is preferred for verification purposes. This requires a careful determination of the volume or weight of the individual batches and chemical analysis of samples taken from the material. Proper homogenization and sampling of the material is highly important for these techniques. Also, sample transport and storage require special attention. Correct analytical results can be obtained only if representative samples are taken from well homogenized material, and only if these samples have retained their representative character until the time the analysis is performed. The characterization of standard materials used for calibration of the non-destructive assay technique also requires very accurate chemical analysis.

To fulfil the chemical analytical demands of IAEA safeguards, the Agency operates its own Safeguards Analytical Laboratory (SAL) at Seibersdorf, near Vienna [3]. This laboratory has been staffed and equipped to cope with a maximum sample load of about 2 000 samples per year. The laboratory is able to analyse all kinds of nuclear material for its thorium, uranium or plutonium content, to perform mass-spectrometric analyses for uranium and plutonium isotopes, to determine radio-nuclides by γ-ray spectroscopy and to make analyses for impurities in nuclear materials by emission spectroscopy. Wet chemical techniques such as oxidimetry and controlled potential coulometry, as well as gravimetry and mass-spectrometric isotope dilution analysis, are used for determining the thorium, uranium and plutonium content of the samples.

Samples taken from dissolved irradiated fuels are analysed exclusively by mass-spectrometric isotope dilution analysis. Because of the extremely high radiation level of these materials, only diluted samples of spent fuel are shipped to the Safeguards Analytical Laboratory. The accurate dilution is done at the safeguarded facility under the observation of Agency inspectors. At the same time, exactly known amounts of U-233 and Pu-242 or

Pu-244 tracers, also called spikes, are added to duplicate samples to allow for calculating element concentrations from measured isotope ratios. In order to reduce the total amount of radioactive materials in the samples, a special micro-technique, called the 'resin bead technique', is under development. This is expected to make possible the determination of isotopic ratios in samples containing no more than $0.00001-0.00005$ mg of uranium and plutonium.

To cope with peak sample loads and to maintain continuous control over the performance of SAL, a certain number of samples are routinely sent for analysis to other recommended laboratories in member states. SAL also participates in various international inter-comparison exercises and international analytical quality assurance programmes. Many difficulties with respect to destructive analyses related to the representativeness of samples, sample shipment and waste disposal have been solved during the past few years; other problems, particularly those related to the timely availability of analytical results, require further attention.

In IAEA safeguards, surveillance and containment [4] to provide information on the movement (or non-movement) of nuclear material in the absence of inspectors are important complementary measures to material accountancy. At present these techniques are used to preserve the validity of previously verified nuclear material data and to provide knowledge of material flow at important points in the fuel cycle. During the past few years, the increasing number of nuclear facilities subjected to IAEA safeguards has led to a significant development of containment and surveillance instruments and devices.

Seals are widely applied by inspectors in order to verify the integrity of containment. Several different types of seal have been developed during recent years. Metallic seals with particular features to detect substitution or tampering are in routine use. Fibre optic seals, which allow the detection of tampering with the sealing wire as well as remote interrogation of seal status, are under development. Pressure-sensitive paper seal labels are used for short-term applications. Ultrasonic seals are undergoing field testing for specific purposes such as underwater application.

Optical surveillance is primarily used to verify the correctness of reports and records regarding spent fuel movements. Various types of single-frame movie cameras and still cameras have been developed for this purpose. Super 8-mm movie cameras with specially fitted quartz timers have proven to be most suitable in many cases.

Closed circuit television systems (CCTV) with remotely controlled cameras are now under intensive field-testing. These CCTV systems have remarkable advantages over optical surveillance systems: for example, pictures can be taken in poor light, radiation fields do not interfere, the picture capacity is significantly higher, the recorded picture can be viewed on site without any delay, and so on. However, the use of these highly sophisticated CCTV systems is presently limited by their high price, the need for maintenance by experts, insufficient reliability in difficult environmental conditions, and their bulk nature.

For special safeguards purposes, several specific monitors have been developed. These include tamper-resistant spent fuel bundle counters used to register the number of spent fuel elements discharged from the core of an on-load refuelled reactor, spent fuel verifiers used for the detection of substitution of spent fuel elements by dummies, reactor power monitors based on a track-etch technique, sensitive neutron detectors used for verification of records on reactor operation, and passive gamma/neutron detectors used to detect the undeclared removal of nuclear material through small openings or ports.

IV. The future

Further development in the field of containment and surveillance is directed towards increased reliability of instruments, particularly camera and CCTV systems; improvement of ease of operation, serviceability and tamper resistance, incorporation of remote status interrogation and long distance picture transmission possibilities; and maximization of cost effectiveness.

Although extensive research and development activities in many scientific laboratories in IAEA member states have led during the past years to a remarkable improvement of IAEA safeguards technology, further substantial effort is necessary to enable the IAEA to implement its safeguards system in a manner which on the one hand avoids hampering the economic and technological development in the field of peaceful nuclear activities and, on the other hand, provides for sufficient international assurance that nuclear material is not diverted from peaceful nuclear activities to nuclear weapons or other nuclear explosive devices and constitutes a credible deterrence to the proliferation of nuclear weapon capabilities. The IAEA will be able to achieve this goal only if it is strongly supported by its member states. Several member states, particularly the United States, have recently entered into comprehensive programmes of support to IAEA safeguards research and development. These programmes not only promote the development of new safeguards methods and techniques, but also aim at overcoming all kinds of implementation difficulties.

References

1. *Safeguards Against Nuclear Proliferation* (Almqvist & Wiksell, Stockholm, 1975, Stockholm International Peace Research Institute), p. 6.
2. 'Non-destructive assay: instruments and techniques for Agency safeguards', *International Atomic Energy Agency Bulletin*, Vol. 19, No. 5, October 1977, pp. 34–37.

3. 'The Safeguards Analytical Laboratory: its functions and analytical facilities', *International Atomic Energy Agency Bulletin,* Vol. 19, No. 5, October 1977, pp. 38–47.
4. 'Surveillance and containment measures to support IAEA safeguards', *International Atomic Energy Agency Bulletin,* Vol. 19, No. 5, October 1977, pp. 20–26.
5. Konnov, Y. and Sanatani, S., *Development of Containment and Surveillance Measures for IAEA Safeguards,* SM 231–119, International Symposium on Nuclear Material Safeguards, 2–6 October 1978, (IAEA, Vienna, 1978).

Paper 13. Safeguards techniques

W. A. HIGINBOTHAM

Square-bracketed numbers, thus [1], refer to the list of references on page 196.

I. Introduction

The safeguards programme of the International Atomic Energy Agency (IAEA) has played a significant role in the containment of the proliferation of nuclear weapons, and ought to play a more important role in the future. Effectiveness of IAEA safeguards to deter diversion is fundamental to any discussion of technical and institutional measures to reduce the danger inherent in proliferation. The preceding paper has described the technologies employed by the IAEA for a variety of applications at different types of facility. This paper discusses the technologies which might be appropriate for a future, larger reprocessing facility. First, however, the relationship of safeguards, in the national context, to safety and to radiation protection is discussed.

II. Nuclear energy programmes and safety measures

Nuclear energy is a clean and economical energy source, which should prove to be of great benefit to mankind. Like all other major energy sources, it is associated with certain risks, several of which are unique to nuclear energy: (*a*) the source material and the fission products released in the production of energy by fission are radioactive; (*b*) a nuclear power reactor could suffer a serious accident and spread radioactivity over a large area; (*c*) nuclear energy processes are similar to the processes that lead to nuclear weapons; (*d*) nuclear fuel cycles involve some materials which are similar to those used in nuclear explosives; and (*e*) these features of nuclear energy might be exploited by individuals, groups or nations to threaten society. All of these problems are of concern to any nation which intends to employ nuclear energy.

Programmes have been developed by individual nations and associations of nations to control the risk of diversion. Generally, the programmes to contain proliferation, to keep radiation exposures below acceptable levels, to make reactors safer, to find ways to dispose of nuclear wastes reliably for all time, and to protect nuclear materials and facilities from subnational adversaries have been designed and implemented separately, at the national and the international levels, although there is considerable overlap between the objectives of the separate programmes and much duplication of effort. In the following, the relationship of nuclear safeguards to these other programmes will be noted.

Reactor safety

So far, nuclear power reactors have a near-perfect safety record, due to the fact that the possibility of serious accidents was recognized from the start, and that great emphasis has been placed on safe design and safe operation. Nonetheless, accidents could occur and it is conceivable that some reactor will have a very serious accident. Although local nuclear opponents have created incidents, and local extremists have in a few cases caused material damage at reactor sites, there has not yet been any serious attempt to sabotage a reactor. The design of reactors for safety makes it unlikely that any but a very knowledgeable group of adversaries would be able to cause a serious reactor accident. Still, that possibility does exist, and any safeguards programme must try to prevent it from occurring. In the USA, the Nuclear Regulatory Commission devotes considerable effort to reactor safety studies, and to developing reactor safety regulations. It also requires physical protection at nuclear power reactors in order to prevent employees or armed invaders from causing a major reactor accident.

Nations which produce or purchase nuclear reactors have a common interest in safe design, safe operation and prevention of sabotage. Designs which reduce the likelihood of accidents, by incorporating features inherent to the reactor proper and by redundant safety systems, also make reactors more resistant to acts of sabotage. Administrative measures taken to prevent sabotage also reduce the chances that inadvertent acts by operators might lead to accidents. If a reactor could be designed so that no serious accident could occur, this would eliminate the possibility of sabotage to threaten the public. Although there is some cooperation between those responsible for reactor safety and those responsible for physical protection, more direct and formal coordination would be beneficial, not only at the national level but also internationally. More widespread cooperation on reactor safety and physical protection should be of special interest to the Third World nations which need to purchase nuclear power reactors but are less well prepared than the supplier states (*a*) to ensure that their reactors incorporate the latest safety features; (*b*) to provide reliable and effective operators, and (*c*) to protect these facilities from local adversaries.

Another issue is that of the common objectives of the safety and safeguards programmes at nuclear fuel processing facilities. The safety objectives are to contain radioactivity, and to protect the workers and the public. The safeguards missions are to prevent diversion of nuclear materials which might be used in a nuclear explosive device or dispersed in a populated area, and to prevent sabotage.

Fuel and waste processing

At fuel fabrication and reprocessing facilities, the safety and safeguards programmes not only have overlapping objectives but also employ many common techniques. Both monitor the operations of the facility to ensure that the materials are contained. Both should be in a position to halt operations and to require an investigation if materials appear to be missing. An important safety consideration is avoidance of accumulations of nuclear material which might sustain a 'critical' reaction. For this purpose, it is necessary to keep careful accounts of the materials transferred from one process step to another. An efficient, highly automated system of measurements and process controls can prevent accumulation of critical amounts and also serve the need for accountancy for safeguards. Safety personnel check incoming and outgoing packages for content and containment, and monitor all low-level emissions, as do safeguards personnel. Both make considerable use of radiation detectors. In fact since safety and safeguards personnel at nuclear facilities do work together, it should be more efficient to recognize the ways in which these two activities could reinforce and supplement each other.

III.   National and international safeguards

Although there is considerable overlap, national and international safeguards objectives and activities are not the same. National systems are designed to prevent theft, diversion or sabotage by subnational adversaries. The nation is able to require and to enforce physical protection measures, to provide police protection, and to pursue and to arrest. The general structure of national systems is probably best decribed in the IAEA documents *The Physical Protection of Nuclear Materials* [2] and the *Report on the Advisory Group Meeting on States' Systems of Accounting for and Control of Nuclear Material* [3]. An update of the latter is in preparation.

The objective of the IAEA, on the other hand, is "the timely detection of diversion of significant quantities of nuclear material from peaceful nuclear activities . . . , and deterrence of such diversion by the risk of early detection" [4]. The IAEA is supposed to detect diversion, not to prevent it.

The goals of the IAEA, and a description of the system that it has developed to achieve these goals, are described in the *IAEA Safeguards Technical Manual* [5].

The Agency's criterion of a 'significant quantity' of nuclear material is the amount that might be needed for a fission explosive: 8 kg of plutonium, 8 kg of U-233, 25 kg of U-235 contained in highly enriched uranium, or 75 kg of U-235 contained in low-enriched or natural uranium. The Agency has also concluded that 'timely' detection should be related to the time that it might take, after a diversion had occurred, for a nation to convert the diverted material to a form suitable for manufacture of a nuclear explosive device. Thus detection of diversion of a significant quantity of plutonium or highly enriched uranium should occur within days or weeks; of plutonium in highly radioactive spent fuel, in from weeks to months; and of 75 kg of U-235 contained in natural or low-enriched uranium, in a period of a year or less [6].

In order to design its safeguards system, the IAEA postulates the strategies a nation might attempt for diversion from a particular facility. This is a most important safeguards technique for the design of national systems as well. INFCIRC/66 states only in a general way what sort of conclusion the Agency should report. INFCIRC/153 is more explicit: "the technical conclusion of the Agency's verification activities shall be a statement, in respect of each material balance area, of the amount of material unaccounted for over a specific period, and giving the limits of accuracy of the amounts stated". It has been assumed by those drafting the official IAEA documents that the operators of nuclear facilities would have high-quality measurement and material-accounting programmes, and would provide accurate reports to the local state and to the IAEA. On the basis of this assumption, the IAEA has planned to monitor the facility accountability procedures, to audit its records and to 'verify' the credibility of the facility records by duplicating some of the measurements, and by more qualitative measurements of additional items by non-destructive means. For this purpose the Agency would employ recognized sampling strategies and statistical analysis methods to determine whether or not it should accept the facility values for material unaccounted for (MUF), and its uncertainty.

To illustrate the technologies which may be needed for the safeguarding of future, large, bulk processing facilities, and how the technologies might be employed, safeguards plans for future reprocessing and mixed-oxide fabrication facilities are described below.

IV.   Reprocessing plant safeguards

Future reprocessing facilities will process very large amounts of plutonium.

For example, a reprocessing plant designed to process 5 tonnes a day of spent fuel, would produce 40 to 50 kg per day of plutonium as plutonium nitrate, which in turn would be converted to plutonium dioxide (PuO_2), probably within the same commercial complex. The IAEA considers two possible diversion strategies and combinations thereof. At one extreme is the abrupt diversion of a significant quantity (8 kg of plutonium, in this case). At the other extreme is a low level of diversion over an extended period. In order to counter the first diversion strategy, it would be necessary to detect an abrupt diversion within a very few days. In order to counter the second strategy, a combination of accountancy, containment and surveillance would need to be employed. With the technologies to be described, it should be possible to achieve both objectives. Only three major areas will be considered here: the receiving—storage area, the input accountability area and the chemical separations area. Other areas are the plutonium nitrate to dioxide conversion process, the plutonium storage areas, the waste treatment areas, the uranium nitrate conversion process, and the analytical laboratory together with its sample lines.

The design of such future facilities may significantly affect the ease or the difficulty which the IAEA will experience in attempting to achieve its goals. The effects of the IAEA safeguards on the operations of the facility should also be considered. If the containment is poorly defined, or equipment is inconveniently located or difficult to observe, both the Agency personnel and the facility personnel will find it difficult to determine the cause of discrepancies in accounting, or to assess equipment failures or accidental losses. Agency inspectors would need to verify certain features of the design information before operation of the equipment. At a reprocessing plant this would be difficult. Consequently, the IAEA should be consulted at an early stage of the design, regarding the physical features, the arrangements for verifying items in storage, the design of accountability vessels, and so on.

The receiving—storage area is similar to the spent-fuel storage pool at a reactor, except that there will be much more activity. Both operators and IAEA inspectors will observe receipts and record identity of spent fuel assemblies and of those assemblies transferred from the pool for disassembly and dissolution. The IAEA procedures will involve item accounting, containment and surveillance. Improved seals are being developed to aid in item identification and surveillance. The IAEA has developed automatic cameras and television monitors for surveillance of such areas. The Japanese are exploring the feasibility of applying tamper-resistant motion recorders on the cranes needed to move about the fuel assemblies, at the Tokai-Mura reprocessing facility.

By its nature, the receiving—storage area will have substantial containment features, and the heavy, radioactive fuel elements can be transferred only with difficulty. The techniques for identifying receipts, checking the location of assemblies stored in the pool, and identifying the 10 to 30 assemblies transferred to the disassembly cell each day will need to be improved; the Agency inspectors should be provided with terminals and a data link. There is little opportunity for a protracted diversion in this material balance area (MBA). The combination of item accounting, containment and surveillance should promptly detect diversion of one or several fuel assemblies.

The assemblies will probably be transferred to the mechanical disassembly cell in batches as received from one reactor at a time, so that the contained uranium and plutonium can be measured, after dissolution, and credited to the particular reactor. This is the most important measurement at the 'back-end' of the fuel cycle, since the plutonium content of spent fuel can be estimated only approximately on the basis of original uranium content and reactor burn-up. This plutonium measurement becomes the starting-point for all subsequent plutonium measurements—at the output of the reprocessing plant, after conversion and at several points in the mixed-oxide fuel fabrication process. Unfortunately, this is at present the least accurate of these measurements.

The headers are mechanically removed from the fuel assemblies, the rods are chopped into pieces and the oxide pellets are dissolved in nitric acid. The undissolved pieces of cladding are discarded as waste. Non-destructive assay (NDA) methods, which have been developed to assay these pieces for any retained plutonium, will be tested at Tokai-Mura for IAEA use.

The dissolver solution is transferred to an input accountability vessel where the volume or weight of the liquid is determined, and samples are drawn for assay. The overall accuracies for the plutonium in a reactor batch (up to 300 kg Pu) are today in the range of 0.5−1 per cent. For future facilities it should be possible to improve this substantially. This would require that accountability vessels be carefully designed for accuracy and for ease of calibration. The devices used to measure the liquid weight (or volume) should be improved. It is important that the contents be thoroughly homogenized before the samples are withdrawn. Finally, the analytical techniques could be made more accurate and less tedious. Some techniques under study are: (a) accountability vessels with favourable geometric shape and rigidity for more accurate volume calibration; (b) pressure transducers with digital read-out to replace manometers for measurement of liquid level and specific gravity (manometers are difficult to read, while transducers are less subject to human error and may be linked to data systems); and (c) uranium and plutonium concentrations determined by isotopic dilution mass spectrometry. Alternatives are X-ray fluorescence for rapid assay of uranium and plutonium concentration, γ-ray spectrometry for uranium and plutonium isotopic composition, and automated analytical techniques for quantitative separation of uranium, and plutonium from fission products, and the assay of the relevant constituents.

At least two techniques have been developed which may be employed by the IAEA to confirm the validity of the input measurement at a reprocessing plant. One has been used by operators at reprocessing plants and reactors to confirm the measurement described above. In this technique it is necessary to have accurate data on the original uranium content and enrichment of the fuel assemblies, before irradiation in the reactor. Then using the ratio of plutonium to uranium in the dissolver, and the burn-up (for example, at 30 000 MWd/t, 3 per cent of the U and Pu would be fissioned), the amount of plutonium in a batch of spent fuel from a reactor can be computed. Accurate volume determinations are not important. Minor corrections must be made for loss to neptunium or trans-plutonium elements. The second technique, which has been especially studied in Europe, requires the accurate measurement of uranium and plutonium isotopic ratios (and possibly of fission products and heavier elements), and some empirical data on each class of reactor. Since the consumption and production of these isotopes are a function of the radiation history of fuel in a reactor, such data may be used to confirm the information provided on reactor burn-up and to detect substitutions. These supplementary techniques predict what should be measured directly at the input accountability vessel *and* whatever may have been discarded with the pieces of cladding.

Since reprocessing plants will not be cleaned out for physical inventories more frequently than two to four times per year, it will be necessary to develop methods to ensure that 8 kg of plutonium will not be diverted between physical inventories. Although it would be theoretically possible to employ containment and surveillance measures for this purpose, it would be more convincing to use shorter-term accounting techniques as well. In the shorter term, the relative accuracy of accounting would be poorer than for the longer period, but the absolute uncertainties should be much smaller. Containment and surveillance at a reprocessing plant would be difficult to enforce, whereas a combination of these measures, with near-to-real-time accounting, should present major risks of detection to a potential diverter.

The chemical processing area must be designed for remote operation and be surrounded by heavy shielding. The process vessels are connected to each other and to the outside by a maze of pipes, and many control pipes and wires must be run into the equipment. Consequently, it may be very difficult for the IAEA to understand the containment and where to apply its surveillance instruments.

As mentioned above, it will be most important that the Agency inspect such facilities during their construction and before they start to operate with radioactive fuels. It has been suggested that IAEA inspectors participate with the designers in the projected check-out procedures which determine whether the pipes are properly connected. Tamper-resistant surveillance devices have been developed to monitor valves and the flows in pipes, to check that ports have been opened, to monitor the presence of personnel, and so on. The design and testing of a surveillance system for such a process area have not yet been attempted, although this is technically feasible. The experiments should be performed.

Assuming that a containment—surveillance system can be designed which will ensure that the nuclear materials are contained within the process area, the near-to-real-time accounting system would employ the measurements at the input accountability vessel described above, those at the uranium and plutonium nitrate product load-out areas, the accountability measurements of the radioactive liquid waste streams, and methods to verify the plutonium and uranium passing through the process vessels. The content of product and waste streams would be based on measurements of the volume (or weight) in accountability vessels, and on the assay of samples, as before. Due to the radioactivity and other contamination of the waste streams, measurement of the plutonium concentration is difficult. On the other hand, since the absolute amounts are small, high accuracy is not required. The uranium and plutonium products should be measured accurately, since they are pure solutions with little radioactivity. Also, the concentration of uranium and plutonium here can be automatically measured by γ-ray absorptiometry (the absorption by the heavy metal of γ-rays of selected energy), and the isotopic content can be continuously measured by γ-ray spectrometry.

There would, in fact, be three relatively independent systems to compare: the containment, the input−output accountability, and the observation of the plutonium in the processing equipment.

The combination of the measurements of feed and withdrawals would today have an accuracy of about 5 per cent and a 24-hour delay at a reprocessing plant, reducing to 1 per cent as time went on. At future facilities, it should be possible to reduce the one-day balance error to about 2 per cent and the long-term one to about 0.5 per cent. In either case, the short-term uncertainty would be dominated by uncertainty as to the in-process inventory which could vary by 50 to 100 kg, due to normal operations or to diversion, in a reprocessing facility handling five tonnes of spent fuel per day.

After removal of the fission products, which occurs about half-way through the process, it is possible to install and to maintain a variety of instruments to monitor the level and concentration of liquids in the process vessels. At the head end it will be necessary to employ methods which would not be affected by the radioactivity or require replacement. For efficient operation, the plant management will also wish to have good data on the materials in process and on the flows. It is technically possible to provide this area of a reprocessing plant with the instruments and the sample lines so that the in-process plutonium could every hour be accounted for to 1 or 2 kg. Any sudden diversion or any inadvertent loss would be promptly detected. With the input−output measurement data, the Agency could achieve its quantitative objective (8 kg Pu) and its time objective of a few days. It would have almost continuous assurance that the facility was being operated as declared. And in the case of any doubt, it would be able to check its accountancy data with its in-process data and with its containment −surveillance data, either to resolve the issue or to pinpoint a local area which should be investigated.

The Agency has also set as an objective the detection of gradual loss of

8 kg of plutonium during a year. Since it would be difficult for the IAEA to have a detection sensitivity of 0.5 per cent of the throughput (7 500 kg Pu in six months), it would not be possible to achieve this objective by accountancy. Rather, the detection of a small, continuing diversion would need to be based on the technique described above and on the fact that the continuing presence of inspectors should make quite high the probability of detection of a protracted diversion or of many repeated small diversions. Agency inspectors should carefully verify the physical inventory process, and take advantage of that opportunity to recalibrate instruments, and to reassess the containment/surveillance features and instruments. During operations, the material balance for the plutonium and uranium should be compared; and, since the isotopic composition of the plutonium changes frequently, additional verification of the operations can be performed.

This plan presumes that IAEA inspectors would be on duty at all times to monitor the receiving, storage and shipment areas, to attend to instruments, to observe measurements, and to take samples. In order to analyse samples, the Agency will need to have a local analytical laboratory, and a local safeguards office equipped with modern data storage and processing equipment. The system as a whole must be designed to ensure the independence of the Agency's verification activities.

V.   Safeguards for conversion and fabrication facilities

The plan for IAEA safeguards at plutonium conversion and mixed-oxide fuel fabrication facilities would follow a similar philosophy, employing a combination of accounting, containment and surveillance measures. The latter facilities can be more accessible for inspection, and the necessary measurement techniques are more highly developed and more accurate. Sophisticated systems for coordinated surveillance and near real-time accounting are being developed for national safeguards application at such facilities [7]. These techniques should be equally effective for the IAEA. However, it will be necessary to determine which features the Agency would actually find useful, and to ensure that the Agency can verify the reliability and tamper resistance of each sensor, data link and computer element. The IAEA could technically meet its safeguards objectives. However, much of the technology will need to be tested and refined under realistic conditions; IAEA safeguards strategy will need to be developed; and suitable staff would need to be trained.

VI.   Conclusions

It would be a mistake to concentrate only on the sensitive facilities enumerated above and to ignore the mundane problems associated with natural and low-enriched fuel-processing facilities and reactors. There will always be many more reactors than enrichment or reprocessing plants. Even if reprocessing and breeders are postponed, by the turn of the century there will be hundreds of reactors in many nations, producing spent fuel containing plutonium. Although the high radioactivity of spent fuel may serve as a barrier to diversion and proliferation, it is not an absolute barrier and it is a barrier which will become less effective in time.

The IAEA safeguards effort which has been suggested above for enrichment and reprocessing facilities would be substantial—perhaps a crew of 10 to 30 professionals residing near each one. However, this high concentration would occur at only a few places. The number of reactors to be monitored will, on the other hand, be large. The Agency already observes the change of fuel assemblies at light water reactors (perhaps for two weeks each year), monitors the spent fuel storage pool several times a year, and visits on-line fuelled reactors even more often. Several member states are working with the IAEA to develop improved monitoring equipment, seals and techniques. These should make the IAEA procedures less burdensome and more effective. Still, it would appear to be necessary for the Agency to send inspectors to a light water reactor whenever the head is removed to transfer fuel and whenever fuel is to be shipped to a reprocessing plant or to remote storage. Sophisticated surveillance and data-transmission equipment would be needed if more frequent visits are to be avoided.

To summarize:

1. Verification of design information is the starting-point for the IAEA safeguards implementation plan. If possible, the Agency should be consulted at the design stage, and the design should take into consideration the needs of the Agency, to assist both the IAEA and the facility operator.

2. Since Agency inspectors as well as the operators can make mistakes, the Agency system should be of the highest quality, and it should, if possible, contain self-checking features.

3. There will inevitably be some 'false alarms'. Effective measures should be mutually developed for the resolution of such incidents.

4. It is most important that the Agency be permitted and encouraged to experiment with its techniques and strategies in order to evaluate the techniques and to optimize its procedures, in the interest of effectiveness and of minimizing its impact on the facilities.

References

1. von Baeckmann, A., see Paper 12 of the present volume (pp. 179–86).
2. *The Physical Protection of Nuclear Materials,* IAEA-INFCIRC/235, September 1975.
3. *Report on the Advisory Group Meeting on States' Systems of Accounting for*

and *Control of Nuclear Material*, IAEA-AG-26 (Brno, Czechoslovakia, July 1975).
4. *The Structure and Content of Agreements between the Agency and States Required in Connection with the Treaty on Non-Proliferation of Nuclear Weapons*, IAEA-INFCIRC/153, May 1971.
5. *IAEA Safeguards Technical Manual*, Part A, IAEA-174 (IAEA, Vienna, 1976).
6. Hough, G., Shea, T. and Tolehenkov, D., Technical Criteria for the Application of IAEA Safeguards, IAEA Symposium, Vienna, 2–6 October 1976 (IAEA, Vienna, 1978).
7. Keepin, G. R. and Lovett, J. E., *The Potential Value of Dynamic Materials Control in International Safeguards*, IAEA-SM-231/133.

Chapter 9.   Exporting policies

Paper 14.   Applications of US non-proliferation legislation for technical aspects of the control of fissionable materials in non-military applications

W. H. DONNELLY

Square-bracketed numbers, thus [1], *refer to the list of references on page 221.*

I.   Introduction

This chapter deals with the effectiveness of bilateral, multinational and international controls with attention to safeguards, the multinational fuel cycle and exporting policies. The purpose of this paper is (*a*) to briefly analyse US legislation intended to reduce the risks of the further spread, or proliferation, of nuclear weapons, or the ability to make them, associated with the civil use of nuclear power; and (*b*) to comment upon implications of that legislation for the question of proliferation.

II.   The ideal use of nuclear energy

US legislation dealing with the relation between nuclear power and nuclear weapons implies a vision of an ideal future use of nuclear power. By reference to this legislation and some reading between its lines—which is always an uncertain undertaking—it is possible to arrive at the following features of a nuclear power utopia.

Ideally, international commerce in the domestic production of separated plutonium, highly enriched uranium and separated uranium-233 would be avoided, at least for the next few decades. World nuclear power generation would be confined to natural or slightly enriched uranium, with terminal storage of unreprocessed spent fuel in international facilities, or at least in national facilities under international auspices and inspection. Plutonium would not be used for fuel. Nuclear supplier and user nations would have agreed upon common restrictions on nuclear trade, with no transfers of plant and equipment for enrichment or reprocessing, and they would impose sanctions on nations that violate non-proliferation commitments. All nations would have ratified the Treaty on the Non-Proliferation

199

of Nuclear Weapons (NPT) and accordingly, except for the nuclear weapon states, all would permit International Atomic Energy Agency (IAEA) safeguards to be applied to all their nuclear activities. Nuclear weapon states would voluntarily place their civil nuclear facilities under IAEA safeguards. Uniform and effective standards for physical security of nuclear materials and facilities would apply throughout the world. Nuclear supplier nations would keep control over what is done with the equipment, materials, fuels and technology that they provide, and user nations would have to acquire both permission to transfer, enrich or reprocess supplied nuclear materials, and approval for storage of spent fuel.

The main incentives for nations to adhere to such non-proliferation commitments would be an assured, reliable supply of nuclear power plants and equipment by supplier states and of nuclear fuel by an International Nuclear Fuel Authority, which might also store spent fuel. Nations would be committed not to build new enrichment or reprocessing plants on a national basis and would place existing facilities under international auspices and inspection. If and when reprocessing were needed, it would probably be provided by a facility under some kind of international management and control. IAEA safeguards would provide timely warning of diversion.

National and international research and development would concentrate on perfecting nuclear fuel cycles that recover more of the potential energy of uranium and thorium resources while not increasing the risks of proliferation beyond that posed by the light water reactor fuel cycle with terminal storage of spent fuel.

On the whole, this ideal nuclear future would emphasize the use of natural uranium or slightly enriched fuel cycles for some decades to come, with emphasis on improved fuel efficiency, and would discourage fuel cycles that involve easily accessible plutonium, U-233 or highly enriched uranium. Commercial deployment of the plutonium breeder and of reprocessing would be deferred into the more distant future when safeguards might be improved enough to offset the proliferation risks associated with present approaches to the fast breeder reactor.

III. Overview of US legislation: the Nuclear Non-Proliferation Act

The centre-piece of US legislation for limiting the risks of the further spread of nuclear weapons associated with civil uses of nuclear power is the Nuclear Non-Proliferation Act of 1978. A summary of this legislation follows, together with identification of other legislation containing non-proliferation provisions.

Findings, policy and purpose

On 10 March 1978 President Carter signed into law the Nuclear Non-Proliferation Act of 1978 [1] and so approved a comprehensive package of policies and measures to reduce or sever the link between civil nuclear power and the spread of nuclear weapons. Its enactment was controversial. Some interests urged direct and vigorous use of US influence or leverage, derived from the supply of enriched uranium by the USA, to pressure other countries into committing themselves to stronger non-proliferation measures. Others warned of undesirable consequences from unilateral US action and cautioned that radical measures could cause the United States to lose out in the world nuclear market and so reduce its ability to influence foreign nuclear decisions which might increase risks of proliferation. The final version represents a compromise and reflects the combined ideas and thinking of the Congress, the President and his Administration.

The act consists of an assessment of the dangers of proliferation, a statement of US policy, a statement of government programmes in support of this policy, and six parts, or titles, which deal respectively with assurance of nuclear fuel supply, strengthening IAEA safeguards, criteria and organization for control of US nuclear exports, revision of US agreements for nuclear cooperation and other negotiations, provision of alternatives to nuclear energy for other countries, and administration.

The legislative assessment of the risks

The Nuclear Non-Proliferation Act opens with a finding, or assessment, by the Congress of the grave threat of proliferation. Here Congress finds and declares that:

... the proliferation of nuclear explosive devices or of the direct capability to manufacture or otherwise acquire such devices poses a grave threat to the security interests of the United States and the continued international progress toward world peace and development. Recent events emphasized the urgency of this threat and the imperative need to increase the effectiveness of international safeguards and controls on peaceful nuclear activities to prevent proliferation . . . [2].

National legislative policy

In response to the urgent assessment of the proliferation situation, Congress outlined in the act a fourfold policy: (a) to provide international nuclear fuel supply assurance and more effective international control over nuclear commerce, including the establishment of international sanctions; (b) to confirm the reliability of the United States as a supplier of nuclear reactors and fuels to nations that agree with its non-proliferation policies; (c) to encourage ratification of the Non-Proliferation Treaty; and (d) to cooperate with other nations in finding suitable technologies for energy production, particularly alternatives to nuclear power.[1]

Purposes of the act

The stated purpose of the act is to promote this fourfold non-proliferation policy in four principal ways: (a) to establish a more effective framework for international cooperation in energy and to ensure that world development of peaceful nuclear power and nuclear exports do not contribute to proliferation; (b) to make the United States a reliable supplier of nuclear reactors and fuels to nations that adhere to its non-proliferation policies; (c) to provide incentives for other nations to join in such international cooperation and to ratify the NPT; and (d) to ensure effective US controls over its nuclear exports.[2]

Contents of the act

Five of the six titles to the Nuclear Non-Proliferation Act contain the operating provisions (a) to advance the policies of the act through a combination of national and international incentives; (b) to strengthen the safeguards system of the International Atomic Energy Agency; and (c) to establish stronger national and international constraints so that it will be more difficult for nations or terrorist organizations (subnational groups) to seize nuclear materials that could be used to make nuclear weapons or explosives.

---

[1] Section 2(a) of Public Law 95-242 states that it is the policy of the United States to: (a) actively pursue through international initiatives mechanisms for fuel supply assurances and the establishment of more effective international controls over the transfer and use of nuclear materials and equipment and nuclear technology for peaceful purposes in order to prevent proliferation, including the establishment of common international sanctions; (b) take such actions as are required to confirm the reliability of the United States in meeting its commitments to supply nuclear reactors and fuel to nations which adhere to effective non-proliferation policies by establishing procedures to facilitate the timely processing of requests for subsequent arrangements and export licences; (c) strongly encourage nations which have not ratified the Treaty on Non-Proliferation of Nuclear Weapons to do so at the earliest possible date; and (d) cooperate with foreign nations in identifying and adapting suitable technologies for energy production and, in particular, to identify alternative options to nuclear power in aiding such nations to meet their energy needs, consistent with the economic and material resources of those nations and environmental protection.

[2] Section 3 of Public Law 95-242 states that it is the purpose of the act to promote its policies by: (a) establishing a more effective framework for international cooperation to meet the energy needs of all nations and to ensure that the worldwide development of peaceful nuclear activities and the export by any nation of nuclear materials and equipment and nuclear technology intended for use in peaceful nuclear activities do not contribute to proliferation; (b) authorizing the United States to take such actions as are required to ensure that it will act reliably in meeting its commitment to supply nuclear reactors and fuel to nations which adhere to effective non-proliferation policies; (c) providing incentives to the other nations of the world to join in such international cooperative efforts and to ratify the treaty; and (d) ensuring effective controls by the United States over its exports of nuclear materials and equipment and of nuclear technology [3].

The sixth title deals with administration of the act and with reporting to Congress by the President.

IV. Incentives to adhere to US non-proliferation ideas

US incentives

The Nuclear Non-Proliferation Act implicitly recognizes limitations on what the United States can accomplish unilaterally, and therefore provides incentives that go beyond what is possible by unilateral action. To this end, Title I would re-establish the United States as a more reliable supplier of nuclear power plants and fuel for those nations that agree to its non-proliferation ideas.[3] Title I would also reinforce this incentive by establishing international assurance of nuclear fuel supply, backed up by an interim nuclear fuel stockpile, so that nations adhering to certain non-proliferation commitments could be protected from the risk of interrupted fuel supply for reasons other than violation of non-proliferation commitments. This title also provides a statutory basis for US participation in the International Nuclear Fuel Cycle Evaluation.

The act would re-establish the United States as a reliable supplier by the measures listed in appendix 14A. The emphasis on an assured US supply of enrichment takes on special importance because in 1974 the United States cut off taking new enrichment contracts from foreign customers. The statements about timely action seek to reassure foreign customers who might doubt the ability of Federal agencies quickly to carry out the procedures for approval and control of US nuclear exports specified in the act.

International incentives

As indicated above, the act would also offer the incentive of an international approach to meeting nuclear fuel needs. As shown in appendix 14B, the act requires the President to discuss such approaches with supplier and user nations and to negotiate for establishment of an International Atomic Fuel Authority to provide nuclear fuel services, including the storage of nuclear materials and spent fuel. As an interim measure, the President was to report to Congress by early September 1978 on the creation of an interim stockpile of enriched uranium large enough to fuel production of 100 000

---

[3] Because of full commitment of its enrichment capacity, in 1974 the United States stopped accepting new contracts for enrichment services.

megawatt years of electricity, and on the possible foreign participation in the US uranium enrichment facilities.[4]

Re-evaluation of the nuclear fuel cycle

One of President Carter's non-proliferation initiatives of April 1977 was to sponsor an International Nuclear Fuel Cycle Evaluation (INFCE), which had its organizational meeting in October 1977. The INFCE study has become a major international review of ways in which nuclear energy can be used while minimizing the dangers of proliferation. The act gives statutory recognition to this evaluation in its direction to the President to invite all nuclear supplier and recipient nations to: " . . . reevaluate all aspects of the nuclear fuel cycle, with emphasis on alternatives to an economy based on the separation of pure plutonium or the presence of high enriched uranium, methods to deal with spent fuel storage, and methods to improve the safeguards for existing nuclear technology"[5].

Strengthening IAEA safeguards

The Nuclear Non-Proliferation Act attaches importance to a "strengthened and more effective International Atomic Energy Agency and to a comprehensive safeguards system administered by the Agency to deter proliferation" [6]. The act directs the government to seek to act with other nations to strengthen safeguards in five ways. The government is to:

(*a*) continue to strengthen IAEA safeguards by contributing funds, technical resources and other support;

(*b*) ensure that the IAEA has the resources to carry out its safeguards responsibilities;

(*c*) improve the IAEA safeguards systems to ensure the timely detection of a possible diversion of nuclear material usable in nuclear explosives, the timely dissemination of information regarding such diversion, and the timely implementation of internationally agreed procedures in the event of such diversion;

(*d*) ensure that the IAEA receives timely data for administration of an effective and comprehensive international safeguards programme and that the IAEA provides timely notice to the world community of any evidence of a violation of any safeguards agreement to which it is a party; and

---

[4] These international fuel assurances would be limited to nations that adhere to policies designed to prevent proliferation. The act directs the President to seek to ensure that the benefits of such assurances are available to non-nuclear weapon states only if they accept IAEA safeguards on all their peaceful nuclear activities, do not manufacture or otherwise acquire any nuclear explosive device, do not establish any new enrichment or reprocessing facilities, and place any such existing facilities under effective international auspices and inspection [4].

(*e*) encourage the IAEA to provide nuclear supplier nations with data they need to assure themselves of adherence to bilateral commitments to such supply by their client states [6].

As for improving national safeguards systems in other countries, the act directs the US Department of Energy (DOE), in consultation with the Nuclear Regulatory Commission (NRC), to set up training in safeguards and in physical security for nations, and groups of nations, which have developed or acquired nuclear items or may be expected to do so. Any such programme is to include training in "the most advanced safeguards and physical security techniques and technology, consistent with the national security interests of the United States" [7].

V. Non-proliferation controls for international nuclear commerce

The act looks to international agreements to establish conditions and controls for world nuclear commerce that will reduce the risks of proliferation associated with the growing and changing use of nuclear power. It provides policy and direction to the President and to the government promptly to seek such agreement. In addition to previously mentioned non-proliferation commitments for access to the proposed International Nuclear Fuel Authority, the provisions of the act enumerated in the sections below bear directly upon international constraints and controls for nuclear commerce, including the imposition of sanctions.

Conditions for international nuclear commerce and domestic use of nuclear energy

The President is to take "immediate and vigorous steps" to seek agreement from nations to impose five conditions upon nuclear exports, and in addition to agree to three conditions affecting their internal use of nuclear power, as stated respectively in appendices 14C and D.

Sanctions

A notable feature of the act is its emphasis upon unilateral and international sanctions for violations or abrogation of non-proliferation commitments. As noted above, the act specifies that it is the policy of the United States to pursue the establishment of more effective international controls over the transfer and use of nuclear items, including the establishment of common international sanctions [8]. The act then specifies seven actions which would trigger a cut-off of US nuclear exports, as listed in appendix 14E.[5]

Responses to violation of non-proliferation commitments

The act directs the government to negotiate with other nations and groups of nations to adopt general principles and procedures, including common international sanctions, to be followed should a nation violate any material obligation for peaceful uses of nuclear energy or should any nation violate the principles of the NPT, including the detonation of a nuclear explosive by a non-nuclear weapon state [10].

Responses to diversion, theft or sabotage

Similarly, the act directs the government to negotiate for international procedures to be followed in the event of diversion, theft or sabotage of nuclear materials or sabotage of nuclear facilities, and for recovering nuclear materials that have been lost or stolen, or obtained or used by a nation or by any person or group in contravention of NPT principles [11].

Bilateral and unilateral non-proliferation measures

In addition to negotiations of international agreements to decouple nuclear power from the spread of nuclear weapons, the act prescribes a set of independent bilateral and unilateral measures. It directs the President to obtain specific non-proliferation commitments from US nuclear trading partners in his renegotiation of existing bilateral agreements for nuclear cooperation, and makes most of these conditions mandatory criteria for government approvals of requests for nuclear export licences or authority to transfer certain nuclear technologies. It gives legislative standing to and prescribes conditions and criteria for US approvals for certain things other countries may do with US nuclear exports. This last measure can affect and complicate the day-to-day operating decisions of foreign nuclear power organizations. Finally the act requires the establishment or revision of many procedures for government action in the negotiation and renegotiation of agreements for cooperation, decisions on applications for licences or authorizations, and action on foreign requests for approval to do certain

---

[5] However, the act also authorizes the President to waive the cut-off if he determines that it would be "seriously prejudicial to the achievement of United States non-proliferation objectives or otherwise jeopardize the common defense and security". In keeping with the post-Viet Nam resurgence of congressional interest in foreign policy, the act further provides that a waiver by the President shall not become effective if Congress adopts a concurrent resolution of opposition [9].

things with or to US nuclear exports. Notable among these is the mandatory requirement for the US Arms Control and Disarmament Agency (ACDA) to make a Nuclear Proliferation Assessment Statement for agreements for nuclear cooperation, and the discretion to that Agency to make such assessments for foreign requests for approval to do certain things with or to US nuclear exports [12].

Non-proliferation commitments for US nuclear cooperation

The act specifies nine commitments, listed in appendix 14F, to be obtained by the President in his renegotiation of existing agreements for nuclear cooperation with nuclear trading partners of the United States. While many of these conditions are similar to those informally agreed upon among participants in the London talks, the act gives them statutory standing. It provides some flexibility for the President in that he may exempt a proposed agreement for cooperation from any of these nine requirements if he determines that the inclusion of any such requirement would be "seriously prejudicial to the achievement of US non-proliferation objectives or otherwise jeopardize the common defense and security" [13].[6]

The President is also directed to try to include in these agreements provision for cooperation in protecting the international environment from radioactive, chemical or thermal contamination ensuing from peaceful nuclear activities [16].

Concerning existing agreements for cooperation, the act represents a balance between those who argued for unilateral termination of US nuclear cooperation until new agreements could be negotiated and those who argued that the United States should honour its existing agreements. It directs the President immediately to begin. to renegotiate the agreements [17]. It also specifies that these new requirements shall not affect the authority to continue cooperation under existing agreements for cooperation [18].

Unilateral US conditions for approval of exports and transfers

Backing up the requirement that the President obtain upgraded nonproliferation commitments for US nuclear trading partners, the act also imposes new additional conditions for government approval of nuclear exports and it expands government control over export parts for nuclear

---

[6] However, the President must submit agreements for cooperation to the Congress which, in essence, can disapprove an agreement by concurrent resolution [14]. Note also that, should the President waive the requirement for full-scope safeguards for a non-nuclear weapon state, and if Congress does not disapprove his waiver, the first export licence or authorization with respect to that state which is issued 12 months after the congressional review period is to be submitted to Congress for further review [15].

facilities. The act provides two years for renegotiation of agreements for cooperation with international organizations (Euratom and the IAEA), during which time not all the new criteria need be applied. Thereafter, they must be applied unless waived by the President whose authority, in turn, is subject to possible congressional veto.

The act gives statutory standing to six export criteria (see appendix 14G) which are already met by existing agreements for cooperation. These are in addition to the still-standing criterion of the 1954 Atomic Energy Act that an export not be inimical to the common defence and security. Of these criteria, those for prior US approval of retransfers and reprocessing are likely to meet with resistance from the European Atomic Energy Community (Euratom). The act adds a seventh export criterion, taking effect on 10 September 1978: that IAEA safeguards must be maintained for all peaceful nuclear activities in a non-nuclear weapon state [19]. So as not unduly to constrain the President in conduct of foreign relations, the act provides that compliance by international organizations with the export criteria for retransfers and reprocessing are not necessary if renegotiations are in progress [20]. Also, if the negotiations with international organizations last longer than two years, the President is authorized to extend this grace period for 12 months at a time. However, he must first determine that failure to continue cooperation would be "seriously prejudicial to the achievement of U.S. non-proliferation objectives or otherwise jeopardize the common defense and security", and, second, notify the Congress. Congress may disapprove such an extension by a joint resolution which, in turn, goes to the President who may veto it, whereupon Congress would have the opportunity to override the veto [21].

The requirement for full-scope safeguards is tempered by authority to the President to waive its application. Here, the procedure is less complicated. A nuclear export to a non-nuclear weapon state that does not have full-scope safeguards in effect cannot be made unless the President notifies the Nuclear Regulatory Commission (NRC) (for materials or equipment) or the Department of Energy (DOE) (for certain technology) that he has determined that failure to approve an export would be "Seriously prejudicial to the achievement of U.S. non-proliferation objectives or otherwise jeopardize the common defense and security". Also, the export cannot be made until the first decision involved is submitted to the Congress for 60 days, during which time Congress can stop the export by a concurrent resolution. If Congress does so, no further export can be made to that state during the remainder of the two-year term of that Congress, unless the state agrees to full-scope safeguards or the President notifies the Congress that he has determined that "significant progress has been made in getting such agreement" or that US foreign policy interests dictate reconsideration and the Congress does not further disagree by concurrent resolution. If Congress does not object and the export goes forward, the act brings future exports before Congress at yearly intervals for possible disapproval until the state has agreed to full-scope safeguards.

Concerning component parts for nuclear facilities, the act directs the NRC, in consultation with the Secretaries of State, Energy and Commerce and the Director of ACA, to determine which items are "especially relevant from the standpoint of export control because of their significance for nuclear explosive purposes". No such item can be exported unless the NRC issues a general or specific licence for its export after a finding, based on a reasonable judgement of assurances provided and other information available to the Federal Government, that their criteria or equivalents are met. The three criteria require: (a) that IAEA safeguards are applied to the items; (b) that no such item is to be used for any nuclear explosive device; and (c) that no such item is to be transferred without prior US consent.

In addition, the NRC must determine in writing that the issuance of each such general or specific licence will not be inimical to the common defence and security [22].

As a further control, no major critical component of any uranium-enrichment, nuclear fuel-reprocessing, or heavy-water production facility shall be exported under any agreement for peaceful cooperation unless the agreement specifically designates such components as items for export [23].[7]

In the US scheme for control of nuclear exports, the control for transfer of certain nuclear technology is assigned to the DOE. This control is based on a long-standing provision of the 1954 Atomic Energy Act making it unlawful for any person to "directly or indirectly engage in the production of any special nuclear material outside of the United States", except as authorized in an agreement for cooperation or upon authorization by the Secretary of Energy that such activity will not be inimical to the interest of the United States [24]. Under this authority, the DOE requires that a person (which includes firms and other organizations) apply to it for authorization to transfer nuclear technology relating to chemical processing of spent fuel, production of heavy water, or separation of uranium isotopes.[8]

Post-export controls

From the beginning, US agreements for nuclear cooperation have required cooperating countries to acquire US approval before doing certain things with US exports. In the early days of nuclear power, when most of the cooperation was in experimentation, these requirements presented few difficulties. With the growth of working nuclear power plants among US nuclear trading partners, the requirements of existing and renegotiated agreements for cooperation are likely increasingly to interject the US government into some of the day-to-day operating decisions of their nuclear

---

[7] The act defines a major critical component as any component part or group of component parts which the President determines to be essential to the operation of a complete uranium enrichment, nuclear fuel reprocessing, or heavy water production facility.
[8] Compare reference [25].

power industries. The Nuclear Non-Proliferation Act now gives statutory recognition to these requirements for US approvals, which are called 'subsequent arrangements'. Besides specifying procedures and criteria for US action on such requests for US approvals, the act establishes conditions for reprocessing of special nuclear material exported by the United States or produced from US nuclear exports. The Secretary of Energy may not enter into a subsequent arrangement for reprocessing of spent fuel or retransfer of recovered plutonium until he determines that the arrangement will not be inimical to the common defence and security, and reports his reasons to Congress and the report lies before Congress for 15 days. The act then sets out somewhat different conditions, depending upon whether the reprocessing plant is new or existing. For a new reprocessing plant, the Secretary may not enter into any subsequent arrangement for reprocessing or for subsequent retransfer of recovered plutonium in quantities of more than 500 grams unless in his judgement, and that of the Secretary of State, this will not result in a "significant increase of the risk of proliferation beyond that which exists at the time that approval is requested" [26]. Among all the factors in making this judgement, foremost consideration is to be given to whether or not the reprocessing or retransfer will take place under conditions that will "ensure timely warning to the United States of any diversion well in advance of the time at which the non-nuclear-weapon state could transform the diverted material into a nuclear explosive device" [26]. As for existing reprocessing facilities,[9] the Secretary of Energy is to attempt to ensure that such reprocessing or retransfer will take place under conditions "comparable to those which in his view, and that of the Secretary of State, satisfy the standards for reprocessing in a new facility" [27]. The act, however, disclaims any intention permanently or unconditionally to prohibit the reprocessing of spent fuel owned by a foreign nation which has been supplied by the United States [28].

The act gives special attention to subsequent arrangements for the return of spent fuel to the United States. The Secretary of Energy may not enter into such an arrangement involving a direct or indirect commitment of the United States for storage or other disposition of any foreign spent power reactor fuel unless three conditions are met [29].[10] One requires that such a commitment lie before Congress for 60 days of continuous session and not take effect if during this time Congress adopts a concurrent resolution of disapproval. Alternatively, the President can submit to Congress a detailed generic plan for such disposition or storage, to take effect if Congress during a period of 60 days does not disapprove it by concurrent

---

[9] The act defines such a facility as one that has processed power reactor fuel assemblies or has been the subject of a subsequent arrangement before the date of the act (10 March 1978).

[10] Summarized, the three requirements are: (a) that such a US commitment has been submitted to Congress for a period of 60 days of continuous session during which it can be disapproved by concurrent resolution; (b) that the Secretary of Energy has obtained the concurrence of the Secretary of State and has consulted with the Director of ACDA and the Secretary of Defense; and (c) that the Secretary of Energy has complied with, or the arrangement will comply with, all other statutory requirements of the Atomic Energy Act.

resolution. Any generic plan submitted by the President is to include a detailed discussion relating to policy objectives, technical description, geographic information, cost data and justifications, legal and regulatory considerations, environmental impact information and any related international agreement or understandings [29].

The act gives the President some flexibility under certain conditions. A subsequent arrangement can be made for storage or other disposition in the United States of limited quantities of foreign spent power reactor nuclear fuel if the President makes two determinations and notifies Congress [30].[11]

Administration and procedures

The act includes detailed specifications of procedures to be followed by government agencies for action on requests for export licences or authorizations, and for approval of subsequent arrangements. Sprinkled through these procedures are admonitions that the actions must be carried out swiftly, so as to strengthen the reputation of the United States as a reliable nuclear supplier. The act requires inter-agency coordination, authorizes the Executive Branch to block exports, and gives the President power to authorize exports if the NRC is unable to issue a licence or takes too long to do so. ACDA is given a statutory role analogous to its functions for arms exports in the mandatory requirement that it provide a Non-Proliferation Assessment Statement for new or renegotiated agreements for cooperation and its discretionary authority to prepare such assessments for subsequent arrangements, although not for nuclear export licences. The act contains a deceptively simple statement that it is not intended to require the NRC independently to visit, or to prohibit the NRC from independently visiting, other countries in connection with its consideration of IAEA safeguards [31]. In essence this would open the way for the NRC to make such visits. Notable features of these procedures are summarized in appendices 14H, I and J.

US and international sanctions

As indicated above, one purpose of the act is to establish international and unilateral sanctions for actions that might dangerously increase the risks of proliferation. As shown in appendix 14K, the only specific sanctions mentioned in the act are the cut-off of nuclear trade and assistance. Presumably, the directive to negotiate for common international sanctions could include other measures, but no examples or other directions are given.

---

[11] The two determinations by the President are (a) that a commitment for storage or other disposition of such limited quantities is required by an emergency situation; and (b) that it is in the national interest to take such immediate action. In his notice to the Congress, the President is to provide a detailed explanation and justification.

VI. Providing alternatives for nuclear power

The final approach of the Nuclear Non-Proliferation Act to severing the link between nuclear power and the potential spread of nuclear weapons is to help underdeveloped countries find attractive alternatives to nuclear power. Nuclear materials that are not present in a country can be neither diverted by the government nor stolen by subnational groups. With this in mind, Congress drafted the act to direct that the United States endeavour to cooperate with other nations, international institutions and private organizations to assist in the development of non-nuclear energy resources. The government is to cooperate with Third World and industrialized nations in protecting the international environment from contamination from both nuclear and non-nuclear energy activities. It is to seek to cooperate with and aid underdeveloped countries in meeting their energy needs through the development of non-nuclear resources and the application of non-nuclear technologies "consistent with economic factors, the material resources of those countries, and environmental protection" [32]. Additionally, the United States is to encourage other industrialized nations and groups of nations to undertake similar cooperation with underdeveloped countries. In support of these objectives, the act authorized a threefold programme for US cooperation with underdeveloped countries for the purpose of: (a) meeting their energy needs for their development; (b) reducing their dependence on petroleum, with emphasis on use of solar and other renewable energy resources; and (c) expanding the energy alternatives available to such countries.

The programme is to include cooperation in evaluating the energy alternatives of underdeveloped countries, facilitating international trade in energy commodities, developing energy resources and applying suitable energy technologies. The programme is to include energy assessment, both general and for specific countries, and cooperative projects in resource exploration and production, training, research and development. The Department of Energy is to arrange for the exchange of US scientists, technicians and energy experts with those of underdeveloped countries. By 10 March 1979 the President is to report to Congress on the feasibility of expanding this bilateral cooperation into an international cooperative effort which would include the creation of a scientific peace corps [33].

VII. Other legislation affecting US non-proliferation policy

Several other recently completed acts of legislation contain provisions relating to proliferation and its control. Notable among these are studies required by the Export Administration Act Amendment of 1977, the sanctions and study of provisions of the International Security Assistance Act of 1977, the sanctions of the Export–Import Bank Act Extension of 1978, and the Foreign Assistance and Related Programs Appropriations Act of 1978, and the funding for research and studies in the authorization act for the Department of Energy for fiscal year 1978. Summaries of this legislation follow below.

The Export Administration Act Amendment of 1977

On 22 June 1977 President Carter signed into law an extension of the Export Administration Act (Public Law 95-52) which contains several requirements tangentially related to export of US nuclear technology. It requires the Secretary of Commerce to study the transfer of technical data and other information to any country to which exports are restricted for national security purposes, that is, to the Socialist bloc. The act also requires the Secretary of Commerce to study the problem of the export, by publication or other means, of technical data or other information. The Secretary was to report on both by 22 June 1978 (within 12 months), giving an assessment of the impact of the export of technical data on the national security and foreign policy of the United States, and making recommendations for monitoring such exports without impairing freedom of speech, press or scientific exchange.

The International Security Assistance Act of 1977

On 4 August 1977 the President signed the International Security Assistance Act (Public Law 95-92) with two non-proliferation provisions and one relating indirectly to proliferation.

The act amends the Foreign Assistance Act of 1961 by revising a previous amendment of Senator Symington that had established unilateral US sanctions for certain nuclear exports. The new provisions deal separately with sanctions for certain actions relating to enrichment and to reprocessing. Concerning enrichment, the act adds a new Section 669 to cut off funds, under the International Security Assistance Act or the Arms Export Control Act, to any country that delivers nuclear enrichment equipment, materials or technology to any other country, or receives such equipment, materials or technology from any other country, unless two conditions are met before delivery. First, the supplying and receiving countries must have agreed to place all such items upon delivery under multilateral auspices and management when available; and second, the recipient country must have agreed with the IAEA to place all such items and also all nuclear fuel and facilities in the country under IAEA safeguards. The President may continue to furnish prohibited assistance if he determines and certifies

213

in writing to the House and the Senate that the termination would have a serious adverse effect on vital US interests, and that he has received reliable assurances that the country in question will not acquire or develop nuclear weapons or assist other nations in doing so. The bill specifies procedures for Senate action on any joint resolution to terminate or restrict such assistance.

As for reprocessing, the act adds a new Section 670, which deals somewhat differently with this proliferation problem. It also provides for the cut-off of funds specified in Section 699 to any country that delivers nuclear reprocessing equipment, materials or technology to any other country or receives such equipment, materials or technology from any other country (except for the transfer of reprocessing technology associated with the investigation, under international evaluation programmes in which the United States participates, of technologies that are alternatives to pure plutonium reprocessing), or is not a nuclear weapon state under the NPT and detonates a nuclear explosive. The President may continue to furnish such assistance if he determines and certifies in writing to the Congress that the termination of such assistance would be seriously prejudicial to the achievement of US non-proliferation objectives or would otherwise jeopardize the common defence and security.

The act also prohibits the use of any funds under the Foreign Assistance Act of 1961 for fiscal year 1978 to finance the construction of, the operation or maintenance of, or the supply of fuel for, any nuclear power plant under an agreement for cooperation.

The act contains other provisions indirectly related to non-proliferation. It directs the President to conduct a comprehensive study of the policies and practices of the US government with respect to the national security and military application implications of international transfer of technology, in order to determine whether such policies and practices should be changed. The study is to examine (a) the nature of technology transfer; (b) its effect on US technological superiority; (c) the rationale for transfers of technology from the USA to foreign countries; (d) the benefits and risks of such transfers; (e) trends in technology transfers by the USA and other countries; (f) the need for controls, the effectiveness of end-use controls, and possible unilateral sanctions if end-use restrictions are violated; (g) the effectiveness of existing organizations in regulating technology transfers from the United States; (h) the adequacy of existing legislation and regulations; and (i) the possibilities for international agreements with respect to transfers.

The Export–Import Bank Act Extension of 1978

On 26 October 1977 the President approved Public Law 95-143, to extend and amend the Export–Import Bank Act. The extension contains two non-proliferation provisions. First, it requires that an Export–Import Bank loan or financial guarantee for any export involving nuclear power, enrichment, reprocessing, research or heavy water production facilities, shall lie before Congress for a 25-day period before the Bank can give its final approval. Second, it established a system for the Secretary of State to report certain undesirable foreign nuclear actions to the Bank, after which the Bank may not approve any future financial transaction unless the President determines that it is in the national interest and he notifies Congress at least 25 days before the Bank approves the transaction. The triggering actions include violation, abrogation or termination of IAEA safeguards or of a US agreement for nuclear cooperation, or explosion of a nuclear device by a non-nuclear weapon country.

The Foreign Assistance and Related Programs Appropriations Act, 1978, Foreign Assistance Act

On 31 October 1977 the President approved Public Law 95-148 to appropriate funds for foreign assistance and related programmes for fiscal year 1978. Title IV included a non-proliferation restriction on funds appropriated to the Export–Import Bank for fiscal year 1978. It prohibited the use of these funds to finance the export of nuclear equipment, fuel or technology to any country, other than a nuclear weapon state eligible to receive economic or military assistance under the Foreign Assistance Act, that detonated a nuclear explosive after 31 October 1977.

The Department of Energy Act of 1978—Civilian Applications

On 28 February 1978 the President approved the Department of Energy Act of 1978—Civilian Applications, which authorized funds for DOE nuclear research and development for fiscal year 1978. The authorization contained several provisions relevant to non-proliferation policy and actions. Section 101 included $20 million for international spent fuel disposition and $13 million for research, development, assessment, evaluation and other activities at the privately owned Barnwell Nuclear Fuels Plant in South Carolina related to alternative nuclear fuel cycle technologies, safeguards systems, spent fuel storage and waste management; another item provided $5 million for research and development on means to reduce the ability to divert plutonium from its intended purposes and to increase detectability of plutonium if it should be diverted. It also directed the Department of Energy to study the Barnwell plant to determine if it might be used in support of US non-proliferation objectives. The report was due by 28 August 1978. One million dollars was authorized for the study. The act authorized the Department of Energy to undertake studies, in cooperation with other nations, on a multinational or international basis, designed to determine the general feasibility of the expanding capacity of spent fuel storage facilities, and authorized appropriations of $20 million, subject to a restriction that no funds available to the Secretary of Energy shall be directly or indirectly used for the repurchase, transportation or storage of any foreign spent power-reactor nuclear fuel unless the President (*a*) determines that use of funds for these purposes is required by an emergency situation; (*b*) determines that it is in the interest of the US common defence and security to do so; and (*c*) notifies Congress of the determination and action with a detailed explanation and justification, and the matter lies before Congress for 30 days. The act requires the Secretary of Energy, in cooperation with the Secretary of State, to report to the Congress by 25 August 1978, on the effects of President Carter's non-proliferation policy statement of 20 April 1977, upon agreements for cooperation in nuclear research and development.

VIII.   Implications for the control of certain technical nuclear activities

The technical aspects of the control of fissionable materials in non-military applications can focus on 10 discrete technological activities ranging over development of nuclear technologies, supply of fuel cycle services and development of safeguards technology. The Nuclear Non-Proliferation Act of 1978 has notable implications for most of these activities, as summarized in appendix 14L.

To the extent that the act can influence future world development and use of nuclear power, this influence will tend to discourage or defer the more sensitive technologies and encourage placement of national nuclear fuel cycles under international auspices and inspection, or their incorporation into multinational or international organizations. The fuel cycle steps likely to be so affected would include enrichment, reprocessing, storage of sensitive fuel materials (whether in bulk or in fabricated fuel), spent fuel storage, and transport. The principles of the act and its policy to limit the spread of sensitive nuclear materials would tend to discourage or delay commercial use of breeder or other nuclear fuel cycles that present more of a proliferation risk than do nuclear power reactors fuelled with natural or slightly enriched uranium plus terminal storage of unreprocessed spent fuel under international auspices and inspection. However, the act provides no criteria for comparing the proliferation risks of one nuclear fuel cycle against those of another.

In the sections below follows a discussion of the implications of the act for each of the 10 technical activities covered in this volume (see the list in appendix 14L).

Proliferation-resistant fuel cycles

Apart from providing a legislative charter for the International Nuclear

Fuel Cycle Evaluation, the Nuclear Non-Proliferation Act does not touch directly on proliferation-resistant fuel cycles. It shows a strong desire for reliable, early warning of theft or diversion of materials from nuclear fuel cycles—soon enough for effective international response. If early warning cannot be assured by proposed fuel cycles and associated safeguards, the principles of the act would oppose their introduction.

'Timely' warning is described in the act as providing warning of any diversion well in advance of the time at which a non-nuclear weapon state could transform the diverted material into a nuclear explosive device [34].

Enrichment

The act recognizes the need for enrichment and supports further use of the light water reactor using slightly enriched uranium, with terminal storage of unreprocessed spent fuel under international auspices and inspection. It would affect the use of enrichment technology by placing enrichment in an international organization, or under international auspices and inspection. Furthermore, it supports expansion of US enrichment capacity so that the United States can supply more of the world's enrichment needs. The act would discourage international commerce in enrichment facilities by the threat of unilateral cut-off of US nuclear exports. Twelve provisions of the act dealing with proliferation aspects of enrichment are summarized in appendix 14M.

Waste disposal

The act deals with waste disposal tangentially and in the context of spent fuel storage. Its general assumption is that reprocessing is to be avoided, which would make moot questions of the disposal of fission products from nuclear fuels. Concerning spent fuel storage, the act provides for return of some spent fuel to the United States, but only in an emergency and not as a general proposition. Instead, the act favours storage of spent fuel in international facilities or under international auspices and inspection. On the whole, the act would tend to de-emphasize development of waste disposal and substitute development of spent fuel storage.

Transport of nuclear materials

The act supports the International Nuclear Fuel Cycle Evaluation (INFCE). Some INFCE studies are directed at spiking and other ways of making nuclear fuel dangerous and difficult to steal or divert from transport. It also directs the President in his negotiations for international nuclear export policies to seek agreement that any international shipment of significant quantities of source or special nuclear material, or irradiated materials, shall be conducted under adequate security measures and under international safeguards [35]. The emphasis on adequacy of physical security measures in the parts of the act dealing with conditions for agreements for cooperation and criteria for approval of nuclear exports logically extends to physical security for shipments of nuclear materials. On the other hand, the act's directive to negotiate for creation of an International Nuclear Fuel Authority does not mention international transport services as one of its functions.

For the most part, the act supports the proposition that sensitive nuclear materials moving in international commerce should be in a form difficult and dangerous to try to move or to use, and should be subject to strong requirements for physical security and be carried out under IAEA safeguards.

Breeder reactors

Many provisions of the act tend to discourage or defer commercial use of breeder reactors by discouraging or limiting the reprocessing of nuclear fuels. The act does not specifically mention development of breeders as such. However, its initial finding that the proliferation of the 'direct capability' to manufacture or otherwise acquire nuclear explosive devices poses a grave threat to US security and to world peace, taken in the context of current writings that associate a high proliferation risk with the breeder, would be read by many as intended to discourage and delay commercial use of the breeder.

Appendix 14N summarizes 12 provisions of the act relating to reprocessing. While most of these would impose controls and constraints on reprocessing, one is a specific disclaimer of congressional intention to prohibit permanently or unconditionally the reprocessing of spent fuel owned by a foreign nation which has been supplied by the United States.

Peaceful nuclear explosives

The Nuclear Non-Proliferation Act includes eight provisions relating to nuclear explosives. None exempts peaceful nuclear explosive devices. To the extent it is successful, the act would tend to discourage and penalize development and use of nuclear explosives by a non-nuclear weapon state, including those for peaceful use. The penalties are limited to cut-off of US nuclear exports, and to possible recall of exported nuclear materials and equipment, and derived materials. On the other hand, the act makes no mention of the use of peaceful nuclear explosives by nuclear weapon states and provides no policy for US responses to possible requests from non-nuclear weapon states for peaceful nuclear explosive services. Appendix 14O summarizes the eight provisions of the act relating to nuclear explosives.

IAEA safeguards

Provisions of the act affecting nuclear safeguards are summarized in appendix 14P. They underscore the importance attached to effective safeguards and reaffirm past US support for the IAEA's safeguards function. The act adds three goals for improvement of the safeguards system. Each emphasizes timely actions by the Agency which will surely require improvements in safeguards technology.

Fusion breeders, laser fusion and nuclear reactors in space

The act does not specifically mention fusion breeders, laser fusion or nuclear reactors in space. Its ideas, however, could extend to and affect the use of these technologies and their control.

The act's emphasis on control and limitation for reprocessing plants, particularly the goal of placing existing facilities under international auspices and inspection, would probably affect commercial use of the fusion breeder. Once a fertile material, uranium-238 or thorium, has been converted into a fissile material, plutonium or uranium-233 respectively, by exposure to neutrons from a fusion reactor, the fissile materials would have to be recovered by processes closely resembling the reprocessing of spent fuel from fission reactors. A world trend towards placing this part of the nuclear fuel cycle in international or multinational hands probably could affect plans and decisions for the continued development of the fusion breeder. Also, the concepts of the act could cause international transfer of technological information on fusion breeders to be limited to countries which agree to the non-proliferation conditions.

Laser fusion would be little affected by the act. However, application of laser fusion technology probably would feel the influence of controls on world nuclear commerce favoured by the act.

Nuclear reactors for space also would not be explicitly affected by the Nuclear Non-Proliferation Act. Here again, if the fuel for such reactors is plutonium-239 or highly enriched uranium, the act's concepts of international commerce in these materials and products containing them could indirectly affect the manufacture and storage of fuel for such reactors.

All three technologies would be affected in the United States by provisions of the Atomic Energy Act of 1954, as amended, which are intended to protect the public health and safety from risks of nuclear materials. Commercial use of fusion breeders and laser fusion in the United States probably would have to be licensed by the Nuclear Regulatory Commission, with that licensing including extensive analysis of the associated environmental impacts as required by the National Environmental Policy Act. As for nuclear reactors for space, whether or not their construction and use would be subject to NRC licences, exhaustive analysis of possible effects on public health and safety and upon the environment would be required, which would probably cause the engineers to give increased attention to these matters. In turn, this could influence design, reliability, costs and timing of such reactors.

For all three, the philosophy evident in the Nuclear Non-Proliferation Act would cause each to be treated as a potential new risk for the spread of nuclear weapons or the materials from which to make them, and would influence the conditions for their domestic use and for related international trade.

IX. Overview

The initiators of the Nuclear Non-Proliferation Act of 1978 expect it to influence the attitudes of other nations towards their future development of nuclear technology, their commercial use of nuclear power and the controls they might impose individually and collectively to keep the associated risks of proliferation within limits acceptable to society. To the extent that the act is successful, it will influence this nuclear future in directions that will keep proliferation risks no greater, and ideally lower, than those of fuel cycles using uranium or slightly enriched uranium with terminal storage of unprocessed nuclear fuels. It would strengthen the technologies for safeguards systems, particularly to assure early warning and action in the event of theft or diversions of nuclear materials. It would restrain international trade and cooperation in those nuclear technologies and products that would make dangerous nuclear materials more readily available, such as reprocessing, laser enrichment and centrifuge enrichment. If new ways of producing plutonium from uranium-238 and uranium-233 from thorium are successfully developed, such as the fusion breeder or perhaps laser fusion processes, such new plants and equipment would be placed in international hands, or at least under international auspices and inspection. All of these effects would tend to keep sensitive nuclear materials and the means to produce them out of the hands of non-nuclear weapon states, which would be compensated for these limitations by assurance of a reliable supply of nuclear fuels to those states that agree to the desired non-proliferation commitment.

Overall, the act will tend to apply a constant, persistent pressure upon those responsible for the future development and use of nuclear energy to be ever mindful that what they do can affect the further spread of nuclear weapons, and to make sure that what they do will not increase these risks.

References

1. Public Law 95-242, US Statutes, Vol. 92, p. 120, US Code, chapter 22, para. 3201.
2. Public Law 95-242, Section 2, US Statutes, Vol. 92, p. 120.
3. US Statutes, Vol. 92, Section 3, pp. 120−21.
4. Public Law 95-242, Section 104(d), US Statutes, Vol. 92, p. 123.
5. Public Law 95-242, Section 105, US Statutes, Vol. 92, p. 123.
6. Public Law 95-242, Section 201, US Statutes, Vol. 92, p. 124.
7. Public Law 95-242, Section 202, US Statutes, Vol. 92, p. 124.
8. Public Law 95-242, Section 2(a), US Statutes, Vol. 92, p. 120.
9. Atomic Energy Act of 1954, Section 129, as amended by Public Law 95-242, Section 307, US Statutes, Vol. 92, p. 139.
10. Public Law 95-242, Section 203, US Statutes, Vol. 92, p. 124.
11. Public Law 95-242, Section 203, US Statutes, Vol. 92, p. 125.
12. Atomic Energy Act of 1954, Sections 131 (a)(1) and 123(a), as amended by Public Law 95-242, Section 303(a) and 401, US Statutes, Vol. 92, p. 127.
13. Atomic Energy Act of 1954, Section 123''a, as amended by Public Law 95-242, Section 401, US Statutes, Vol. 92, p. 143.
14. Public Law 95-242, Section 401, US Statutes, Vol. 92, p. 144.
15. Atomic Energy Act of 1954, Section 128''b, as amended by Public Law 95-242, Section 305, US Statutes, Vol. 92, p. 137.
16. Public Law 95-242, Section 407, US Statutes, Vol. 92, p. 147.
17. Public Law 95-242, Section 404(a), US Statutes, Vol. 92, p. 147.
18. Public Law 95-242, Section 405(a), US Statutes, Vol. 92, p. 148.
19. Atomic Energy Act of 1954, Section 128''a, as amended by Public Law 95-242, Section 306, US Statutes, Vol. 92, p. 137.
20. Atomic Energy Act of 1954, Section 126''a (2), as amended by Public Law 95-242, Section 304(a), US Statutes, Vol. 92, p. 133.
21. Atomic Energy Act of 1954, Section 128''b(2) and (3), as amended by Public Law 95-242, Section 306, US Statutes, Vol. 92, pp. 137−38.
22. Atomic Energy Act of 1954, Section 109(b), as amended by Public Law 95-242, Section 309(a), US Statutes, Vol. 92, p. 141.
23. Public Law 95-242, Section 402(b), US Statutes, Vol. 92, p. 145.
24. Atomic Energy Act of 1954, Section 57(b), as amended by Public Law 95-242, Section 302, US Statutes, Vol. 92, p. 126.
25. Unclassified Activities in Foreign Atomic Energy Programs, Part 810.7, Title 10, US Code of Federal Regulations.
26. Atomic Energy Act of 1954, Section 131''b(2), as added by Public Law 95-242, Section 303(a), US Statutes, Vol. 92, p. 128.
27. Atomic Energy Act of 1954, Section 131''b(3), as added by Public Law 95-242, Section 303(a), US Statutes, Vol. 92, p. 129.
28. Atomic Energy Act of 1954, Section 131''d, as added by Public Law 95-242, Section 303(a), US Statutes, Vol. 92, p. 129.
29. Atomic Energy Act of 1954, Section 131''f(2), as added by Public Law 95-242, Section 304(a), US Statutes, Vol. 92, p. 131.
30. Atomic Energy Act of 1954, Section 131''f(c), as added by Public Law 95-242, Section 304(a), US Statutes, Vol. 92, p. 130.
31. Atomic Energy Act of 1954, Section 126(a)(2), as amended by Public Law 95-242, Section 304(a), US Statutes, Vol. 92, p. 133.

32. Public Law 95-242, Section 501, US Statutes, Vol. 92, p. 148.
33. Public Law 95-242, Section 503, US Statutes, Vol. 92, p. 149.
34. Atomic Energy Act of 1954, Section 131''b(2), as added by Public Law 95-242, Section 303(a), US Statutes, Vol. 92, p. 128.
35. Public Law 95-242, Section 403(c), US Statutes, Vol. 92, p. 147.

Appendix 14A

Improving the reliability of the United States as a nuclear supplier: provisions of the Nuclear Non-Proliferation Act of 1978

US policy. "The United States, as a matter of national policy, shall take such actions and institute such measures as may be necessary and feasible to assure other nations and groups of nations (Euratom) that may seek to utilize the benefits of atomic energy for peaceful purposes that it will provide a reliable supply of nuclear fuel to those nations and groups of nations which adhere to policies designed to prevent proliferation." (Section 101, US Statutes, Vol. 92, p. 121)

US capacity to supply. "The United States shall ensure that it will have available the capacity on a long-term basis to enter into new fuel supply commitments consistent with its non-proliferation policies and domestic energy needs." (Section 101, US Statutes, Vol. 92, p. 121)

"The Secretary of Energy is directed to initiate construction, planning and design, construction, and operation activities for expansion of uranium enrichment capacity, as elsewhere provided by law." (Section 102, US Statutes, Vol. 92, p. 122)

"The President shall promptly undertake a study to determine the need for additional United States enrichment capacity to meet domestic and foreign needs and to promote United States non-proliferation objectives abroad. The President shall report to the Congress on the results of this study within twelve months . . . " [March 10, 1979] (Section 103, US Statutes, Vol. 92, p. 122)

Timely action ". . . The Commission [Nuclear Regulatory Commission] shall, on a timely basis, authorize the export of nuclear materials and equipment when all the applicable statutory requirements are met." (Section 101, US Statutes, Vol. 92, p. 121.)

" . . . the Secretary [of Energy] as well as the Nuclear Regulatory Commission, the Secretary of State, and the Director of the Arms Control and Disarmament Agency are directed to establish and implement procedures which will ensure to the maximum extent feasible, consistent with this Act, orderly processing of subsequent arrangements and export licenses with minimum time delay." (Section 102, US Statutes, Vol. 92, p. 122)

Appendix 14B

International assurances of reliable nuclear supplies in the Nuclear Non-Proliferation Act of 1978

Negotiation of international approaches. " . . . The President shall institute prompt discussions with other nations and groups of nations, including both supplier and recipient nations, to develop international approaches for meeting future worldwide nuclear fuel needs." (Section 104(a), US Statutes, Vol. 92, p. 122)

Negotiation for an International Nuclear Fuel Authority. "In particular, the President is authorized and urged to seek to negotiate as soon as possible with nations possessing nuclear fuel production facilities or source material, and such other nations and groups of nations, such as the IAEA, as may be deemed appropriate, with a view toward the timely establishment of binding international undertakings providing for:" (Section 104(a), US Statutes, Vol. 92, p. 122)

An International Nuclear Fuel Authority. " . . . the establishment of an international nuclear fuel authority with responsibility for providing agreed upon fuel services and allocating agreed upon quantities of fuel resources to ensure fuel supply on reasonable terms in accordance with agreement between INFA and supplier and recipient nations."

INFA conditions. " . . . a set of conditions . . . under which international fuel assurances under INFA auspices will be provided to recipient nations, including conditions which will ensure that the transferred material will not be used for nuclear explosive devices."

International facilities. " . . . feasible and environmentally sound approaches for the siting, development, and management, under effective international auspices and inspection, of facilities for the provision of nuclear fuel services including the storage of special nuclear materials."

Spent fuel storage. " . . . the establishment of repositories for the storage of spent nuclear reactor fuel under effective international auspices and inspection."

Credit for spent fuel. "The establishment of arrangements under which nations placing spent fuel in such repositories would receive appropriate compensation for the energy content of such spent fuel if recovery of such energy content is deemed necessary or desirable . . . "

Sanctions. "Sanctions for violation of the provisions of or for abrogation of such binding international undertakings."

An interim stockpile. "The President shall submit to Congress not later than six months after the date of enactment of this Act proposals for initial fuel assurances, including creation of an interim stockpile of uranium enriched to less than 20 percent in the uranium isotope 235 (low-enriched uranium) to be available for transfer pursuant to a sales arrangement to nations which adhere to strict policies designed to prevent proliferation when and if necessary to ensure continuity of proposals for the transfer of low-enriched uranium up to an amount sufficient to produce 100,000 MWe years of power from light water nuclear reactors, and shall also include proposals for seeking contributions from other supplier nations to such an interim stockpile pending the establishment of INFA." (Section 104(b), US Statutes, Vol. 92, p. 123)

Foreign enrichment participation. The President is to report to Congress by September 10, 1978, on the desirability of and options for foreign participation, including investment, in new U.S. uranium enrichment facilities. (Section 104(c), US Statutes, Vol. 92, p. 123)

Appendix 14C

Desired conditions for international nuclear trade specified in Section 403 of the Nuclear Non-Proliferation Act of 1978

Section 403 of the act directs the President to seek agreement from all nations that no nuclear items will be transferred unless there is agreement to the following stipulations:

(1) *No nuclear explosives.* "[N]o nuclear materials and equipment and no nuclear technology in, under the jurisdiction of, or under the control of any non-nuclear-weapon state, shall be used for nuclear explosive devices for any purpose or for research on or development of nuclear explosive devices for any purpose, except as permitted by Article V, the Treaty";

(2) *Full-scope safeguards.* "IAEA safeguards will be applied to all peaceful nuclear activities in, under the jurisdiction of, or under the control of any non-nuclear-weapon state";

(3) *Physical Security.* "[A]dequate physical security measures will be established and maintained by any nation or group of nations on all of its nuclear activities";

(4) *Retransfers.* "[N]o nuclear materials and equipment and no nuclear technology intended for peaceful purposes in, under the jurisdiction of, or under the control of any nation or group of nations shall be transferred to the jurisdiction of any other nation or group of nations which does not agree to stringent undertakings meeting the objectives of this section";

(5) *No nuclear explosive assistance.* "[N]o nation or group of nations will assist, encourage, or induce any non-nuclear-weapon state to manufacture or otherwise acquire any nuclear explosive device."

Appendix 14D

Desired limitations upon domestic use of nuclear power specified in Section 403 of the Nuclear Non-Proliferation Act of 1978

Sections 403(b) and (c) of the act direct the President to seek agreement from other nations to the following three limitations upon their domestic use of nuclear power:

(1) *Enrichment, reprocessing, fabrication, and stockpiling.* "No source or special nuclear material within the territory of any nation or group of nations, under its jurisdiction, or under its control anywhere will be enriched . . . no irradiated fuel elements containing such material which are to be removed from a reactor will be altered in form or content, and no fabrication or stockpiling involving plutonium, uranium 233, or uranium enriched to greater than 20 percent in the isotope 235 shall be performed except in a facility under effective international auspices and inspection, and any such irradiated fuel elements shall be transferred to such a facility as soon as practicable after removal from a reactor consistent with safety requirements. Such facilities shall be limited in number to the greatest extent feasible and shall be carefully sited and managed so as to minimize the proliferation and environmental risks associated with such facilities. In addition, there shall be conditions to limit the access of non-nuclear-weapon states other than the host country to sensitive nuclear technology associated with such facilities."

(2) *Fuel storage.* "Any facilities within the territory of any nation or group of nations, under its jurisdiction, or under its control anywhere for the necessary short-term storage of fuel elements containing plutonium, uranium 233, or uranium enriched to greater than 20 percent in the isotope 235 prior to placement in a reactor or of irradiated fuel elements prior to transfer as required in subparagraph (1) shall be placed under effective international auspices and inspection."

(3) *Physical security.* "Adequate physical security measures will be established and maintained with respect to all nuclear activities within the territory of each nation and group of nations, under its jurisdiction, or under its control anywhere, and with respect to any international shipment of significant quantities of source or special nuclear material or irradiated source or special nuclear material, which shall also be conducted under international safeguards."

Appendix 14E

List of actions triggering cut-off of US nuclear exports specified in the Nuclear Non-Proliferation Act of 1978

Section 307 of the act adds a new section 129 to the Atomic Energy Act of 1954, as amended, which specifies four actions that would trigger cut-off of US nuclear exports to non-nuclear weapon states and three actions that would cut off US nuclear exports to any nation or group of nations.

Triggering actions for cut-off to non-nuclear weapon states:

(1) Detonation of a nuclear explosive device.

(2) Termination or abrogation of IAEA safeguards.

(3) Material violation of an IAEA safeguards agreement.

(4) Engaging in "activities involving source or special nuclear material and having direct significance for the manufacture or acquisition of nuclear explosive devices", and failure to take steps "which, in the President's judgment, represent sufficient progress toward terminating such activities."

Triggering actions for cut-off to all nations:

(1) Material violation of a US agreement for cooperation on terms of supply for US items.

(2) Assisting, encouraging, or inducing any non-nuclear weapon state to engage in " . . . activities involving source or special nuclear material and having direct significance for the manufacture or acquisition of nuclear explosive devices" and failing to take steps which in the President's judgement " . . . represent sufficient progress toward terminating such assistance, encouragement, or inducement."

(3) Entering into an agreement for the transfer of reprocessing equipment, materials, or technology to a non-nuclear weapon state, except in connection with the International Nuclear Fuel Cycle Evaluation or pursuant to an international agreement or undertaking to which the United States subscribes.

Appendix 14F

Conditions for US agreements for nuclear cooperation required by the Nuclear Non-Proliferation Act of 1978

Section 123''a of the Atomic Energy Act of 1954, as amended by Section 401 of the Nuclear Non-Proliferation Act of 1978, requires the President to obtain agreement to nine conditions in negotiation or renegotiation of agreements for cooperation in peaceful uses of nuclear energy. These are summarized as follows:

(1) *Safeguards for exports.* A guaranty that safeguards will be maintained for all nuclear items exported and for all nuclear material used in or produced through the use of exported nuclear materials and equipment, irrespective of the duration of other provisions in the agreement or whether the agreement is terminated or suspended for any reason.

(2) *Full-scope safeguards.* For a non-nuclear weapon state, maintenance of IAEA safeguards for all nuclear materials in all its peaceful nuclear activities.

(3) *No nuclear explosives.* A guaranty that no nuclear materials and

equipment or sensitive technology transferred and no special nuclear materials derived from them will be used for any nuclear explosive device, or for related research and development, or for any other military purpose.[1]

(4) *US right of return.* Except for agreements with nuclear-weapon states, a US right to require return of any nuclear materials and equipment transferred and of any special nuclear materials derived from them if the cooperating state detonates a nuclear explosive device or terminates or abrogates an agreement providing for IAEA safeguards;

(5) *US consent for transfers.* A guaranty that any material or any Restricted Data transferred or any nuclear facility transferred, or any special nuclear materials produced through such transfers will not be transferred to unauthorized persons or beyond the jurisdiction or control of the cooperating party without the consent of the United States.[2]

(6) *Physical security.* A guaranty that adequate physical security will be maintained for nuclear material transferred or derived from nuclear transfers.

(7) *Prior US approval for reprocessing, enrichment or alterations.* A guaranty that no nuclear material transferred to a country or derived from US exports will be reprocessed, enriched, or otherwise altered without prior approval of the United States.

(8) *Advance US approval for storage of nuclear materials.* A guaranty that no plutonium, U-233 or highly enriched uranium transferred or derived from US exports will be stored in any facility that has not been approved in advance by the United States.

(9) *Non-proliferation commitments for derived items.* A guaranty that special nuclear materials and nuclear facilities produced or constructed through the use of any sensitive nuclear technology transferred will be subject to all the preceding non-proliferation commitments.

Appendix 14G

List of statutory criteria for approval of US nuclear exports

Immediate criteria. Section 127 of the Atomic Energy Act of 1954, as

---

[1] Sensitive nuclear technology is defined by Section 4(a)(6) of the Nuclear Non-Proliferation Act to mean any information (including information incorporated in a production or utilization facility or an important component part thereof) which is not available to the public and which is important to the design, construction, fabrication, operation or maintenance of a uranium enrichment or nuclear fuel reprocessing facility or a facility for the production of heavy water, but shall not include Restricted Data...

[2] Restricted Data is defined in Section 11(y) of the Atomic Energy Act of 1954 as amended to mean all data concerning (a) design manufacture, or utilization of atomic weapons; (b) the production of special nuclear material; or (c) the use of special nuclear materials in the production of energy, but shall not include data declassified or removed from the Restricted Data category ...

amended by Section 305 of Public Law 95-242 (US Statutes, Vol. 92, p. 136), lists six criteria for approval of US nuclear exports, as summarized below:

(1) *IAEA safeguards*. IAEA safeguards are to apply to the nuclear export, to previous exports, and to materials derived from them.

(2) *No nuclear explosives*. No proposed or past nuclear export or materials derived from them will be used for any nuclear explosive device or for related research and development.

(3) *Physical security*. Adequate physical security will be maintained for proposed exports and nuclear materials derived from them.

(4) *Prior approval of retransfers*. Prior US approval must be obtained for retransfer of US nuclear exports or materials derived from them.

(5) *Prior approval of reprocessing*. Prior US approval must be obtained for the reprocessing of exported nuclear materials, or such materials derived from them, and also for alteration of spent fuel elements from a US reactor.

(6) *Sensitive nuclear technology*. No sensitive nuclear technology shall be exported unless the foregoing conditions shall apply to any derived nuclear materials or equipment.

Full-scope safeguards, a delayed criterion. Section 128 of the US Atomic Energy Act of 1954, as amended by Section 306 of Public Law 95-242 (US Statutes, Vol. 92, p. 137), imposes a seventh criterion, to apply after 10 September 1979: Non-nuclear weapon states must have all their peaceful nuclear activities under IAEA safeguards. Section 128''b provides for waiver of this criterion and for certain congressional action.

Common defence and security. Section 102 of the Atomic Energy Act of 1954, as amended, specifies that no licence for a nuclear reactor, including those for exports, may be issued if, in the opinion of the Commission, the issuance would be "inimical to the common defense and security or to the health and safety of the public". Section 123''b of the act requires a Presidential finding for an agreement for cooperation that it will promote, and will not constitute an unreasonable risk, to the common defence and security.

Appendix 14H

Notable procedural requirements for negotiation and renegotiation of agreements for nuclear cooperation

Negotiation. The Secretary of State negotiates agreements with the technical assistance and concurrence of the Secretary of Energy and in consultation with the Director of ACDA.

Submission to the President. After consultation with the NRC, the agreement is submitted to the President jointly by the Secretaries of State and Energy, accompanied by their views and recommendations and those of the NRC and ACDA.

Nuclear Proliferation Assessment Statement. The Director of ACDA provides the President with an unclassified Nuclear Proliferation Assessment Statement regarding the adequacy of safeguards and other control mechanisms and the peaceful use assurances contained in the agreement for cooperation to ensure that any assistance supplied will not be used to further any military or nuclear explosive purpose.

Presidential action. The President approves and authorizes execution of the agreement and makes a written determination that the performance of the proposed agreement "will promote and will not constitute an unreasonable risk to the common defense and security".

The President then submits the proposed agreement, together with his approval and determination, to the Congress.

Congressional action. If a proposed agreement does not involve large nuclear reactors—that is, less than five megawatts thermal output—the agreement must lie before Congress for 30 days of continuous session before taking effect. Congress can waive any or all of the waiting period.

If the proposed agreement involves more powerful reactors, the agreement must lie before Congress for 60 days of continuous session and shall not take effect if during this time Congress adopts a concurrent resolution stating that it does not favour the agreement.

Following submission of a proposed agreement, Congress may request the views of the Departments of State and Energy and of ACDA as to whether the safeguards and other controls of the agreement provide an adequate framework to ensure that any contemplated exports will not be inimical to or constitute an unreasonable risk to the common defence and security.

Source: Section 123 of the Atomic Energy Act of 1954, as amended by Section 401 of Public Law 95-242, US Statutes, Vol. 92, pp. 142–45.

Appendix 14I

Notable procedural requirements for NRC licensing of US nuclear exports

Part I.   Executive Branch actions

Executive Branch veto. The NRC may not license a nuclear export without notice by the Secretary of State that it is the judgement of the Executive Branch that the proposed export "will not be inimical to the common defense and security".

Procedures. The Secretary of State is responsible for procedures for preparation of the Executive Branch judgement.

Timing. The Executive Branch judgement is to be completed within 60 days of receipt of an application unless the Secretary of State in his discretion authorizes additional time because it is in the national interest to do so.

Part II.   NRC actions

NRC judgement. The NRC also may not license a nuclear export until it finds, based on a reasonable judgement of the assurances provided and other information available to the Federal Government, that the nuclear export criteria or their equivalent are met.

Single findings. The NRC is authorized to make a single finding for more than a single application where the same country is involved in the same general time frame for items of similar significance for nuclear explosives and under reasonably similar circumstances.

Abbreviated review. The NRC is authorized to find that there is no "material changed circumstance" associated with a new application from that of the last application which was approved by the Commission using all required procedures, and such a finding shall be deemed to satisfy the requirement for findings by the Commission.

Timely action. The NRC is to give timely consideration to licence requests.

Presidential action. If the NRC does not issue a licence because it is unable to make statutory determinations required, or if it does not act within 120 days after the Executive Branch delivers its judgement, the President can authorize the export, but he must determine that withholding of the export would be "seriously prejudicial to the achievement of U.S. non-proliferation objectives, or would otherwise jeopardize the common defense and security".

Congressional review. If the President authorizes such an export, before doing so he must submit his Executive order to Congress together with his explanation and the matter must remain before Congress for 60 days of continuous session. Congress can block the export by concurrent resolution during this period.

NRC procedures. The NRC is directed to promulgate regulations for export licence action and for associated public participation.

Source: Atomic Energy Act of 1954, Section 126, as amended by Public Law 95-242, Section 304(a), US Statutes, Vol. 92, pp. 131–35.

Appendix 14J

Notable procedural requirements for Department of Energy action on subsequent arrangements

Negotiation. The Secretary of Energy negotiates subsequent arrangements. The Secretary of State has the leading role in any negotiations of a policy nature for storage or disposition of spent fuel or approvals for transfer of nuclear materials, facilities or technology.

Consultation. The Secretary of Energy is to obtain the concurrence of the Secretary of State and consult with the ACDA Director, the NRC and the Secretary of Defense.

Determination. The Secretary of Energy must make a written determination that such an arrangement will not be "inimical to the common defense and security".

Timing and public notice. A subsequent arrangement is not to take effect until 15 days after its publication in the *Federal Register* together with the Secretary of Energy's determination. Subsequent arrangements for retransfer or reprocessing of nuclear materials from the United States also must lie before Congress for 15 days of continuous session.

Nuclear Non-Proliferation Assessment Statement. The ACDA Director has discretion to prepare a Nuclear Proliferation Assessment Statement, if in his view a proposed subsequent arrangement "might significantly contribute to proliferation", regarding the adequacy of the safeguards and other control mechanisms and the application of the peaceful use assurances of the agreement to ensure that assistance to be furnished will not be used to further any military or nuclear explosive purpose.

Reprocessing. The United States will give timely consideration to required requests for prior approval for reprocessing, to transfer of recovered plutonium, and for alteration of spent fuel. It will try to expedite such consideration when the actions are set out in a bilateral or international agreement.

Procedures. The Secretary of Energy is to establish procedures for consideration of requests for subequent arrangements.

Source: Atomic Energy Act of 1954, Section 131, as added by Public Law 95-242, Section 303(a), US Statutes, Vol. 92, pp. 127–29.

Appendix 14K

Summary of provisions in the Nuclear Non-Proliferation Act of 1978 relating to sanctions against certain proliferation actions

Policy. It is the policy of the United States to "actively pursue . . . the establishment of more effective international controls over the transfer and use of nuclear materials and equipment and nuclear technology for peaceful purposes in order to prevent proliferation, *including the establishment of common international sanctions*". (Emphasis added) (US Statutes, Vol. 92, p. 120)

International negotiations. The United States shall seek to negotiate with other nations and groups of nations to adopt general principles and procedures, *including common international sanctions*, to be followed in the event that a nation violates any material obligation with respect to the peaceful use of nuclear materials and equipment or nuclear technology, or in the event that any nation violates the principles of the NPT, including the detonation by a non-nuclear weapon state of a nuclear explosive device. (US Statutes, Vol. 92, p. 124)

Termination of US nuclear exports. The act forbids nuclear exports to any non-nuclear weapon state that is found by the President to have: detonated a nuclear explosive device; or terminated or abrogated IAEA safeguards; or materially violated an IAEA safeguards agreement; or "engaged in activities involving source or special nuclear material and having direct significance for the manufacture or acquisition of nuclear explosive devices, and has failed to take steps which, in the President's judgement, represent sufficient progress toward terminating such activities". (US Statutes, Vol. 92, p. 138)

The act forbids nuclear exports to any nation or group of nations, which includes nuclear weapon states, that is found by the President to have:

(1) materially violated the terms of an agreement for cooperation or the terms of supply for exports not supplied under an agreement for cooperation; or

(2) "assisted, encouraged, or induced any non-nuclear-weapon state to engage in activities involving source or special nuclear material and having direct significance for the manufacture or acquisition of nuclear explosive devices, and has failed to take steps which, in the President's judgment, represent sufficient progress toward terminating such assistance, encouragement, or inducement"; or

(3) "entered into an agreement . . . for the transfer of reprocessing equipment, materials, or technology to the sovereign control of a non-nuclear-weapon state" except in connection with INFCE. (US Statutes, Vol. 92, p. 138)

A condition for US agreements for cooperation. US agreements for cooperation in peaceful uses of nuclear energy with non-nuclear weapon states are to include a "stipulation that the United States shall have the right to require the return of any nuclear materials and equipment transferred . . . and any special nuclear material produced through the use thereof if the cooperating party detonates a nuclear explosive device or terminates or abrogates an agreement providing for IAEA safeguards". (US Statutes, Vol. 92, p. 142).

Appendix 14L

Summary of notable implications of the Nuclear Non-Proliferation Act of 1978 for control of certain technological activities in the non-military applications of nuclear energy

1. *Development of proliferation-resistant fuel cycles*. The act supports study and analysis through INFCE provision.

2. *Enrichment of uranium*. It favours ban on international trade in enrichment plants, equipment and technology; would put existing enrichment plants under international auspices and inspection, and perhaps combine them into an International Nuclear Fuel Authority. It would have non-nuclear weapon states agree not to build new enrichment plants.

3. *Waste disposal*. No direct treatment. By discouraging reprocessing, it would reduce need for waste disposal for separated fission products and transuranic materials. It encourages storage of spent fuel under international auspices and inspection, and perhaps as a function of an international organization.

4. *Transport of nuclear materials*. It supports study and analysis through INFCE. It would assure adequate physical security and put international shipments under IAEA safeguards.

5. *Breeder reactors*. No direct treatment. By discouraging reprocessing, and by putting reprocessing plants under international auspices and inspection, it could tend to discourage, delay or redirect development of breeders.

6. *Safeguards technology*. It promotes improvement of safeguards technology to assure early warning and action. It makes assurance of early warning a principal consideration for US approval for reprocessing of its nuclear exports abroad.

7. *Peaceful nuclear explosives*. It would ban development and use of nuclear explosives of any kind, peaceful or otherwise, by non-nuclear weapon states, cut off nuclear cooperation and trade to states that do not comply, and perhaps require return of nuclear exports to such states.

8. *Fusion breeders*. No direct effect. Results of INFCE could influence future development so as to limit proliferation risks. Depending upon success of development, international commerce in fusion breeder equipment might be controlled or banned, and their use placed under international auspices and inspection, or direct international operation.

9. *Laser fusion*. No direct effect. Results of INFCE could influence future development to limit proliferation risks. Depending on success of technological development, international commerce in laser fusion equipment might be controlled or banned, and their use placed under international auspices and inspection, or direct international operation.

10. *Nuclear reactors in satellites*. No direct effect.

Appendix 14M

Summary of provisions of the Nuclear Non-Proliferation Act of 1978 relating to enrichment of uranium

Policy. It is the policy of the United States to take such actions as are required to confirm the reliability of the United States in meeting its commitments to supply nuclear reactors and fuel to nations which adhere to effective non-proliferation policies by establishing procedures to facilitate the timely processing of requests for subsequent arrangements and export licences. (US Statutes, Vol. 92, p. 120)

Capacity. The United States shall take such actions as may be necessary and feasible to assure other nations and groups of nations that it will provide a reliable supply of nuclear fuel to those who adhere to non-proliferation policies of the act. (US Statutes, Vol. 92, p. 121)

The United States shall ensure that it will have available the capacity on a long-term basis to enter into new fuel supply commitments consistent with its non-proliferation policies and domestic energy needs. (US Statutes, Vol. 92, p. 121)

The Secretary of Energy is to take steps for expansion of US uranium enrichment capacity. (US Statutes, Vol. 92, p. 122)

The President shall promptly undertake a study of the need for additional US enrichment capacity to meet domestic and foreign needs and to promote US non-proliferation objectives abroad. The President is to report the results by 10 March 1979. (US Statutes, Vol. 92, p. 122)

International supply. The President is promptly to start with other nations and groups of nations discussions of international approaches to world-wide nuclear fuel supply, including creation of an International Nuclear Fuel Authority. He is to negotiate with nations possessing nuclear fuel production facilities or source materials, and others, to establish binding international undertakings to provide for, among other things, the siting, development and management under effective international auspices and inspection of facilities for provisions of nuclear fuel services. (US Statutes, Vol. 92, p. 122)

US stockpile. The President is to report to Congress by 10 March 1979, with proposals for initial fuel assurances, including the creation of an interim stockpile of enriched uranium to be available for transfer to nations which adhere to strict non-proliferation policies when and if necessary to ensure continuity of their nuclear fuel supply. The proposal is to consider transfer of low enriched uranium up to an amount sufficient to produce 100 000 megawatt years of electrical energy from light water reactors, and shall also include proposals for contributions from other supplier countries. (US Statutes, Vol. 92, p. 122)

Conditions of access. Access to international sources of nuclear fuel is to be limited to non-nuclear weapon states that, among other things, do not establish any new enrichment facility and place any existing facilities under effective international auspices and inspection. (US Statutes, Vol. 92, p. 123)

Termination of US nuclear exports. One triggering action by a non-nuclear weapon state for cut-off of US nuclear exports is to engage in activities "involving source or special material and having a direct significance for the manufacture of nuclear explosive devices", and to fail to take steps which, in the President's judgement, represent sufficient progress toward terminating such activities. (US Statutes, Vol. 92, p. 138)

A condition of agreements for cooperation. The Secretary of State in renegotiating agreements for nuclear cooperation is expected to get guarantees from US nuclear trading partners that no exported nuclear materials, or materials derived from US nuclear exports, will be enriched without prior US approval. (US Statutes, Vol. 92, p. 143)

A limitation on US exports. No source or special nuclear material shall be exported for enrichment or for reactor fueling to any nation or group of nations that has entered into a new or amended agreement for cooperation except pursuant to such agreement. (US Statutes, Vol. 92, p. 145)

International agreement. The President is to negotiate with other nations and groups of nations for agreement on common export policies. One is that no source material will be enriched except in a facility under effective international auspices and inspection. Such facilities shall be limited in number to the greatest extent feasible and shall be carefully sited and managed so as to minimize the proliferation and environmental risks. In addition, there shall be conditions to limit the access of non-nuclear weapon states other than the host country to sensitive nuclear technology associated with such facilities. (US Statutes, Vol. 92, p. 146)

Appendix 14N

Summary of provisions of the Nuclear Non-Proliferation Act of 1978 relating to reprocessing of nuclear fuels

Compensation for energy value of spent fuel. The President is to seek arrangements for nations that place spent fuel in international repositories to receive appropriate compensation for its energy content, "if recovery of such energy content is deemed necessary or desirable". (US Statutes, Vol. 92, p. 122)

No new facilities. International fuel assurances are limited to those nations that, among other things, agree not to establish any new reprocessing facilities and to place existing facilities under international auspices and inspection. (US Statutes, Vol. 92, p. 123)

Consideration of reprocessing requests. The United States will give timely consideration to all requests for prior approvals for reprocessing of nuclear materials to the "maximum extent feasible" and will attempt to expedite such considerations when the terms for reprocessing are set forth in an agreement for cooperation or in some other international agreement to which the United States is party and subject to congressional review. (US Statutes, Vol. 92, p. 128)

Conditions for US approval of reprocessing requests:

(1) The Secretary of Energy may not approve the retransfer of nuclear materials exported by the United States or derived from its exports for reprocessing or for the subsequent retransfer of recovered plutonium in quantities greater than 500 grams until he determines that it will not be inimical to the common defence and security, reports his reasons to Congress and a period of 15 days of continuous session has elapsed. (US Statutes, Vol. 92, p. 128)

(2) For new reprocessing facilities, the Secretary of Energy may approve reprocessing or subsequent retransfer of resulting plutonium in quantities greater than 500 grams only if in his judgement and that of the Secretary of State, such reprocessing or retransfer will not result in "a significant increase of the risks of proliferation beyond that which exists at the time that approval is requested". In this judgement, "foremost consideration" will be given to whether or not the reprocessing or retransfer will take place under conditions that will ensure "*timely warning* to the United States of any diversion well in advance of the time at which the non-nuclear weapon state could transform the diverted material into a nuclear explosive device". (Emphasis added) (US Statutes, Vol. 92, p. 128)

(3) For existing reprocessing facilities, the Secretary of Energy shall attempt to ensure, in entering into any subsequent arrangement for reprocessing or subsequent retransfer of recovered plutonium in quantities greater than 500 grams, that this will take place under conditions comparable to those which, in his view and that of the Secretary of State, satisfy the standards for reprocessing in a new facility. (US Statutes, Vol. 92, p. 128)

No prohibition. The act is not intended to prohibit, permanently or unconditionally, the reprocessing of spent fuel owned by a foreign nation, which has been supplied by the United States. (US Statutes, Vol. 92, p. 129)

An export criterion. One of the six criteria for decisions on US nuclear exports is that no material to be exported and no special nuclear material produced through its use will be reprocessed without prior US approval. (US Statutes, Vol. 92, p. 136)

Termination of US nuclear exports. One triggering action by a non-nuclear weapon state for cut-off of US nuclear exports is to engage in activities "involving source or special material and having direct significance for the manufacture of nuclear explosive devices", and to fail to take steps which, in the President's judgement, represent sufficient progress toward terminating such activities. (US Statutes, Vol. 92, p. 138)

Another triggering action is for any nation or group of nations to agree to transfer reprocessing equipment, materials or technology to a non-nuclear weapon state, except in connection with INFCE, unless this condition is waived by the President. (US Statutes, Vol. 92, p. 139)

A condition of agreements for cooperation. The Secretary of State, in renegotiating US agreements for nuclear cooperation, is expected to acquire a guarantee from US nuclear trading partners that no material transferred and no material used in or produced through use of US nuclear exports will be reprocessed without prior US approval. (US Statutes, Vol. 92, p. 143)

International agreement. The President is to negotiate with other nations and groups of nations for agreement on common export policies. One is that no source or special nuclear material will be reprocessed except in a facility under effective international auspices and inspection. Such facilities shall be limited in number to the greatest extent feasible and shall be carefully sited and managed so as to minimize the associated proliferation and environmental risks. In addition, there shall be conditions to limit the access of non-nuclear weapon states other than the host country to sensitive nuclear technology associated with such facilities. (US Statutes, Vol. 92, p. 146)

Appendix 14O

Summary of provisions in the Nuclear Non-Proliferation Act of 1978 relating to nuclear explosives

No nuclear explosives. The nuclear fuel assurances provided in the act are limited to those non-nuclear weapon states that do not manufacture or otherwise acquire any *nuclear explosive device.* (US Statutes, Vol. 92, p. 123)

Nuclear Non-Proliferation Assessment Statements. The Director of the Arms Control and Disarmament Agency is to prepare a Nuclear Non-Proliferation Statement for new or renegotiated agreements for cooperation. In his discretion, he may prepare such statements for US approvals provided for in the agreements (subsequent arrangements) that in his view might "significantly contribute to proliferation". The statement is to address the adequacy of safeguards and other control mechanisms and the application of peaceful uses assurances of agreements to ensure that assistance furnished will not be used to further any military or *nuclear explosive* purpose. (US Statutes, Vol. 92, p. 127)

An export criterion. One of the six criteria for US decisions on nuclear exports is that no export will be used for any nuclear explosive device or for related research and development. (US Statutes, Vol. 92, p. 136)

Termination of US nuclear exports. One triggering action by a non-nuclear weapon state for cut-off of US nuclear export is to detonate a nuclear explosive device. (US Statutes, Vol. 92, p. 138)

Another triggering action is for a state to assist, encourage or induce any non-nuclear weapon state to engage in activities involving source or special nuclear material and having direct significance for the manufacture or acquisition of nuclear explosive devices, and has failed to take steps which, in the President's judgement, represent sufficient progress toward terminating such action. (US Statutes, Vol. 92, p. 138)

A condition for agreements for cooperation. The Secretary of State in renegotiating agreements for nuclear cooperation is expected to get a guarantee that no nuclear export or derived materials will be used for any nuclear explosive device or for related research and development. (US Statutes, Vol. 92, p. 146)

Another condition is a right for the United States to require the return of any nuclear materials and equipment exported and any derived special nuclear material if a non-nuclear weapon state detonates a nuclear explosive device. (US Statutes, Vol. 92, p. 142)

International agreement. The President is to negotiate with other nations and groups of nations for agreement on common export policies. One policy is that nuclear transfers will be made only to nations that agree that no nuclear materials and equipment and no nuclear technology within any non-nuclear weapon state will be used for nuclear explosive devices for any purpose or for related research and development. (US Statutes, Vol. 92, p. 146)

Appendix 14P

Summary of provisions of the Nuclear Non-Proliferation Act of 1978 relating to safeguards technology

Congressional finding. "Recent events emphasize . . . the imperative need to increase the effectiveness of international safeguards and controls on peaceful nuclear activities to prevent proliferation." (US Statutes, Vol. 92, p. 120)

Policy. "It is the policy of the United States to actively pursue . . . the establishment of more effective international controls over the transfer and use of nuclear materials and equipment and nuclear technology for peaceful purposes in order to prevent proliferation . . . " (US Statutes, Vol. 92, p. 120)

The United States is committed to a "strengthened and more effective International Atomic Energy Agency and to a comprehensive safeguards system administered by the Agency to deter proliferation." (US Statutes, Vol. 92, p. 124)

US actions. The USA shall seek to act with other nations to:

— continue to strengthen the safeguards programme of the IAEA and contribute funds, technical resources and other support to assist the IAEA in effectively implementing safeguards;

— ensure that the IAEA has the resources to carry out its safeguards functions (as defined in the IAEA statute); and

— improve the IAEA safeguards system (including accountability) to ensure:

(1) the timely detection of a possible diversion of source or special nuclear materials which could be used for nuclear explosive devices;

(2) the timely dissemination of information regarding such diversion; and

(3) the timely implementation of internationally agreed upon procedures in the event of such diversion. (US Statutes, Vol. 92, p. 124)

Paper 15. Nuclear exporting policies

B. SANDERS

Square-bracketed numbers, thus [1], *refer to the list of references on page 249.*

I. Introduction

It is difficult to give a clear picture of the present world position with respect to the international exchange of nuclear materials, equipment and know-how. It is in fact hardly meaningful to speak of 'export policies' as if there existed a set of clearly defined, uniformly applied policies. Nor is the 'export' side of the picture its only decisive aspect. The situation is, rather, characterized on the one hand by a lack of clarity in, and of unity among, the export policies of the various nuclear supplier countries, and on the other hand by efforts on the part of actual or would-be importers of nuclear items to influence developments in line with their thinking.

It is therefore tempting to ignore the differences between the various approaches and to speak, more generally, in terms of the 'position of the suppliers' or the 'attitude of the industrialized states', as opposed to the 'approach of recipient states' or the 'position of Third World nations'. To the extent that one can at all define collective stances and speak of particular groups of states as advancing specific approaches and reflecting particular interests, the nuances between various positions within such groups must not be overlooked. Nevertheless, this paper attempts to indicate some common trends among the approaches of those states that appear to be the principal actors on the international nuclear scene.

II. Export and import policies

The industrialized states capable of exporting nuclear materials, equipment and technology, and as such involved in formulating an 'export policy', although important protagonists, are no longer the sole determinants of the nuclear export dialogue. Perhaps the principal novelty of recent years is that a number of actual or potential recipients of such exports are making

241

concerted efforts to influence the conditions of supply and are beginning to define their own positions towards the suppliers' approaches. This situation has the makings of an incipient conflict, but, as tensions increase, so do attempts from both sides to reach a consensus. This may be yet another new element in the situation.

All countries concerned have two basic interests in common: to develop a nuclear energy potential where this is required, and to prevent this development from advancing the proliferation of nuclear military capabilities. The difference is really one of emphasis. The industrialized states which seek to restrict the export of nuclear items for the sake of reducing proliferation risks are at the same time interested in developing a nuclear market abroad. Many Third World nations, on the other hand, emphasize their interest in full and ready access to such a market; they may not attach as high priority to the prevention of nuclear weapon proliferation as do some of the nuclear supplier states, but this wish for access to various nuclear supplies cannot be interpreted as an indiscriminate rejection of the non-proliferation aim.

The nuclear suppliers

The gap in the respective supplier and recipient perceptions is wide and involves diverse important political as well as technical, strategic, economic and psychological considerations. In very general terms, a group of states capable of exporting nuclear material, facilities, equipment or technology have sought to establish certain constraints in respect of their nuclear exports, in order to reduce the possibility of further nuclear weapon proliferation. These states (Belgium, Canada, Czechoslovakia, France, the Federal Republic of Germany, the German Democratic Republic, Italy, Japan, the Netherlands, Poland, Sweden, Switzerland, the USSR, the United Kingdom and the United States) have held consultations on a common policy in this regard, in what has become popularly known as the 'London Suppliers Club'. In January 1978, they sent letters to the International Atomic Energy Agency (IAEA), setting forth the principles and detailed guidelines they would henceforth follow in their export of nuclear material, equipment and technology [1].[1] The basis of this joint policy is a Trigger List of items that should be exported only on such conditions as: (*a*) the recipient's assurance explicitly to exclude uses which would result in any nuclear explosive device; (*b*) effective physical protection of the items supplied; and (*c*) the application of IAEA safeguards strictly in accordance with the Agency's rules pertaining to the duration and coverage of those safeguards. The suppliers' policy also provides that 'sensitive' facilities (specifically, installations handling or producing material that can be used

---

[1] See also reference [2], containing Australia's announcement that it will apply criteria which satisfy those guidelines and have certain additional requirements.

in a nuclear explosive) which utilize technology directly supplied or derived from other supplies should also be placed under IAEA safeguards. The guidelines call for restraint in the transfer of 'sensitive' facilities, technology and materials usable in nuclear weapons. They further include some constraints in regard to retransfer, the use of supplied facilities for the enrichment of uranium above a certain level, and the separation of plutonium through the reprocessing of irradiated fuel. Several of the states involved (the German Democratic Republic, Poland and the USSR) have further indicated that they support the principle that Trigger List items should be exported only if all nuclear activities in the recipient state are under IAEA safeguards.[2] In addition, the guidelines encourage alternatives to national enrichment and reprocessing plants, such as multinational fuel centres. The supplier states undertake to keep in contact and consult each other on the implementation of the guidelines and particularly on action to be taken in case of a violation of the understandings accepted by recipients. Some variations in approach among members of the London Club will be mentioned below.

The nuclear recipients

A number of Third World countries are opposed to any restriction on the export of nuclear technology. They regard such restrictions as running counter to the commitment which industrialized states, in particular nuclear weapon states, have accepted in Article IV of the Non-Proliferation Treaty (NPT), and they consider restrictions on exports as a reflection of an attempt by the industrialized states to keep the Third World in subservience.[3] The attitudes of Third World countries are also not homogeneous and are influenced by their political situation, their state of nuclear development, and their security perceptions.

India is one of the most outspoken of these countries, in its dual capacity as a leader of the non-aligned nations and as having reached a high level of development in nuclear technology. Its attitude towards such restrictions —which have affected provisions for the US-supplied nuclear power station at Tarapur, and the installation at Rajasthan which was set up with Canadian assistance—is reflected in various publications which blame the NPT for having perpetuated the technological monopoly of the nuclear weapon powers (for example, in such phrases as "a bizarre new world of nuclear aristocracy ruling the non-nuclear serfdom" [3]). India claims that under the NPT, all nuclear benefits are conferred on the nuclear weapon

---

[2] A similar provision is included in the US Nuclear Non-Proliferation Act of 1978 and forms part of the Canadian nuclear export policy.
[3] This brief description is derived from remarks made on 18 May 1978 by Mr D. A. V. Fischer, Assistant Director General for External Relations with the IAEA, at a Panel on Exploring the Relationships of Export Policy and Non-Proliferation of the international conference on regulating nuclear energy organized by the Atomic Industrial Forum.

powers and all the burdens on the non-nuclear weapon nations, and sees the measures adopted by suppliers as constituting a "technological bondage" which no self-respecting nation could accept.[4]

Similar opinions are heard from Yugoslavia which, although generally professing to be in agreement with the concept of IAEA safeguards on peaceful nuclear activities, resists any additional measures of control which go beyond those directly inherent in the NPT.[5] These views were presented with considerable force during the May–June 1978 United Nations General Assembly Special Session on Disarmament, in particular in the discussion of the final document of that session. The document contains relatively little about the question of proliferation of nuclear weapons [5] and it clearly shows the extent to which, in the interest of reaching consensus, account has been taken of the approach of Third World nations. Brazil also played an important part in the discussions which led to the final text and particularly to the subdued manner in which non-proliferation measures in general and the NPT in particular have been referred to.

The supplier/recipients

A third category of states involved in the debate is formed by those countries which are both industrially developed and actual or potential suppliers of nuclear items, but which are still dependent on other states for some of the nuclear materials, equipment or technology they need for their nuclear fuel cycle. An example of such a state is Japan, which seeks to reduce its dependence on uranium supplies from abroad by promoting the use of breeder facilities and which has a recycle programme that relies on plutonium extracted from spent reactor fuel. It therefore has difficulty in accepting restrictions, such as those sought by the USA, on the reprocessing of irradiated uranium fuel.[6]

III. Safeguards measures

Although the awareness of the risks of nuclear proliferation has particularly grown in the past five years or so, partly as a result of the Indian nuclear explosion in May 1974, the idea that various restrictions should be imposed on the use of nuclear energy for peaceful purposes is by no means new. Already in the mid-1940s, the Baruch Plan introduced the concept of physical international control combined with inspection. Since then, a whole

---

[4] The Indian attitude is excellently reflected in reference [3].

[5] See, for example, reference [4]. One may assume that the Yugoslav attitude has been influenced by delays it encountered in obtaining US supplies of equipment and material for its first nuclear power station. These delays apparently occurred when Yugoslavia found diffi-culty in accepting US requirements regarding reprocessing and retransfer, reportedly made after the original supply conditions had been agreed on.

[6] See, for example, reference [6].

spectrum of measures has been proposed to reduce the danger of proliferation inherent in the spread of nuclear technology and materials, once it was obvious that this spread could not be prevented altogether. The measure initially chosen to accompany the export of nuclear equipment and fuel was the application by the supplier of 'safeguards' (that is, measures to ascertain how nuclear materials and installations have been used, so as to detect, within a given time, the diversion of a given quantity of nuclear material). Such bilateral safeguards began to be replaced in the 1960s by international safeguards, in particular those of the International Atomic Energy Agency. Because ample information about this development has been published elsewhere, it is not described here. Such safeguards, whether bilateral or multilateral, were connected only with specific supplies and did not cover such nuclear activities as the recipient states were capable of undertaking by their own means.

The NPT of 1968 provided that the entire nuclear effort of states party to that treaty should be covered by safeguards. A similar provision was included in the Treaty on the Denuclearization of Latin America (Treaty of Tlatelolco of 1967).

In the early 1970s several suppliers, as well as a number of other states supporting a strict non-proliferation régime, came to the conclusion that such partial safeguards would not in themselves be adequate. Consequently, moves were made to strengthen the safeguards régime of the IAEA through provisions extending the duration of safeguards for the lifetime of supplied items or installations and materials derived therefrom. At the same time, measures were devised to ensure that so-called sensitive technologies, when exported, should remain under control in the foreseeable future. Attempts were also made to achieve 'full-scope safeguards' in states not party to either of the two aforementioned treaties. Especially in the framework of the London Suppliers Club, an attempt was made to draw up common guidelines with regard to these requirements.

A new element arising from those discussions was the introduction of certain constraints with regard to exports of nuclear items. Hitherto, it had been generally assumed that as long as there was adequate provision for international safeguards, virtually any nuclear item could and would be exported to states ready to undertake not to use such imports for any military purpose, or for the manufacture of an explosive device for any purpose whatsoever. However, a group of major exporters agreed for the first time to exercise restraint in the transfer of sensitive facilities, technology and weapon-usable materials, to apply special controls with regard to enrichment technology and to require the suppliers' consent for the retransfer of certain items—all measures serving to restrict certain exports and, logically, leading to some continuing involvement of suppliers. In one form or another, this approach has been reflected in the legislation of several of the supplying countries concerned.

In the USA, the Nuclear Non-Proliferation Act was signed in March 1978 (see preceding paper). The act contains a set of 'denials' in respect of certain items which the USA believes create such a large risk that their export should be avoided altogether, wherever possible. It also includes a set of 'controls' in terms of safeguards and conditions for physical security in regard of exported items, backed up by sanctions. The act provides that no US nuclear export may be transferred to any other nation without prior approval of the USA and that no fuel exported from the USA may be reprocessed without such approval. New agreements for cooperation must include the provision, as a condition for a continuing US nuclear supply, that IAEA safeguards cover all nuclear materials and equipment regardless of whether or not these have been supplied by the United States.[7]

The Canadian nuclear export policy sets similar requirements with regard to retransfers, controls on enrichment and reprocessing, and sanctions. In the case of Canada, the reintroduction of the bilateral element in the safeguards régime is particularly noticed. Thus Canada requires 'standby safeguards' to take effect if and when the NPT and its concomitant safeguards régime run out or are abrogated. Canada similarly exercises this right in recipient states not party to the NPT, with respect to agreements covering Canadian supplies. Further, Canada aims to verify that its requirements with regard to restrictions on reprocessing, retransfers and the use of technology are being adhered to. It also continues to press for the adoption of 'full-scope safeguards' and does not wish to make any nuclear exports to states that have not ratified the NPT or otherwise accepted such safeguards [8].

France has actively participated in the London discussions and subscribes to the guidelines that have resulted from them, but its approach appears to diverge in several respects from that of other participants. While not a party to the NPT, and in fact highly critical of that instrument, France has repeatedly demonstrated its concern about nuclear proliferation and its support for international safeguards. In the past, however, it appears to have felt that such safeguards should constitute an adequate guarantee for the prevention of nuclear proliferation. This is reflected by its agreement in 1975 to supply reprocessing plants to the Republic of Korea and to Pakistan. Agreements for safeguards in respect of both plants were concluded between the states concerned and the IAEA [9, 10]. Apparently under pressure from the USA, however, South Korea seems to have decided not to go in for reprocessing and the arrangements with France have consequently lapsed. With regard to Pakistan, press reports often mentioned the concern of a number of governments and particularly that of the USA about this intended supply, but both parties repeatedly affirmed their intention to go through with the agreement. Recently, however, it was reported that France decided to withdraw its sale of reprocessing equipment to Pakistan [11–13]. According to these reports, France advised the government of Pakistan that it could only supply a plant of a type whose end-product contained a mixture of uranium and plutonium unfit for the

---

[7] See the publication of the US Library of Congress, Congressional Research Service [7]. Over the years, this Service has produced an invaluable collection of material on the question of nuclear proliferation.

production of nuclear weapons; this offer was refused by Pakistan.

The new French approach is also demonstrated by the joint declaration issued on 22 June 1977 by General Secretary Leonid Brezhnev and President Giscard D'Estaing on the non-proliferation of nuclear weapons, in which they recall that both the French and the Soviet governments were in favour of "limiting the transfer of nuclear materials suitable for the production of nuclear weapons or other nuclear explosive devices, as well as equipment and technology with which to produce these materials" [14]. However, France has made it clear that it would not exert pressure on any non-nuclear weapon state to refrain from using any given nuclear technology if it could develop this through its own efforts or in cooperation with another state.

The Federal Republic of Germany has not shown that it has reversed the policy which permitted it in 1975 to conclude an agreement with Brazil for the export of all elements in the nuclear fuel cycle, including reprocessing and enrichment facilities. There are no indications so far of any intention on the part of either side to reconsider this arrangement, which has been severely criticized by a number of supplier states. The government of the Federal Republic of Germany, however, announced on 17 June 1977 that it would not, for the present, permit the export from its country of certain sensitive nuclear items, but that its previous commitments in this regard would be honoured.

Australia and Sweden are among the states that have announced restrictive export measures similar to those of other suppliers; in a recent agreement with Finland, the former has stipulated, *inter alia*, that its consent is required before the uranium it has supplied undergoes high enrichment or reprocessing. It also provides for the possibility of suspending shipments in case of a failure to adhere to any of the terms of the agreement or to the safeguards provided for [15].

The states most capable of supplying nuclear material, equipment, facilities and know-how, although varying in approach to the conditions they would set for such supplies, thus show the collective inclination to restrict their exports to a greater degree than before. They are confronted by a group of potential recipient states which in principle seek access to equipment and technologies that might further their nuclear development, as they perceive its need. The latter states show an increasing tendency to combine forces in an attempt to break the suppliers' monopoly. It remains to be seen whether they can do so. If one or more non-nuclear weapon states eventually succeed, perhaps in a joint venture, to make themselves independent of the nuclear supplies of the present exporters, this would probably be greatly injurious to the non-proliferation régime as we now know it.[8]

---

[8] A cogent plea for international nuclear cooperation under adequate safeguards, in the interest of both non-proliferation and economic development, is made in an article by Meyer-Wöbse [16]. The author advances the thesis that a suppliers' boycott against Third World non-nuclear weapon states would be counter-productive, because there are several among them capable of important nuclear activity. If those states should turn to each other for mutual assistance, they could soon establish a high level of nuclear independence, and they would then be unlikely to accept international safeguards.

An impasse seems to have been reached. In the words of the communiqué issued after the summit meeting of Heads of State and Government in London in May 1977:

Increasing reliance will have to be placed on nuclear energy to satisfy growing energy requirements and to help diversify sources of energy. This should be done with the utmost precaution with respect to the generation and dissemination of material that can be used for nuclear weapons. Our objective is to meet the world's energy need and to make peaceful use of nuclear energy widely available, while avoiding the danger of the spread of nuclear weapons.

We are also agreed that, in order to be effective, non-proliferation policies should as far as possible be acceptable to both industrialized and developing countries alike [17].

IV.   The INFCE

To help solve this conundrum, the United States took the initiative for a study of less proliferation-prone nuclear fuel cycles. This resulted in the International Fuel Cycle Evaluation (INFCE) which has been under way since October 1977. It has been agreed that the evaluation should be carried out with mutual respect for each country's nuclear position and decisions; it will not jeopardize their fuel cycle policies, international cooperation, agreements or contracts for peaceful use of nuclear energy, provided that agreed safeguards measures are applied. In assessing whether INFCE will be able to further a solution to the present problem, it must be remembered that even if it can come up with new technical approaches, considerable time for research and development is needed before such approaches are available on an industrial scale. No doubt various alternative fuel routes are available. In connection with the Pakistani situation, mention has already been made of co-reprocessing methods yielding material unsuitable for weapon production. Recent press reports [18] divulge that France, the Federal Republic of Germany and the USA are discussing the possible joint development of a non-proliferation-prone uranium enrichment process.[9] But an early breakthrough, in the sense of the discovery of a series of alternative processes acceptable to all concerned and available in the relatively short term, is very unlikely.

Rather than looking for a single solution which would meet the various desiderata of all concerned—exporters and importers, and within each of these two groups the diverse requirements of the various members—it would seem to be unavoidable to strive for the realization of a package of several partial measures which could satisfy diverse requirements.

---

[9] This process was discussed in reference [19a].

The basic requirement remains the strengthening both of international cooperation in the peaceful uses of nuclear energy and of international safeguards. It might be productive, also from the point of view of the suppliers, to meet the wish of Third World nations for the former, which is now taking the form of a call for an international conference on peaceful uses of nuclear energy. Such a conference might be held after the results of the INFCE have been obtained. The suppliers should demonstrate their readiness to function as reliable sources of nuclear fuel and equipment, adhering to contracts of supply once they have been concluded. The fullest possible use should be made of existing international institutes such as the IAEA, and consideration should be given to such joint enterprises as multinational and regional fuel cycle centres. The possibility provided in the statute of the IAEA to make, under the aegis of that organization, arrangements for the international management and storage of plutonium and spent fuel should be seriously considered and, wherever possible, realized. Unilateral constraints and verification activities which tend to irritate recipient states should be avoided and be replaced by international measures; where such measures are necessary to verify the observance of undertakings that go beyond safeguards, the IAEA should be given the means and the task to do so. Exercises such as the INFCE should be promoted and followed up wherever they contain a promise of practical solutions. The nuclear activities of all states, whether safeguarded or not, should be publicized by the IAEA and the UN as a confidence-building measure and consideration should be given to the publication of all international transfers in the nuclear field. States in a position to do so should offer other states facilities for the disposal and storage of nuclear waste; such a measure might be used as an inducement to submit all nuclear activities to international safeguards or to refrain from reprocessing [19b].

There is obviously no panacea, given the present circumstances, but through these and similar measures, international cooperation might be promoted and the sting might be taken out of the present disagreements between groups of countries which fundamentally share the same concern: that of promoting nuclear energy without similarly promoting nuclear weapon proliferation.

References

1. IAEA document INFCIRC/254.
2. IAEA document INFCIRC/254/Add.1.
3. Poulose, T. T., ed., *Perspectives of India's Nuclear Policy* (Young Asia Publications, New Delhi, Stockholm), esp. pp. 136–60.
4. Brezaric, J., 'Against nuclear monopoly', *Review of International Affairs*, Vol. 39, No. 669, 20 February 1978.
5. UN document A/S-10/2, paragraphs 66–71.
6. Kawakama, Koichi, 'The nuclear fuel cycle in Japan', *Bulletin of the Atomic Scientists*, Vol. 34, No. 6, June 1978, pp. 17–18 (reprint from *Atoms in Japan*, September 1977).

7. Donnelly, W. H., ed., *Nuclear Proliferation Factbook* (US Library of Congress, Congressional Research Service, Environment and Natural Resources Policy Division, Washington, D.C., 23 September 1977).
8. IAEA document INFCIRC/243.
9. IAEA document INFCIRC/233.
10. IAEA document INFCIRC/239.
11. *International Herald Tribune*, 24 August 1978.
12. *Le Monde*, 25 August 1978.
13. *Wall Street Journal*, 25 August 1978.
14. Press Release No. 122, 23 June 1977, Mission of the USSR to the United Nations.
15. *New York Times*, 21 July 1978.
16. Wöbse, Gerhard, 'Nucleare Zusammenarbeit in der Dritten Welt' ['Nuclear cooperation in the Third World'], *Aussenpolitik*, First quarter 1978, pp. 63−72.
17. *UN Disarmament Yearbook*, Vol. II (United Nations, New York, 1977), p. 142.
18. 'Talks progress on non-weapons A-fuel', *International Herald Tribune*, 5 September 1978.
19. *Proceedings of the International Conference on Nuclear Power and its Fuel Cycle* (IAEA, Salzburg, May 1977).
    (a) —, Vol. 3, Fréjacques, C., *et al.*, 'Evolution des procédés de séparation des isotopes de l'uranium en France', Paper No. IAEA-CN-36/257, pp. 203 ff.
    (b) —, Vol. 1, Eklund, S. (closing remarks at Salzburg Conference), p. 799.

Chapter 10. Multinational and international controls

Paper 16.   A nuclear fuel supply cooperative: a way out of the non-proliferation débacle

A. R. W. WILSON

I.   Introduction

Heightened international concern is affording the world perhaps its last chance to do more than temporize with the risk that civilian nuclear energy developments will lead to a proliferation of nuclear weapons. Opportunities for concerted international action to restrain proliferation will decrease as nations become increasingly seized with the urgency of coping with energy shortages.

Thirty years ago, when civilian applications of nuclear energy were the preserve of a few, it was possible to contemplate their internationalization. But having failed to grasp that opportunity, the world is committed to coping with the dangers inherent in national control of materials and facilities which can be diverted to nuclear weapon production.

II.   The inadequacy of safeguards

With civilian nuclear activities shielded by national sovereignty, the international community has sought to avoid their misuse through mutual restraint and reassurance. States have been encouraged to become parties to treaties which incorporate undertakings not to manufacture nuclear weapons, and a system of international safeguards[1] has been set up to provide reassurance that the undertakings are being honoured. Unfortunately, these non-proliferation treaties have been unable to attract universal membership. Some states object in principle to the discriminatory nature of

---

[1] Technically the term 'safeguards' refers to the accountancy, inspection and physical control measures used to verify that nuclear materials, facilities and technology intended for civilian purposes have not been diverted to any proscribed purpose. However, in general usage it is often taken to mean both the undertaking and the verification measures.

their provisions and the intrusions into national sovereignty which they entail. The growth of unsafeguarded nuclear activities in those states not members of non-proliferation treaties has been restrained by the safeguards which most nuclear supplier states have attached to their nuclear exports. Importing states have been required to give an undertaking that they will not divert the imported nuclear supplies to any nuclear explosive purpose and to accept international safeguards on the imported nuclear supplies to verify that the undertaking is being respected.

Non-proliferation treaties and export safeguards have been barely adequate to contain proliferation in a world where only a few non-nuclear weapon states have acquired an enrichment or a reprocessing facility, an essential adjunct for the production of nuclear weapons. Now that many states are looking to the possibility of obtaining enrichment or reprocessing facilities, international concern is rising. Even safeguarded national enrichment or national reprocessing facilities introduce an increased risk of diversion. More importantly, they could represent the first step in the state's eventual acquisition of an unsafeguarded facility. Export safeguards can reduce the risk of directly imported enrichment and reprocessing facilities being misused and they may be able to slow down the transfer of technology, but they cannot prevent a state building up the competence to design and construct its own facilities.

Concern over the adequacy of existing non-proliferation arrangements goes beyond the possible acquisition of unsafeguarded facilities or the possibility of diversion within safeguarded facilities. It is now realized that existing arrangements offer little protection against a state using enrichment and reprocessing facilities acquired under safeguards agreements and stockpiles of enriched uranium and plutonium built up within safeguards agreements to manufacture rapidly nuclear weapons following abrogation[2] of those agreements.

III. Recent measures to strengthen the non-proliferation régime

The main nuclear exporting states have reacted to the increased concern over the implications for non-proliferation of civilian nuclear developments by agreeing to exercise discretion in the export of sensitive technologies and by attaching a variety of new safeguards conditions to nuclear exports. Such measures are palliatives rather than cures for the weaknesses of the current non-proliferation régime. While they may buy time, in the longer term they

---

[2] Withdrawal is of less concern since most safeguards agreements require notice of withdrawal and thus allow time for political measures designed to remedy the circumstances prompting it.

could prove counter-productive by reason of the political polarizations which they are engendering. Many Third World states see them as an attempt by the industrialized world to retain for itself the full benefits of nuclear energy, some Non-Proliferation Treaty (NPT) states argue that they amount to an unwarranted qualification of the provisions of that treaty, and several industrialized states object to them as an unjustified intrusion into their national sovereignty. What is needed is an arrangement which dissuades states from stockpiling enriched uranium and plutonium, which discourages them from building unsafeguarded enrichment and reprocessing plants and which reduces the risk of safeguarded plants being misused following the abrogation of safeguards agreements.

The INFCE

The International Nuclear Fuel Cycle Evaluation (INFCE) now in progress provides an international framework within which these problems can be explored. The INFCE is intended to be a wide-ranging international study of issues involved in making nuclear energy for peaceful purposes widely available, while minimizing the danger of the proliferation of nuclear weapons. A central issue of the INFCE is whether there is scope for furthering this objective through the adoption of alternative nuclear fuel cycles.

Unfortunately, the INFCE is unlikely to provide a technological solution to the proliferation problem. The fuel cycles in current use have been adopted on the basis of deliberate technical and economic choices. The probability of achieving an international consensus on the technical and economic viability of any alternative fuel cycle is remote, no matter how much more attractive it might be from a non-proliferation viewpoint. Even if there is a satisfactory low proliferation-risk fuel cycle, there is little prospect that nations which have committed major resources to existing systems would agree to adopt it.

IV.   Multilateral institutional arrangements

Since the INFCE is unlikely to arrive at a technological solution, its main contribution to minimizing the danger of the proliferation of nuclear weapons may be the opportunity it affords the international community to explore new institutional arrangements. The institutional concept which over recent years has attracted the most interest as a possible approach to the nuclear weapon proliferation problem is the idea of multinationally controlled centres for the enrichment and chemical reprocessing of nuclear fuel. Subject to their widespread acceptance as the sole legitimate means of meeting nuclear fuel cycle needs, and to their early establishment, multi-nationally controlled nuclear fuel cycle centres, combined with appropriate controls on the stockpiling of nuclear materials, might serve as an effective means of containing nuclear weapon proliferation risks. Unfortunately, *prima facie*, it seems unlikely that these conditions can be met. Many non-nuclear weapon states consider themselves already technically disadvan-taged *vis-à-vis* the nuclear weapon states. They are unlikely readily to be persuaded to forgo the right to engage in national enrichment or reprocessing programmes, particularly when the NPT itself enshrines the sovereign right of parties "to develop research, production and use of nuclear energy for peaceful purposes without discrimination". The establishment of the complex of multinationally controlled fuel cycle centres which would be needed reasonably to satisfy the full spectrum of national needs, before any further significant spread of enrichment and reprocessing technology, would seem to be precluded by the magnitude of the organization and capital resource problems likely to be involved.

V.   The idea of a nuclear fuel supply cooperative

No doubt the idea of multinationally controlled fuel cycle centres will be explored and developed within the INFCE. This paper proposes an alter-native fuel-supply arrangement which can satisfy current proliferation concerns without introducing the problems of surrender of national sovereignty and implementation delay inherent in the multinational fuel cycle centre approach. The concept is based on the assumption that if a new arrangement is to attract widespread support, it must be one which compels the involvement of states by reason of their immediate self-interest. To this end, the concept aims to establish a cooperative of states dedicated to meeting the nuclear fuel and services requirements of member states, while minimizing the attendant risks of nuclear weapon proliferation.

The cooperative would take the form of a group of states tied by indivi-dual master agreements to a central statutory body. The master agreement would bind the members to the purposes of the cooperative, regulate both the conduct of their own nuclear programme and their interaction with the activities of others, and oblige them to enter into specific safeguards agree-ments with the IAEA. Conclusion of a master agreement would entitle a state to membership of the central statutory body which would function as the governing body of the cooperative. The central statutory body could be set up by a few founder states, each of which would then be obliged to conclude a master agreement with the statutory body.

States members of the cooperative would pledge themselves to cooperate in measures to ensure the availability of adequate and reliable supplies of uranium and enrichment and reprocessing services. So that the member states themselves would draw the full benefit of these efforts, they would undertake not to export nuclear materials, services, facilities or technology to non-member states. The benefits of membership in the cooperative could be enhanced by measures such as fuel-assurance arrangements, nuclear insurance pools, nuclear import—export banks, and centralized radioactive waste depositories.

Nuclear weapon proliferation would be restrained through members undertaking in the master agreement to deny themselves any access to the services of enrichment and reprocessing facilities, whether indigenous or foreign, beyond that necessitated by their power programme, and to accept safeguards developed to verify observance of this undertaking on all relevant nuclear activities. These special-purpose safeguards would be applied by the IAEA under a standard bilateral agreement with the member state of the cooperative.

In practice, member states of the cooperative would undertake not to make use of enrichment or chemical reprocessing services offered by non-member states, and member states operating enrichment plants would be bound to restrict production to contracts authorized by the IAEA on the basis of its assessment of the nuclear fuel requirement of the customer state.[3] Similarly, member states operating reprocessing plants would be bound to return to the customer only that portion of the recovered plutonium which the IAEA judged necessary for the latter's nuclear power programme and to deposit the balance of the plutonium with the IAEA for safe keeping. The Agency would be empowered to apply safeguards to all enrichment and reprocessing plants under the control of members with a view to verifying that these undertakings were honoured. It would also apply safeguards to the enriched uranium and plutonium retained by the customer state to the extent necessary to ensure that it was incorporated into reactor fuel as soon as possible and irradiated to the point where it became worthless as weapon material without further chemical reprocessing.

Enrichment and reprocessing would remain under national control and member states would be free to construct new capacity at their discretion. However, to the extent found practicable, while operated under national control, all new plants would be sited on international territory dedicated for nuclear purposes. Commercial contracts would be negotiated and concluded between companies and organizations and subject to the obligations assumed by the states concerned.

---

[3] Although, for non-proliferation purposes, it is only necessary to prevent the production of highly enriched uranium, the concept envisages the control of all enriched uranium production because the ready availability of low-enriched uranium would make the use of a covert enrichment plant for weapons production a more practicable proposition.

The master agreement

The master agreements would be based on a model agreement developed to cover all of the various nuclear activities in which member states may engage. The preamble to the master agreement would acknowledge the purpose of the undertakings as being:

(a) to promote cooperative measures to ensure member states access to adequate and reliable supplies of uranium and enrichment and chemical reprocessing services at realistic prices, and

(b) to promote confidence that civilian nuclear activities are not being used for nuclear weapon purposes by denying member states access to enrichment and chemical reprocessing services, both domestic and foreign, beyond that necessitated by their nuclear power programmes.

Under provisions in the model agreement, contracting non-nuclear weapon states would undertake:

(a) not to export uranium to non-member states,

(b) to cooperate in measures to ensure the availability of adequate and reliable supplies of uranium at realistic prices,

(c) not to export enrichment or reprocessing plants or enrichment or reprocessing technology to non-member states,

(d) not to avail themselves of enrichment or reprocessing services for plutonium or uranium-233 separation offered by non-member states,

(e) not to provide enrichment or reprocessing services to non-member states,

(f) to prohibit the use of nationally controlled enrichment plants for any production other than for contracts authorized by the IAEA on the basis of that organization's assessment of the customer's nuclear fuel requirements,

(g) to deposit with the IAEA for storage all plutonium and uranium-233 excess to the immediate nuclear fuel requirement of the state concerned,

(h) to require that nationally controlled reprocessing plants return to the customer, whether domestic or foreign, only that portion of the recovered plutonium or uranium-233 which the IAEA approves as necessary for the customer state's nuclear power programme and deposit the balance of the plutonium or uranium-233 with the IAEA for storage on international territory or the territory of nuclear weapon states members of the nuclear fuel cooperative,

(i) to accept IAEA safeguards on all nationally controlled enrichment and reprocessing plants with a view to verifying that the preceding undertakings are honoured,

(j) to accept IAEA safeguards on all enriched uranium and plutonium under the control of the member state to the extent necessary to ensure that it is incorporated into reactor fuel on a timely basis and irradiated to the point where it becomes worthless as nuclear weapon material without further chemical reprocessing, (*k*) to locate new enrichment and reprocessing plants at international sites when so requested by the cooperative,

(*l*) to cooperate in measures to ensure the availability of adequate supplies of enrichment and reprocessing services at realistic prices, and

(*m*) to cooperate with the IAEA in the provision of a buffer stock of enriched uranium to be held under IAEA control on international territory or the territory of nuclear weapon states members of the nuclear fuel cooperative and used to meet unforeseen shortfalls in enriched uranium availability.

To avoid changes in their nuclear status necessitating renegotiation of membership, the model agreement would provide for non-nuclear weapon states undertaking to observe all of these conditions.

Some modification of the conditions would be necessary to accommodate the membership of nuclear weapon states. Such states might be expected to assume the general conditions of membership in respect to their civilian nuclear activities and their international nuclear trade, but to retain the right to utilize nationally controlled enrichment and reprocessing plants for national nuclear weapon programmes.

The inclusion of a general undertaking by non-nuclear weapon states not to engage in the development or production of nuclear weapons is not suggested because it is not functionally necessary to establish the non-proliferation régime and because it would emphasize the. differences in obligation accepted by nuclear weapon member states and non-nuclear weapon member states. It will be noted that, unlike current non-proliferation régimes, the proposed cooperative rests on an undertaking not to misuse enrichment or reprocessing facilities, rather than upon an undertaking not to divert nuclear material to the manufacture of nuclear weapons. However, acceptance of the membership obligations of the cooperative would not involve any conflict with existing safeguards arrangements and, if the parties to safeguards agreements so desired, these could continue in operation concurrently with the cooperative controls.

As a consequence of the preceding obligations, non-nuclear weapon member states would be prevented from using reprocessing and enrichment facilities for other than the immediate purposes of their peaceful nuclear power programmes. This would be accomplished without denying them the opportunity to operate national enrichment and reprocessing facilities where it could be shown that these were required to meet programme needs. States not willing to accept the obligations of membership would be denied access to the potentially most competitive and assured supplies of uranium, to most established enrichment and reprocessing services and to the most advanced technology.

It was pointed out in section II above that an important concern with any non-proliferation arrangement which relies on voluntary restraints is the possibility that a state may abrogate the safeguards agreement and utilize stocks of enriched uranium and plutonium and national enrichment and reprocessing facilities to achieve rapidly a nuclear weapon capability. The nuclear fuel cooperative concept precludes the stockpiling of enriched uranium and plutonium under the control of member states and thus effectively avoids the dangers of stockpiles being diverted to nuclear weapon production. In the longer term, it also offers a solution to the misuse of national enrichment and reprocessing facilities following the abrogation of safeguards agreements, since it looks to a situation where nationally controlled plants will be sited on international territory and come under international controls in the event of the operating state abrogating its membership agreement. Meanwhile, it faces member states contemplating abrogation with the certainty that they would thereby immediately cut themselves off from all of their established external sources of uranium and nuclear-fuel services. By extending the contractual arrangements to prevent national stockpiling of natural uranium, as well as the stockpiling of enriched uranium and plutonium, it would be possible to limit even further the opportunities for member states to misuse nationally operated facilities sited on their territory following abrogation. However, to do so may make membership less attractive to states and thereby reduce the effectiveness of the arrangements.

Implementation of the concept would require its acceptance in principle by the international community, perhaps within the INFCE, the establishment of the controlling statutory body by a group of founding states, the negotiation and conclusion of master agreements between individual states and the controlling statutory body, and safeguards agreements between the individual states and the IAEA. The concept would be viable only if it could rapidly attract wide membership. If the main nuclear exporting and importing states could be induced to accede to membership of the cooperative in the formative stages, most other states with nuclear power programmes could be expected to follow suit since they would otherwise cut themselves off from most foreign nuclear suppliers. Fortunately, the main nuclear exporting and importing states are politically committed to non-proliferation and could be expected to display sympathy for the objectives of the cooperative. Hesitations arising out of their concerns over the commercial implications of entering into membership of the cooperative without knowing whether its membership would eventually embrace most states might be overcome by providing for conditional membership. Conditional membership would convert to unqualified membership in the event of the cooperative attracting appropriate support.

The advantages of the proposal

The following are some of the attractive features of the concept:

(*a*) the non-proliferation régime which it seeks to establish is firmly based on mutual benefit rather than on a discriminatory surrender of national sovereignty;

(*b*) it does not attempt to deny member states the right to engage in enrichment or reprocessing when these activities can be justified on the basis of a programme need;

(c) it does not involve the construction of new facilities or the establishment of new international funds;

(d) it allows commercial arrangements to be settled direct between buyer and seller;

(e) it can be implemented rapidly and sequentially (states can join in as they judge it to be their interest);

(f) it offers an opportunity to avoid the current problem of over-lapping national safeguards requirements;

(g) it allows the safeguarding to be concentrated on the critical enrichment and reprocessing steps in the fuel cycle; and

(g) it offers greater safeguards against the dangers of abrogation than current arrangements.

On some current non-proliferation problems the concept offers only a compromise solution. Of necessity, it has to discriminate between nuclear weapon states and non-nuclear weapon states; in common with existing arrangements, it relies upon member states revealing the existence of all national enrichment and reprocessing facilities, and it requires some surrender of national sovereignty. But the criterion against which its usefulness must be assessed is not how far it falls short of the ideal, but rather the extent to which it offers an opportunity to reduce the risks of nuclear weapon proliferation while facilitating the availability of nuclear energy for peaceful purposes.

Note

This paper is based solely on the personal views of the author. It does not necessarily present the policies of the Australian government.

Paper 17.   A preliminary evaluation of the technical aspects of INFCE

U. FARINELLI

I.   INFCE: technical versus political content

The International Fuel Cycle Evaluation (INFCE) was set up by the Organizing Plenary Conference in October 1977. Its terms of reference specify that its work should be completed in two years; we are thus in the middle of this exercise, and it should be possible to make an initial judgement on how INFCE is proceeding and what can be expected from it. Is INFCE going as well as expected? Is it performing its task? What results will be obtained, and will they be in the direction that was hoped? Does it or will it generate useful technical results?

Unfortunately, such a judgement is not easy to make; it is even more difficult if one wants to confine it to the technical aspects without taking into account the political ones. Although INFCE was set up as a purely technical body to provide the necessary technical information on which the various countries would subsequently base their decisions (decisions that will be essentially political in nature), it has not functioned as such, at least not until now. In most of the working groups, the technical discussions have been much fewer than the lengthy procedural disputes permeated with fundamental differences of a political nature. It is interesting to discuss why this has happened.

The questions which concern INFCE bear on some basic political issues: military security, energy security, economic and environmental impacts of the choices and their influence on the rate and quality of development, and so on. It is difficult to keep these questions in the background and concentrate on the technical facts, even more so since countries rather than individuals are represented at INFCE. The choice of country representatives was made by the respective governments and, although most of the participants are scientists (others are diplomats and foreign affairs officials), it was obvious that they would be chosen to represent their nations' predominant view and would be given political instructions.

But there are other factors that make unbiased technical discussion difficult. The novelty of INFCE is the introduction of a new dimension in the consideration of fuel cycles: the resistance to nuclear weapon proliferation. This new dimension is difficult to quantify; as we shall see below, it is often difficult to agree on whether a certain solution is more or less proliferation-resistant than another, let alone how *much* more or less. This difficulty is enhanced by the fact that quantitative and qualitative judgements on proliferation are often based on classified information and on practical experience that is available only to those countries that already have nuclear weapons. This tends to create a bias among the countries that have voluntarily abstained from considering military applications; this bias adds to the apprehension caused by some of the proposed solutions that in fact discriminate between nuclear weapon states and non-nuclear weapon states in regard to installations and technologies related to peaceful applications.

An even more difficult problem is the setting up of guidelines for evaluating the various fuel cycles from all the different points of view. The emphasis on different aspects varies substantially from one country to another, and it is therefore impossible to arrive at some sort of 'weighted evaluation' that takes into account all the different aspects: the relative weights for, say, the economics, fuel availability and technological difficulty of a certain cycle are vastly different from one country to another. In practice, each participant in the technical discussion tends to start from a predetermined viewpoint (be it non-proliferation or economics) on which his choice of a fuel cycle is based, and to present those arguments that support those cycles that are 'good' in his judgement and that show weak points in the others. This can hardly be considered a free and objective technical discussion.

This situation is further complicated by the way in which INFCE works. There is no standing international group to make the actual calculations and evaluations. All the basic work is done in the individual countries. The reports of the eight working groups are based mainly on the collection and compilation of the contributions provided by the various countries (which reflect in all important aspects their official viewpoints). Even the duties involved in compiling and editing the final report are often divided among the different countries, and the report is supposed to record the points of agreement and disagreement on the different subjects. This contributes to making the discussions in the initial phase more and more based on procedures (such as outlines of the final reports, irrespective of their content, and the assignment of the various parts to different participants). In the final phase (which has not yet begun), discussions will undoubtedly concentrate on the form of the final text and on 'sensitive' words, rather than on reaching a broad consensus on a number of technical conclusions. The whole procedure is more like a diplomatic negotiation than international technical cooperation.

II. The new fuel cycles

There was some hope in the beginning that INFCE would provide a much needed occasion to look at nuclear reactors and fuel cycles from a new perspective. Actually, the history of the development of nuclear energy shows a pattern that is very different from an overall optimization or from a global system analysis. Some sort of optimization was performed on the reactor itself, without much attention to the rest of the fuel cycle; this gave rise to specifications for the fuel to be introduced into the reactor, and to predictions of what would be discharged from the reactor. Working back from the specifications for the fuel, one would find those for the fuel fabrication and for uranium enrichment, and these processes would then be optimized separately. This would even be used to assess the necessity for uranium mining. At the other end of the cycle, the reprocessing was studied starting from the characteristics of the unloaded fuel, and was optimized on its own. Finally, the characteristics of the fuel and of the wastes coming out of the reprocessing plant were used as specifications for re-fabrication and waste disposal. No global optimization was attempted, other than in highly theoretical studies.

Could INFCE provide the occasion for the global optimization of the fuel cycle that took into account not only the economic but also the environmental, safety, resource and non-proliferation problems?

The answer is clearly no, and it was predestined to be so by the very fact that the work of INFCE was split into eight working groups (WGs), most of which attempted the traditional distinction between uranium mining (WG 1), uranium enrichment (WG 2), fast reactors (WG 5), spent fuel storage (WG 6), reprocessing (WG 4), and waste disposal (WG 7). Only two groups dealt with the whole fuel cycle: WG 3, on problems of a predominantly political nature (assurances of long-term supply of technology, fuel, heavy water and services); and WG 8, loaded with a miscellany of the once-through fuel cycle, advanced reactor and fuel cycle concepts, and problems of research reactors. The segmentation of the whole process into the other groups was enhanced by the lack of coordination among the different groups, since the TCC (Technical Coordinating Committee) actually functions as a highly political body and meets only rarely. There is no standing technical coordination at the international level.

In my opinion, a new type of approach could hardly be expected from INFCE for the following reasons. In particular, even the view that in the past an optimization was reached at the level of subsystems is too optimistic. During the 1950s and most of the 1960s, a broad investigation into a vast variety of reactor concepts was carried out in many countries. An illusion of that period was that neutronic calculations and engineering conceptual designs (or, at most, experimental reactors of relatively small power output) could provide the answer in terms of 'the best possible reactor'. Time showed that this idea was a loser. Very few reactor concepts have survived, but there is no reason to believe that they are the 'best' concepts. They are those for which enough money has been concentrated to make a real breakthrough into the commercial market. For various reasons —fall-out from military research, industrial competition, marketing, application of know-how gained in other areas, and so on—enough research, development, engineering, experimentation, demonstration and prototype reactors had been funded. And last but not least, there was a dumping of prices for the first installations that allowed a large-scale sale, and therefore initial investments could in the long term be recovered for a large number of systems. Such a situation is even more likely to exist in the future, as we move towards increasingly expensive systems, and in particular if one considers not only the reactor but the whole fuel cycle. Alternative enrichment or reprocessing schemes are even more expensive than reactors for testing and demonstrating on a scale large enough to yield meaningful results.

Moreover, no simple representation of a reactor or a fuel cycle is likely to be usable for optimization or selection purposes, but at most for a first screening for promising concepts. A realistic assessment of the potentialities of a reactor or a fuel cycle concept can be obtained only from a very detailed analysis that includes all the engineering factors. For example, results obtained in the zero- or one-dimensional burn-up calculation of a reactor are essentially meaningless today. In order to obtain useful results for comparison of different concepts, one must consider the actual fuel elements, with all the structural materials; the appropriate control elements; the detailed operation strategy, consistent with the practical limitations that are met; and the safety analysis and requirements, including quality assurance. No such exercise could be carried out within the two-year time horizon of INFCE except as far as small modifications to proven concepts are concerned. Nothing basically new could be expected except for, at most, some indications that one or another concept deserves being further looked into.

III. Proliferation resistance

Even more discouraging in terms of obtaining basically new results from INFCE is the consideration of the new dimension added by INFCE, that is, proliferation resistance.

First, one must consider the time-scale. Most agree that technical remedies to proliferation are a question for only the next 20–25 years—not because there will be no nuclear weapon problem after that, but because the inevitable spread of technologies and know-how, the many routes to weapon proliferation aside from the power reactor fuel cycle, and the pure application of the theory of probability require that some effective political solution be found before then. But very few of the innovative solutions that are proposed can be applied on a large scale for the next 25 years. This is clear even for some of the more realistic and down-to-earth proposals for relatively small modifications in present systems. For instance, the US proposals for better utilization of uranium by the once-through cycle in light water reactors (LWRs) assume that such changes as lowering the pellet density in standard fuel (by adding diluents or using annular pellets) could be deployed on a large scale after the year 2000, and radial or axial natural or depleted uranium blankets around 1995! When one calibrates the realism of the other proposals against these predictions (which are certainly competent, and unbiased, since the USA has every advantage in showing that uranium conservation measures in LWRs can be quickly introduced) what should predictions be about the large-scale deployment of, say, fusion— fission non-proliferating hybrid reactors, or even the more modest schemes calling for the utilization and the recycling of thorium? Certainly these schemes should be considered inadequate for dealing with the time limit we have set for technical non-proliferation measures.

Second, as mentioned above, it is impossible to quantify proliferation resistance, and in many cases even to decide whether one system is more or less proliferation-resistant than another. There is no sharp distinction between weapon-grade, weapon-usable and non-weapon-usable materials. This distinction depends on the degree of technological sophistication which is available (mostly in relation to assembly time) and on the definition of a 'useful' weapon. What if there is a high probability of pre-detonation? And what is the lower limit of nuclear yield that can still be called a nuclear explosion? Is a ton of 'dirty' plutonium more or less proliferant than 4 kg of very pure Pu-238? The distinction between what is adequate protection against proliferation and what is not is even more hazy. At the beginning of the exercise, many at INFCE thought that denaturation (or isotopic dilution) was absolutely safe, that gamma contamination (spiking) was on a lower level of safety, and that chemical processing provided very little protection. Broadly speaking, this is still true in many respects, but there is much overlap. Enriching U-233 in a 20 per cent U-233/U-238 mixture by centrifuge can be relatively easy (due both to the higher mass difference than in the case of U-235 and to the relatively high initial U-233 content)— perhaps less difficult (at least in certain technological contexts) than chemical separation of plutonium from very highly active fission products, or the highly radioactive metallurgy of 'old' plutonium.

Another point of difference is the time integration. One of the strong points of the Civex proposal, based on fast breeders, is that plutonium is burnt as soon as it is available, and there is no accumulation or 'plutonium mine' such as that provided by the indefinite storage of spent fuel elements whose radioactivity decreases with time. Is it safer to have less plutonium circulating around to be burnt, or to limit the total quantity of plutonium that is stored?

Finally, what is actually demanded of proliferation-resistance measures? Many countries have the technical capabilities to fabricate nuclear weapons if they choose to do so. But the easiest way is certainly not through the power reactor cycle. Is the object of non-proliferation the slowing down of the spread of these technical capabilities to still other countries? Or is it the maximization of the time elapsing between the detectable breach of non-proliferation agreements by a country and its attaining the ability to deploy a weapon (which of course assumes that political means exist for using this time to prevent actual deployment or use of the weapon)?

IV.   The basic solutions and their evolution

As we have seen, the necessity of adopting measures that are effective in the short term, the lack of time to explore to any depth new solutions, and the necessity of basing decisions on facts rather than on hopes have prevented INFCE from elaborating any fundamentally new scheme.

To state the problem in somewhat simplified form, the parties have basically retained their initial positions, which are best represented by the opposite views of the USA and of Western Europe plus Japan.

The US position concentrates on the aspects of *military* security. The US view is that the most sensitive aspects, from the point of view of proliferation, are reprocessing and the use of plutonium, so that the once-through cycle in thermal reactors is the only one that gives adequate protection. Since this cycle is the most wasteful in terms of uranium resources, the USA maintains that earlier predictions of nuclear power growth were much too high, and estimates of uranium resources much too low; it would therefore not be necessary for a long time to recycle the plutonium or to resort to fast breeder reactors, and any decision or commitment in those directions could be postponed to well after the year 2000.

The position of Europe (especially France, FR Germany and Italy) and Japan is just the opposite. The main concern here is for *energy* security. There is more optimism about the growth of nuclear power and more pessimism about uranium reserves (and particularly about the free availability of uranium, whatever the reserves may be). The strategy preferred involves the reprocessing of the fuel from light water reactors, in order to recover both uranium and plutonium, and the use of plutonium to start up fast breeders that are assumed to be close to commercialization.

It would be unfair to say that these two positions have not changed during the first year of INFCE. Although the basic strategies have remained the same, many changes have been introduced in each of them, and these changes are not marginal. Many factors have contributed to this evolution, but INFCE can be credited for having catalysed it.

266

The changes are of course in the direction of making each strategy more acceptable to the other party: for the USA, in the direction of a better utilization of the resources and of a more reliable assurance of availability of fuel; and for Europe and Japan, in the direction of hardening the proliferation resistance of reprocessing and plutonium utilization.

For the first objective, a large number of improvements have been proposed and analysed in order to increase fuel utilization in the once-through mode; however, the only one that seems to be applicable on a relatively short time-scale is the use of higher burn-up in LWR fuel, coupled with a greater number of batches in the fuel cycle. By increasing the initial enrichment of the fuel (from about 3 to perhaps 4.1 per cent) and the burn-up from 30 000 MWd/t to somewhat less than 50 000 MWd/t, and by every year changing one-sixth of the core instead of one-fourth (or one-fifth instead of one-third), it would be possible to increase the uranium utilization by about 15 per cent. Even if the technology for such changes became available in a few years and could be backfitted into the existing reactors (it requires modifications only in the fuel element), it is not certain that this scheme would be adopted universally or quickly by the utilities. Large-scale deployment would be expected no earlier than 1995, so that its effects on uranium resources would not be felt until well into the 21st century.

Another improvement in uranium utilization would be obtained by substituting heavy water reactors for light water reactors. Recent studies have shown that the best resource utilization with a CANDU type reactor is obtained with a slightly enriched (about 1.1 per cent) uranium fuel rather than the natural uranium fuel that is typical of the original Canadian design. It is not clear whether it is realistic to assume such a radical shift of reactor type, but serious studies in this direction have been carried out in the USA, and heavy water reactors keep cropping up in long-term strategic studies from sources that are influential in determining US policy. The once-through cycle in such a reactor would allow a reduction of uranium use of about 40 per cent with respect to the LWR once-through cycle.

A final improvement that has been proposed consists in reducing the U-235 content in the tail of the enrichment process from 0.2 per cent to 0.05 per cent. This reduction would allow a further 15 per cent saving of uranium, and could be backfitted by stripping the accumulated depleted uranium; however, it is not clear what process would be used, and whether the economic costs would be acceptable. Actually, the opposite trend had been predicted as a result of the increased cost of the energy required.

The improvement in the assurance of uranium availability would be obtained by an institutional rather than a technical change: the setting up of an international uranium bank—a proposal that is strongly supported by the USA and is gaining universal acceptance, although it can be predicted that negotiations on details of the proposal will be neither easy nor short.

In the opposite strategy, which requires reprocessing and early deployment of fast breeder reactors, improvements have been made towards increasing the proliferation resistance. The weak point of this strategy (from the point of view of proliferation) is the reprocessing plant and the transportation of plutonium-rich fuel from this plant to the power station. Technical fixes that have been proposed include co-processing and spiking. The former consists in not completely separating uranium from plutonium at any stage of reprocessing, and the latter in leaving enough of the fission products in this mix to make it highly radioactive, but not enough to cause too much parasitic neutron absorption. These two changes are really effective only if they are employed together, since the high level of radioactivity interferes with the chemical separation of plutonium from uranium. Such solutions were discarded at first as impractical and too expensive, but closer analysis has shown that this is not necessarily so. The co-processing scheme could be just as convenient as ordinary reprocessing, at least in some circumstances, while the remote refabrication in α- and γ-tight cells demanded by the spiking method could be necessary in any case if one were using old and dirty plutonium that had undergone more than one reprocessing. Institutional non-proliferation measures call for international control over reprocessing facilities. These facilities could be concentrated in a few highly protected areas.

V.   Conclusions

INFCE has not generated (and probably could not generate) radically new ideas or solutions that would provide an answer to proliferation problems. However, it has promoted a much better understanding of the various points of view and of the real nature of the problems, and it has been instrumental in arriving at modifications of the basic strategies that make them more acceptable and more likely to coexist.

The technical conclusions of INFCE, in their carefully weighted wording, will probably contain neither a drastically new message nor any indication of a possible breakthrough towards a panacea on which to concentrate worldwide cooperation, but they may provide the basis for a compromise encompassing co-existence of the main strategies in a form that could slow down proliferation without compromising energy security and development. It would then be up to the governments to use this time to find a political solution that could be effective where technical remedies fail.

Although the ideas and the numbers quoted are by no means classified or restricted, and for the most part have been published in the open literature, they are based essentially on the draft documents presented at INFCE and on personal discussions. These documents will be published in final form at the end of INFCE.

Note

The ideas and evaluations expressed in this paper reflect the personal opinions of the author and do not necessarily correspond to those of his institution or his government.

Paper 18. Nuclear proliferation: arrangements for international control

J. ROTBLAT

Square-bracketed numbers, thus [1], *refer to the list of references on page 285.*

I. Introduction

The intrinsic link between the peaceful and the military aspects of nuclear energy—the fact that it is impossible to generate electricity in a uranium-based reactor without at the same time producing a nuclear weapon material, plutonium—means that eventually either civilization will be destroyed in a nuclear war, or nuclear energy based on fission will have to be abandoned. Indeed, were it not for the rush to make the atomic bomb during World War II and the resultant nuclear arms race, and for the tremendous outlay of capital and skill for the development of reactors for military purposes, it is quite likely that there would not have been the urge to develop fission as a source of energy in preference to other sources, such as solar. Had similar effort and resources been devoted to the latter, we would by now probably be well on the way to satisfying our energy needs with solar energy, in its direct or indirect forms. In the long run this will have to come in any case, and the use of nuclear energy will become only a short episode in the history of mankind.

However, due to the events of recent history, nuclear energy will inevitably play a role as an energy source on a world scale during the next few decades. How big this role will be is impossible to predict. It will certainly be much slower in coming than was envisaged in the euphoria of the 1960s and early 1970s, when, based on unrealistic rates of industrial growth and continuous exponential population increase, it was envisaged that fission energy would not only become the main source for the future, but would even play a major role during this century. However, even the current, more sombre projections still insist that fast breeder reactors will be necessary early in the next century.

Thus, the threat to world security that is intrinsically present in nuclear energy programmes, particularly those with fast breeders, still exists. During the next few decades, the danger of a large number of nations becoming nuclear weapon powers by virtue of their acquisition of nuclear technology and reactors for peaceful purposes will increase. The ensuing risk of a catastrophic nuclear war is real and serious.

The advocates of large-scale nuclear power programmes regard these fears as grossly exaggerated. They point out (a) that no nation has so far acquired nuclear weapons through its nuclear power programme, although several have been in a position to do so; (b) that a nation wishing to have nuclear weapons can achieve this by means other than power reactors; and (c) that the danger of proliferation can be eliminated by strengthening the Treaty on the Non-Proliferation of Nuclear Weapons (NPT) and other existing safeguards systems. The counter-arguments are: (a) the fact that something has not happened in the past is no guarantee that it will not happen in the future, particularly if the impetus and opportunities for it were greatly increased; (b) the fact that certain dangers already exist is no justification for introducing further dangers; and (c) with regard to the NPT and other safeguards, it is argued in this paper that no amount of patching up of their shortcomings will be effective, without tackling the underlying political issues.

The consequences of a nuclear war are so horrible that all efforts to minimize the chances of its occurring are justified. On this premise, the need to control the nuclear fuel cycle by truly international collaboration and by setting up new UN agencies is advocated in this paper.

II. Existing control machinery and proposals to improve it

The Non-Proliferation Treaty

The NPT was brought into being in order to prevent an increase in the number of nations possessing nuclear weapons, since the United Nations considered that such an increase would seriously enhance the danger of a nuclear war. Agreement on the treaty was reached in 1968 and by 1978 two-thirds of all nations were party to it. There can be no doubt that the NPT is one of the most important arms control measures; it was achieved because of the ardent desire of people all over the world to reduce the peril to mankind created by the discovery of nuclear weapons.

Because of its immense importance to world security, one should avoid deprecating it. On the other hand, it is legitimate to ask, 10 years after the treaty was drafted, whether it is likely to achieve its original objective. The answer appears to be in the negative.

To begin, with, too many nations (over 50) still refuse to sign the NPT. Of the six nations which have demonstrated their possession of nuclear weapons by testing them (India claims to have tested a 'device' rather than a weapon, but this distinction is hardly significant), only three are party to the treaty. Among other non-adherents, more than 10 are so-called 'threshold countries', that is, they are engaged in significant nuclear activities and either already have, or could at short notice produce, nuclear weapons. Moreover, parties to the NPT can withdraw from the treaty after giving three months' notice. The worry this represents arises from the loose wording of Article II, which makes it possible for a country to come very close to producing a nuclear weapon without violating the letter of the NPT.

A new danger to the integrity of the NPT is the attempt to shift its emphasis. Articles I and II, which deal with the ban on transfer and acquisition of nuclear weapons, have always been considered the backbone of the treaty, but recently priority has begun to be given to Article IV. This article, which deals with the right of a party to develop the peaceful uses of nuclear energy and to receive help for this purpose from other nations, was originally included as a compensation for countries forgoing their right to acquire nuclear weapons. However, the provision of cooperation in the peaceful uses of nuclear energy is now being interpreted by some as the main aim of the treaty, even overriding the non-proliferation aspect. For example in the report on the Windscale Inquiry, Mr Justice Parker states: "I also find it difficult to see how a party, which has developed reprocessing technology or created reprocessing facilities, would be otherwise than in breach of the agreement if it both refused to supply the technology to another party and refused to reprocess for it" [1]. Should such an interpretation be generally accepted—that the prime objective of the treaty is to help in the development of nuclear energy, and the danger inherent in handing over plutonium is of secondary importance—the NPT will in effect become a treaty for the peaceful uses of nuclear energy, and as such may be instrumental in promoting the very spread of nuclear weapon capability that it was intended to inhibit.

One of the basic weaknesses of the NPT is the inherent division of nations into two classes: the nuclear weapon states, which have all the privileges and no restrictions, except for the undertaking not to transfer weapons, and all other nations, which have to accept certain restrictive obligations. It was the underlying hope of the treaty that this division would be of a temporary nature, and that all nuclear weapons would eventually be abolished. In fact, Article VI calls on all parties to pursue negotiations towards nuclear disarmament. But after 10 years of hardly any progress in this direction, and with the nuclear arms race between the two great powers continuing and even accelerating, it is not surprising that the feeling against this division is growing. India objected from the beginning to the NPT on these grounds and several other nations, which have refused to sign the NPT, are eager to acquire the full nuclear technology which would enable them to make nuclear weapons and also become members of the privileged class. Other nations which have signed the treaty may see in these developments a threat to their own security or prestige, and this may induce them to do the same. So long as the possession of nuclear weapons is deemed to confer extra prestige and status to their owners, the pressures to join the nuclear weapon club will be very strong. The NPT may in fact be used as a means towards achieving this aim, through the acquisition of nuclear technology via nuclear power programmes.

The IAEA safeguards system

Under Article III of the NPT, each non-nuclear weapon state party to the treaty undertakes to accept safeguards, for the purpose of verifying that no diversion of nuclear energy from peaceful to military uses has occurred. These safeguards were to be negotiated with the International Atomic Energy Agency (IAEA), which was to set up a safeguards system for this purpose.

The aim of the IAEA is the advancement of nuclear energy. The origins of the Agency go back to the Atoms for Peace programme of President Eisenhower, who, in a speech to the UN General Assembly in December 1953, proposed the establishment of an international agency which would devote its activities exclusively to the peaceful uses of atomic energy. The statute of the Agency states as its objectives: "The Agency shall seek to accelerate and enlarge the contribution of atomic energy to peace, health and prosperity throughout the world. It shall ensure, so far as it is able, that assistance provided by it or at its request or under its supervision or control is not used in such a way as to further any military purpose" [2].

Since its foundation in 1957, the IAEA has pursued the promotional aspects of nuclear energy most diligently and effectively and still considers these to be the chief task of the Agency. In an address to commemorate the twentieth anniversary of the Agency, its Director-General, Dr Sigvard Eklund, said:

The role of nuclear power is intimately connected with industrial development in both advanced and developing countries. Until now no immediately available energy source, apart from nuclear power and the classical sources, has reached technical maturity. This being the case, nuclear power will continue to play an increasing role in both advanced and developing countries. The present opposition against this new form of energy will hopefully give way in time to a better understanding of the matter when the communities concerned correctly perceive the environmental implications of all the various energy sources at our disposal [3].

Notice that there is no mention of the proliferation implications.

In its organizational structure, safeguards come under one of the five departments of the Agency. In early years, the Department of Safeguards played only a minor role in the Agency's activities, but this has recently changed. With the growth of nuclear power programmes, and particularly after the NPT came into being, an increasing effort has been put into the safeguards aspect. During the past few years, for example, the proportion of the agency's budget allocated to safeguards has nearly doubled. The comprehensive system of safeguards introduced in 1971 is being updated and expanded both in the technical features of safeguards procedures and in the number of agreements reached with nations. Thus, new sophisticated methods are being developed for surveillance and containment; non-destructive assay techniques are being introduced; material accountancy, which is still the fundamental safeguards measure, is being improved; and a computer-based safeguards information and accounting system is being set up [4].

By the end of 1978, the IAEA had concluded safeguards agreements with 75 non-nuclear weapon states (including the European Atomic Energy Community (Euratom)) which have signed the NPT. In addition, the Agency has safeguards agreements with seven non-signatories of the NPT.

Despite this progress, the IAEA safeguards system leaves much to be desired. Controls over the most important parts of the nuclear fuel cycle, reprocessing and enrichment, are still at the stage of preliminary study. The basic weakness of the IAEA safeguards system still lies in the fact that the Agency has no authority to exercise any police-type control or to provide physical security to prevent theft of nuclear materials, these being the responsibilities of the individual states. Since all safeguards are based on an agreement between the Agency and the government of the given state, their effectiveness depends to a very large extent on the goodwill of that state and the efficiency of the state system of accountancy and control. It would be very difficult to eliminate the risk of diversion if the state itself decided to misuse its materials, say by regularly submitting slightly distorted figures in each accounting statement. Commenting on these deficiencies of the IAEA safeguards system, the Fox Report to the Australian government concluded that "these defects, taken together, are so serious that existing safeguards may provide only an illusion of protection" [5].

It is interesting that the same doubts were expressed in almost the same words 20 years earlier by Homi Bhabha of India, when the statutes of the Agency were being discussed. According to Bertrand Goldschmidt, "He [Bhabha] pointed to the illusory nature of strict safeguards and emphasized that any aid in the nuclear field—be it training opportunities or nuclear materials—was potentially military aid since it might allow a country to switch resources to a military programme" [6]. The Agency itself realizes that its safeguards system is still inadequate and would not necessarily detect diversion of nuclear materials by governments.

The Nuclear Suppliers Club

The ease with which India managed to use materials and facilities supplied for a peaceful programme to test in 1974 an explosive 'device' prompted several exporting countries, actual and potential, to meet to discuss ways and means of supplementing the IAEA safeguards. Seven of these nuclear supplier countries—Canada, France, FR Germany, Japan, the UK, the USA and the USSR—started in 1975 to meet in London; hence they came to be known as the 'London Club'. Later they were joined by eight other countries: Belgium, Czechoslovakia, the German Democratic Republic, Italy, the Netherlands, Poland, Sweden and Switzerland.

The earlier meetings were closed and little information emerged, but the secrecy gradually vanished, and the discussions and decisions from later meetings were made public.

In 1976 the members of the London Club agreed on a code of conduct for sales of nuclear materials, equipment or technology, all of which were defined in a so-called 'Trigger List'. The code requires the recipient country to pledge not to use the acquired facilities for the manufacture of nuclear explosives, to accept IAEA safeguards, to provide physical security for the transferred nuclear facilities, and to exercise restraint in retransfer or re-export of these facilities to other countries. At a 1977 meeting the members agreed to consult with each other on action to be taken in the event of a purchasing country breaching an agreement.

How much will such a code of conduct, indeed the existence of the suppliers group itself, help to reduce the danger of proliferation? Among the items to its credit is the fact that its membership includes France, which has so far refused to sign the NPT. The requirement that importing countries must abide by the IAEA safeguards system, even if they have not made an agreement with the Agency, is also helpful. In the end, however, the protection provided by the guidelines cannot be greater than that of the IAEA safeguards system, and this, as was pointed out above, is still inadequate.

There is also an undesirable aspect of the London Club which may have an aggravating effect. As a group of supplier nations, they may appear to the importing nations as a cartel. Although the group was convened with the laudable aim of preventing a lowering of safety standards which free competition between exporters might have produced, in principle the same group could also discuss and agree on prices to be charged for goods provided. The many examples of such practices in other fields justify these fears. The division of nations into exporters and importers, although not the same as the division into nuclear and non-nuclear weapon states, is bound to produce some mistrust, and this in turn may increase the probability of conflict.

Multinational fuel cycle centres

The Suppliers Club code includes the encouragement of multilateral regional facilities for reprocessing and enrichment. The concept of multinational centres has been proposed and discussed in recent years, not only from the economic benefit aspect, but also as a means of reducing the opportunities for diversion [7]. The IAEA has carried out a study on such multinational facilities, with particular reference to plutonium management.

Obviously, it would be much more difficult for a single nation to divert nuclear materials from a multinational facility than from a plant under its own control. From this point of view, the concept of multinational fuel cycle centres is a step in the right direction. If, however, as has been generally assumed, it envisages the setting up of several such centres rather than one global centre under truly international control, then the disadvantages may well outweigh the advantages. The divergence between East and West and between industrialized and Third World countries is likely to be exacerbated, and competition between centres may result in a lowering of standards. Conflicts of loyalties may also ensue.

An example of such a conflict is the current dispute between the European Economic Community (EEC) and the UK over the proposed purchase by the latter of uranium from Australia. The UK had to sign a safeguards agreement with Australia that the uranium would not be transferred to other countries which are not covered by such an agreement. However, this is in direct contradiction to the Euratom requirement that fissile materials should flow freely among all its members.

The US energy policy

One of the sources of disagreement among members of the London Club is the problem of the export arrangements made by some of them. Thus, the contract between FR Germany and Brazil for the latter to be supplied with the whole gamut of nuclear facilities and technologies was regarded with strong reservation by other members, particularly by the United States.

The realization of the peril in making plutonium available to many nations, even with all the existing safeguards, has grown in recent years in the USA. The danger was seen to lie predominantly in reprocessing plants and fast breeder reactors. This concern found expression in the energy policy announced in April 1977 by President Carter, subsequently enacted in the Nuclear Non-Proliferation Act of 1978, which was passed by Congress in March 1978.

The USA, the largest nuclear exporting country, with a virtual monopoly on the manufacture and supply of enriched fuel in the non-communist world, has the greatest possibility to influence the future development of nuclear energy, and—arising out of this—the greatest responsibility for preventing its misuse. Hence, the flurry of activities, congressional hearings and studies by governmental and non-governmental groups which took place in recent years, most of it triggered by India's test explosion in 1974. The result was the decision by the Carter Administration to stop the reprocessing of spent fuel and to defer further work on demonstration and commercial use of the fast breeder. Although the reasons for these decisions were clearly stated, they were received with hostility in other countries, which preferred to interpret them as a means of strengthening US hegemony and as interference in the economy of other countries. Much of this criticism was inspired by the nuclear industry and obviously had ulterior motives; nevertheless it had to be met and shown to be unjustified, by supplementing the domestic legislation with corresponding proposals in the international arena.

The International Nuclear Fuel Cycle Evaluation

At a seven-nation summit meeting, held in London in May 1977, President Carter proposed the setting up of the International Nuclear Fuel Cycle Evaluation (INFCE) programme. This proposal was accepted, and the organizing conference of the INFCE was convened in Washington in October 1977, with the participation of delegates from 42 countries, including some of those which have not signed the NPT, such as Argentina, Brazil and India.

The main purpose of the INFCE is to investigate the practicality of proliferation-resistant fuel cycles, but in fact it covers the whole spectrum of nuclear power programmes, as is evident from the titles of the eight working groups set up by the Technical Coordinating Committee of the INFCE. These are: (a) fuel and heavy water availability; (b) enrichment availability; (c) assurance of long-term supply of technology, fuel and heavy water services in the interest of national needs consistent with non-proliferation; (d) reprocessing, plutonium handling and recycling; (e) fast breeders; (f) spent fuel management; (g) waste management and disposal; and (h) advanced fuel cycle and reactor concepts. A plenary session of the INFCE was held in November 1978 in Vienna, with 56 countries participating, and completion of the work is likely to take at least another year.

The setting up of the INFCE is an important advance in attempts to reduce the danger of proliferation in the world; for the first time it brought together exporting and importing countries, both signatories and non-signatories of the NPT. But how much can it really achieve? At best it may come up with proposals to develop and give priority to proliferation-resistant fuel cycles based on thorium, and for the international administration of the more proliferation-sensitive nuclear facilities. Such proposals, if generally accepted, would considerably reduce the opportunities for nations, or subnational groups, to acquire materials for nuclear weapons. This would be a very significant achievement in providing a greater degree of security in the short term. In the long term, however, it may have the reverse effect. By making nuclear energy, including breeder reactors and more advanced technology, highly respectable on the international arena, it would encourage more nations to embark on nuclear energy programmes. Unless accompanied by concrete proposals for the development of alternative sources of energy, the final outcome of the INFCE might well be much wider usage of nuclear energy in the world. Since no system of safeguards can be 100 per cent efficient, the probability of a nuclear war indirectly arising from nuclear power programmes may in fact be increased.

III.   The energy problem of the Third World

The most effective way of reducing the risk of nuclear weapon proliferation is to reduce the number of nations that will base their energy needs on nuclear energy. This can be achieved most easily with countries which have not yet started nuclear programmes—as it happens these are almost exclusively in the Third World. Of the 36 countries with commercial nuclear power reactors, either in operation or under construction, only 11 are in the Third World. Of the remaining countries, 90 per cent are underdeveloped. This means that for the great majority of nations the options are still open. Unfortunately, this may not be the case for long, since there are strong pressures for these countries to embark on nuclear power programmes. These lures come from two main sources: the IAEA and the nuclear industry.

The Agency has conscientiously performed its task of promoting the peaceful uses of nuclear energy, and for a number of years has been espousing the cause of nuclear power. It has to be held largely responsible for spreading the myth that nuclear energy is essential for the economic and industrial growth of Third World countries. Fed on its own enthusiasm, the Agency made highly optimistic forecasts about the growth of nuclear power in many countries over the next few decades. Such forecasts have (often intended) self-fulfilling properties and act as an incentive for a country to reach the target allocated to it, particularly when nuclear power is presented as synonymous with progress. It then becomes a matter of prestige for one country not to be seen to lag behind others in technological achievements. Another aspect of the Agency's activities, organizing training courses for scientists from Third World countries, produces the same effect. After returning home, these scientists, often belonging to a country's élite, exert strong pressure on their governments to start nuclear power programmes.

While the motives of the IAEA may be considered altruistic, the motives of the nuclear industry are clearly more mundane, that is, profit-making. The capital investment in nuclear reactors is very high, and a company which manufactures them will try to recoup its capital outlay by selling as many reactors of the same type as possible, whether or not this is in the interest of the client. Export is indeed the avowed aim of the nuclear industry, and high-powered salesmanship techniques are employed. Attractive offers, in the form of long-term loans, often subsidized by the governments of the exporting countries, will be made to prospective customers. Governments themselves may not be averse to making a profit and earning foreign currency, as was the case with the reprocessing plant at Windscale which is owned by the British government.

These two factors—the IAEA promotion campaign and tempting offers from the nuclear industry, which, when combined with the possibility of the facilities and technology being useful for making nuclear weapons, is a prospect undoubtedly alluring to some governments—are likely to lead to the widespread use of nuclear energy, unless effective countermeasures are taken.

There are several reasons why the spread of nuclear reactors to underdeveloped countries should, in their own interest, be avoided. One reason is that nuclear energy is probably the most undesirable form of energy for them. Even the IAEA admits that the competitive position of nuclear power is weaker in underdeveloped countries because of the large initial investment required, because of the unacceptability of large reactors which are much more economical than small ones, and because the cost of nuclear power plants is higher than in the industrialized countries [8]. Neither the economic structure nor the state of industrialization calls for, or makes possible, the utilization of the highly capital-intensive, technologically sophisticated and centralized character of nuclear power. In most of the underdeveloped countries, there is no grid system to distribute the electricity generated in large reactors, and the necessary technical infrastructure will be absent for many years. What these countries need now are small generating units which can be easily installed for individual communities to suit local requirements. Such dispersed sources of energy need less capital, have shorter construction times and greater flexibility, and, in the case of failure, would cause much less damage than would a breakdown in a huge central station.

For many countries, solar energy, which fulfils all these criteria, would be economical even at the present time. Instead, as a result of propaganda, the underdeveloped countries may find themselves equipped with the potential to make nuclear weapons decades ahead of their ability to utilize the reactors for peaceful purposes. Apart from distorting a country's industrial development, the acquisition of nuclear power may have a crippling effect on its economy. The payment of interest on the loan to buy the extremely expensive reactors will be a heavy burden on the population, and may lead to a lowering of the standard of living well before any benefits have accrued from the investment.

A second main reason for discouraging the Third World from undertaking nuclear power programmes is the danger of a new colonialism, that is, a new type of dependence of the underdeveloped countries on the industrialized ones. To keep the reactor going, spent fuel will have to be replaced by fresh fuel elements. With the proposed arrangements whereby enrichment and reprocessing will be carried out by the supplier country, a potential threat of withholding the replacement of fuel will ensue, and a country whose industry is based on nuclear energy could be held to political ransom. Incidentally, this applies not only to underdeveloped countries. It could also happen in the socialist countries, if the Soviet Union were the only one to possess enrichment and reprocessing plants, and equally in the West, if the supplier countries retain their monopoly on enrichment and reprocessing plants, ostensibly as safeguards against proliferation. To avoid this, countries able to afford it will insist on acquiring complete fuel cycle facilities, as Brazil has already done, thus making them virtually nuclear weapon states. In other countries, the effect of this neo-colonialism will be increased mistrust, continual tension and further polarization of political ideologies. All these factors contribute to a greater likelihood of conflict, and thus war, with the possible use of nuclear weapons.

This brings us to the third reason for avoiding the use of nuclear energy in the Third World countries: the increased risk of a nuclear war. The total nuclear power capacity of the underdeveloped countries will for a long time be well below that in the developed countries, but from the point of view of proliferation it is not the power output but the number of countries involved that matters. The more countries which have the potential to make nuclear weapons, the greater is the likelihood of these weapons being used in a conflict, the probability of a conflict increasing faster than the number of countries. The Third World countries constitute nearly 80 per cent of all states. Furthermore, many of these states, having only recently gained independence, have not yet achieved a high degree of stability and, more than other countries, are subject to tribal strife, military coups and other upheavals. Rivalry, or traditional enmity between nations of different religious and ethnic backgrounds, enhances fears of attack and compels them to arm themselves, as evidenced by the huge growth in the conventional arms trade in the Third World in recent years, amounting to an increase of 15 per cent per year [9]. If one of these countries acquired—or is rumoured to have acquired—nuclear weapons, this would be a strong incentive to its neighbours to do the same, and the nuclear disease will spread like an epidemic.

IV.   The need for global action

It follows from the previous section that in order to prevent the further proliferation of nuclear weapons, three additional types of action need to be taken: (*a*) countries which have not yet embarked on a nuclear power programme should be encouraged to meet their energy needs from other sources; (*b*) the pressure on countries to acquire nuclear reactors should be reduced by taking the nuclear energy industry out of the domain of competition and profit-making enterprises; and (*c*) countries which already have nuclear reactors should be provided with fuel cycle arrangements that are free from the threat of new hegemonies.

The initiatives taken hitherto, including the INFCE or the proposals for various types of multinational agreement, are not sufficient to meet these objectives. Their achievement will require an immense effort on a truly international scale. However, the dilemma facing mankind—on the one hand, to ensure that nations will not be deprived of energy, the life-blood of modern society; and on the other hand, to prevent a nuclear catastrophe—is so awesome that it justifies taking steps which would otherwise be considered extreme. Such steps include the setting up of the two new international institutions described below.

A World Energy Organization

In many ways, energy is as vital to the community as food and health are to the individual. As with these two other elements, the absence of energy or its misuse could lead to consequences extending beyond national frontiers, so that energy is indeed a world problem.

In the cases of food and health, this was recognized many years ago by the establishment of the Food and Agriculture Organization (FAO) and the World Health Organization (WHO), which are now among the specialized agencies of the United Nations. For energy, there also exists a specialized UN agency, the IAEA, but this body is concerned with only *one* form of energy: nuclear. This imbalance needs to be remedied by setting up an international agency to deal with all forms of energy, a World Energy Organization (WEO).

This WEO might be a specialized agency of the United Nations, with functions analogous to those of the WHO or the FAO, combined with some of the functions of the World Bank. Its terms of reference might be: to promote and encourage research on various forms of energy other than nuclear (in order not to overlap with the IAEA), and on their utilization; to supply member states with information about developments in the energy field; to advise member states on the forms of energy most suitable for their needs, and on ways of implementing their energy programmes; and to provide financial help towards their realization.

As is envisaged here, the WEO would be a different body from the world energy institutes which have been suggested in the past, although it might collaborate with such institutes or enter into contractual arrangements with them. The main point is that the WEO should have the financial means, in conjunction with the World Bank, to make loans to individual countries for their energy development.

A country wishing to start a new energy programme or to update an existing one would first apply to the WEO for advice. The WEO would institute an inquiry into the most suitable form of energy for the country, and the scope and timetable for its implementation. If the advice from WEO is accepted, the country would be entitled to ask for a loan to carry out its programme. The loan would have generous conditions, so that it would not impose a heavy burden on the economy of the country.

In some cases the advice may be to purchase a nuclear reactor. Due to the military implications of nuclear energy, a loan for this purpose would have stringent conditions attached, relating to the size, type and safety aspects of the reactor. An important condition should be that the reactor is purchased only through the medium of the WEO. In this way, the scope for private companies to sell reactors directly would be greatly reduced. Indeed, with the WEO having virtual control on prices to be charged for reactors, the incentive for free enterprise would soon disappear, and further reactors would be built only on WEO contracts.

An International Nuclear Fuel Agency

The setting up of the WEO is likely to reduce drastically the number of countries choosing the nuclear path. However, there still remains the problem of the more than two dozen non-nuclear weapon states which already have, or soon will have, nuclear reactors and nuclear technology. To deal effectively with the proliferation dangers arising from this, new arrangements for the international operation of certain aspects of the nuclear fuel cycle, coupled with a significant extension of the obligations under the NPT, will be necessary.

These international operations relate primarily to those aspects of the nuclear fuel cycle where diversion to weapon materials can occur most easily, that is, enrichment, fuel fabrication and reprocessing of spent fuel. At present, enrichment plants do not constitute a significant threat, since the uranium-235 content of the fuel used in reactors is too low for weapons. But with the likely introduction of new enrichment techniques such as ultra-centrifuge, jet nozzle or lasers, the production of weapon-grade uranium may become much easier and for this reason the operation of enrichment plants must also be controlled.

It is proposed that the operation of enrichment, fuel fabrication, reprocessing and waste disposal plants be permitted to be carried out only by, or under licence of, a new international body, the International Nuclear Fuel Agency (INFA). However, to be effective, the controlling function of the INFA would have to extend back to the very beginning of the nuclear fuel cycle, the uranium and thorium ores. The INFA would buy such ores from the prospecting countries, and these would undertake not to sell nuclear fuel-bearing ores to anyone else without the authority of the INFA.

The enrichment and fuel-fabricating plants, operated under the licence and strict control of the INFA, would manufacture fuel elements for different types of reactor, and store them in the INFA bank. After the first charging of a reactor, fresh fuel elements would be sold to individual countries, which would be obliged to return spent elements, as well as all fissile-bearing discharge materials. These would be stored by the INFA, either with or without chemical processing. When necessary, reprocessing of spent fuels would be carried out in plants operated by or under licence of the INFA.

These means would ensure that no weapon-grade materials from civilian programmes would be in existence anywhere outside the control of the international authority. The supply of fuel for reactors would be assured by a body owned by the international community and would thus not be subject to political or economic discrimination.

For these measures to be effective, nations would have to undertake not to acquire fuel elements from any source other than the INFA; not to process their spent fuels either themselves or through other countries; and not to acquire, make or store plutonium, or uranium above 10 per cent enrichment, except for such special purposes as research, and then only with INFA approval.

An important feature of these proposals is that these undertakings would apply to all nations, whether or not they possess nuclear weapons. The five nations now overtly possessing nuclear weapons would have to undertake to separate completely their military and civilian uses of nuclear energy. No exchange of materials or facilities between the weapon and peaceful programmes would be permitted, either within the country or with other countries. As far as the civilian use of nuclear energy is concerned, the nuclear weapon powers would have exactly the same obligations and be subject to exactly the same controls and inspections as are all other countries. In this way, the division of the world into two classes, from the point of view of nuclear energy, would disappear.

The above obligations covering the nuclear fuel cycle might most easily be implemented by adding appropriate protocols to the NPT.

With regard to the organizational aspect of the INFA, two courses are available: (a) it is set up as an independent body; or (b) it is attached to the IAEA. At first sight, the second option recommends itself for choice, since in international negotiations it is much easier to add something to an existing establishment than to start a new one. However, other considerations militate against this.

The operation of the INFA would require a safeguards system which in many ways would be different from, and additional to, the present IAEA system. Thus, there would be the need to police the plants operating under INFA licence and to guard all shipments of materials to and from reactors and other plants. To include all this in the organization of the IAEA would make the size of the Department of Safeguards out of proportion to other departments, and would completely distort the structure of the IAEA.

There is still another reason for moving away from the IAEA. Its two tasks—the promotion of nuclear energy and the prevention of its misuse—are to a large extent contradictory; safeguarding involves an element of constraint which does not go hand in hand with promotion. The desirability of separating these two functions, by allocating them to different bodies, has been recognized in other places. For example, in the USA the two similar functions previously held by the Atomic Energy Commission were later allocated to two separate bodies: the Energy Research and Development Administration and the Nuclear Regulatory Commission. On these grounds it would seem logical to remove the Department of Safeguards from the IAEA and make it part of the INFA, which would then have only one task—the prevention of proliferation. This scheme would also ensure that the INFA would not have to start from scratch, but would begin with a budget and trained staff.

A compromise solution would be to put the INFA under the overall umbrella of the IAEA, but to modify the organizational structure of the latter by making the present Department of Safeguards, together with the proposed INFA, an autonomous branch of the IAEA with a separate budget and staff.

V. Concluding remarks

The prospects of reaching agreement on substantive changes to the NPT, and on setting up new international organizations, may appear so slim as to make the whole idea unrealistic. But the haunting nightmare of a nuclear war compels us to reach for the seemingly impossible. In fact, the chances of success may not be so hopeless. Due mainly to the initiative of the US government, the peril of proliferation is now the topic of international discussion, and schemes to tackle it, including an international nuclear fuel bank, have been proposed and are more likely to be considered seriously by other governments.

When ideas of international nuclear fuel management were first mooted, some critics pointed out their similarity to the original Baruch Plan, and on this basis predicted their failure [10]. However, quite apart from the fact that the world situation in 1978 is very different from that in 1946, the main difference between these proposals and the Baruch Plan is that the proposals made here do not directly involve the control of nuclear weapons, and it is this control that has hitherto proved an unsurmountable difficulty.

Ultimately, the spectre of a nuclear war will be laid to rest only by nuclear disarmament and this, as part of the programme of general and complete disarmament, must remain our chief goal. But the removal of the threat of horizontal proliferation will enhance the chances of dealing with vertical proliferation, and thus the chances of the survival of mankind.

References

1. *The Windscale Inquiry*, Vol. 1, Report and Annexes, 26 January 1978 (HMSO, London, 1978), p. 18.
2. International Atomic Energy Agency Statute, 23 October 1956 (United Nations, New York, 1956).
3. Eklund, S., 'Foreword', *International Atomic Energy Agency 1957–1977* (IAEA, Vienna, September 1977), p. 1.
4. *Non-Proliferation and International Safeguards*, Public Information Booklet IAEA 575 (IAEA, Vienna, 1978).
5. *Ranger Uranium Environmental Inquiry. First Report* (Australian Government Publishing Service, Canberra, 1976), p. 147.
6. Goldschmidt, B., 'The origins of the International Atomic Energy Agency', *IAEA Bulletin*, Vol. 19, No. 4, August 1977, p. 12.
7. Chayes, A. and Lewis, W. B., *International Arrangements for Nuclear Fuel Reprocessing*, Proceedings of a Pugwash Symposium (Ballinger Press, Cambridge, Mass., 1977).
8. Woite, G., 'Capital investment costs of nuclear power plants', *IAEA Bulletin*, Vol. 20, No. 1, February 1978, p. 11.
9. *World Armaments and Disarmament, SIPRI Yearbook 1978* (Taylor & Francis,

London, 1978, Stockholm International Peace Research Institute), p. 223.

10. Epstein, W., *The Last Chance* (Free Press, New York, 1976), p. 281.

Part IV

Introduction

B. JASANI

Square-bracketed numbers, thus [1], *refer to the list of references on page 292.*

Of all the variety of applications of nuclear energy, the peaceful applications of nuclear explosions (PNE) and the use of reactors for generating electric power are the most controversial. This is because both have far-reaching implications for the question of the proliferation of nuclear weapons. The controversy regarding the PNE arises because it is virtually indistinguishable from a test explosion for a nuclear weapon. In the case of the power-generating reactors, the controversy arises because they can produce fissile material needed for nuclear weapons and thus contribute to horizontal proliferation, that is, proliferation among a greater number of nations. This aspect of nuclear reactors was discussed more fully in Parts I and II. However, the contribution of nuclear power generators to vertical proliferation, that is, both the increase in the number of nuclear weapons within nuclear weapon states as well as the increase in the quality of such weapons, has received much less attention.

Both these technologies are therefore considered in this Part.

I. Peaceful nuclear explosions

Despite certain technical, safety and political problems, international interest in the peaceful applications of nuclear explosives has been maintained to some extent. The 1963 Limited Test Ban Treaty and the 1968 Non-Proliferation Treaty, and, more recently, the 1976 US–Soviet bilateral Peaceful Nuclear Explosions Treaty as well as the present discussions on a comprehensive test ban treaty have served to focus attention on this very controversial technology.

Article V of the NPT states that the benefits of PNE technology must be made available to non-nuclear weapon states party to the treaty, under "appropriate international observation and through appropriate international procedures". The PNE is attractive because the technology may offer a relatively cheap source of concentrated energy for use in the development of natural resources and in large-scale civil engineering projects.

Basically PNE technology can be used in two ways: (*a*) for underground explosions carried out relatively close to the Earth's surface for creating reservoirs, for constructing canals or for uncovering mineral deposits; and (*b*) for contained explosions carried out deep underground for gas stimulation, for copper leaching or for creating underground cavities for storage purposes. As can be seen from Paper 19, a new application related to seismological studies has been considered since 1971.

Emphasis seems to be shifting towards geophysical applications rather than engineering ones. Applications involving nuclear explosions, however, must be carried out without hazards to populations, private property or the environment and, of course, in accordance with the provisions of the 1963 Limited Test Ban Treaty.

A number of countries have been interested in PNEs, but the most extensive experiments have been carried out in the United States and the Soviet Union. The extents of the PNE programmes of these and other countries are shown in table IV.1.

The USA. Between 1956 and 1974, the US Atomic Energy Commission (presently known as the Department of Energy) spent about $160 million in PNE-related studies which included several tests (table IV.1). About two-thirds of this was spent on excavation studies, but since the Limited Test Ban Treaty, contained underground PNE experiments have been carried out. By the beginning of 1975, some 53 PNE-related tests were carried out, of which three were for gas stimulation. Several scientific experiments were also carried out.

The Soviet Union. The Soviet Union is known to have published results of some 14 PNE experiments, five related to gas stimulation, gas-fire blow-out, oil stimulation, oil-shale mining and underground storage. In addition to these known tests, it is estimated that by the end of 1974, the Soviet Union had conducted at least 19 additional tests in support of its PNE programme. Several other planned projects have been described (table IV.1).

France is the third nation with a nuclear explosion programme related to PNE experiments. By the end of 1974, some 13 contained experimental explosions had been carried out at In Ekker in the Sahara.

India has carried out one underground nuclear explosion related to PNE experiments.

The United Kingdom has maintained an interest in PNEs, and other countries such as Australia, Egypt and Venezuela have expressed interest in having applications performed in their countries.

In order to make Article V of the NPT more meaningful and thereby strengthen the NPT, a number of questions have to be answered, particularly by the second NPT Review Conference to be held in 1980. Despite

Table IV.1. Underground and peaceful nuclear explosions carried out by various countries

Applications	China Actual	China Proposed	France Actual	France Proposed	India Actual	India Proposed	UK Actual	UK Proposed	USA Actual	USA Proposed	USSR Actual	USSR Proposed
Excavation												
Experiments	0	0	0	0	0	0	0	0	7	0	4	–
Highways	0	0	0	0	0	0	0	0	0	1[a]	0	0
Harbours	0	0	0	0	0	0	0	0	0	2[a]	0	0
Canals	0	0	0	0	0	0	0	0	0	1[a]	0	1
Water resources	0	0	0	0	0	0	0	0	0	0	1	1
Overburden removal	0	0	0	0	0	0	0	0	0	0	0	1
Contained												
Experiments	0	0	13	–	1	–	0	0	Many	–	Many	–
Gas and oil stimulation	0	0	0	0	0	0	0	0	3	0	3	2
Gas-fire blowout	0	0	0	0	0	0	0	0	0	0	2	0
Oil shale	0	0	0	1	0	0	0	0	0	1	0	0
Storage	0	0	0	0	0	0	0	0	0	1	3	2
Geothermal energy	0	0	0	0	0	0	0	0	0	1	0	1
Mining	0	0	0	0	0	0	0	0	0	2	1	1
Total PNEs	**0**	**0**	**13**	**1**	**1**	**–**	**0**	**0**	**53**	**10**	**14**	**9**
Total underground explosions	**1**	**0**	**13**	**1**	**1**	**–**	**5**	**0**	**366**	**10**	**129**	**9**

[a] Inactive.

some two decades of work, the technology is still in the development stage. The IAEA has accumulated a significant amount of technical data and operational experience. An international Ad Hoc Advisory Group on Nuclear Explosions for Peaceful Purposes has also been established by the IAEA. Paper 20 describes some of the conclusions of the group. Under these studies, health and safety hazards were considered and were found to be both predictable and manageable. However, the economic aspects are less clear.

Moreover, difficult problems, such as how to treat excavation experiments, since they may violate the Limited Test Ban Treaty, and the fact that a PNE device cannot be distinguished from a weapon, so that the development of PNEs in a country contributes to its nuclear weapon capability, and how to treat PNEs with respect to the comprehensive test ban treaty, remain unsolved. Guidelines drawn up by IAEA expert panels for appropriate international observation and for procedures to deal with requests for PNE services, and the Ad Hoc Advisory Group's work and services made available to all interested member stated of the Agency and states party to the NPT, certainly help to strengthen the NPT by making Article V of the treaty more meaningful. However, a suggestion has been made to extend these services also to non-NPT states so that they would have no excuse for developing their own PNE programmes.

II. Reactors in satellites

Since the beginning of the space age in late 1957, several thousands of man-made space objects have re-entered the Earth's atmosphere and burned up. Of the several hundred pieces of debris that have landed on the surface of the Earth, none has caused any personal injury or damaged any property, but on two occasions contamination of the Earth's surface with radioactive materials has occurred.

Whereas international controls have been established to check horizontal proliferation of nuclear weapons resulting from the civilian uses of nuclear reactors, no such controls exist to check the vertical proliferation resulting from the uses of power generators using radioactive materials or nuclear reactors.

Artificial Earth satellites circling round the Earth in near and geosynchronous orbits are increasingly being used to enhance the performance of Earth-bound weapons carrying their lethal conventional or nuclear warheads. The increased accuracy of delivery of such weapons to their targets with the aid of satellites not only may provide incentives to develop new types of weapon but also is changing the war-fighting doctrines, such as the current counterforce and flexible response doctrines, which would emphasize limited nuclear war-fighting capabilities at various levels.

The premature re-entry, on 24 January 1978, of the Soviet Cosmos 954 satellite carrying a nuclear reactor has focused attention on the military uses of artificial Earth satellites. This particular satellite was an ocean surveillance satellite which used a reactor fuelled with highly enriched uranium. A nuclear power reactor is used on board such satellites to provide electrical energy for electronic equipment such as radars. Details of such nuclear power sources are given in Paper 21.

The earliest satellites needed no more than a watt or two of power to transmit their sensor-readings back to Earth. Today's satellites need more power because they perform many complex tasks. Moreover, the present development of anti-satellite systems makes the possibility of space warfare more likely. This, in turn, may give impetus to the development of nuclear power generators to, for example, replace solar panels, thus increasing the survivability of satellites by reducing the probability of being detected by radars. Some $43 million have been proposed in the USA for the fiscal year 1979 budget by the Department of Energy (DOE) for research on nuclear reactor and isotopic power sources for space applications [4]. The DOE may invest $2 million in a study for a 100-kW space reactor fuelled with plutonium [5]. Applications include a power system for the Global Positioning System and the Air Force Satellites Communications System. The heat created by decaying radioisotopes can be converted into electricity either by dynamic conversion using a turbogenerator (dynamic radioisotope generator) or by static conversion in which non-moving energy conversion devices such as thermoelectric elements are used (radioisotope thermoelectric generator—RTG). The latter types become heavy and less efficient for power levels greater than a kilowatt. For such power requirement, dynamic radioisotope generators have to be used. Such a device will be tested on board a satellite in 1982—83 [5].

Nuclear power sources are being used in the United States on military satellites such as the navigation, meteorological and communications satellites. The Soviet ocean-surveillance satellites are known to carry nuclear reactors as a power source. At least four accidents are known to have occurred involving nuclear power sources deployed on satellites and two of these were military satellites. In both cases, contamination of the Earth's surface and the atmosphere resulted.

In future, nuclear propulsion for rockets may be used and at present nuclear power is being used on board satellites, and warships and submarines are propelled by nuclear reactors. What can be done to prevent such activities, be it in outer space, in air, on the surface of the Earth or ocean surface or below it, to check vertical proliferation? A technical study group, including the USSR and the USA, was set up in 1978 within the United Nations which could lead to international sanctions against nuclear reactors but not the plutonium RTG power sources being used in artificial Earth satellites. A ban on the use of nuclear power sources in satellites is of fundamental importance. However, before such a decision is reached, there are a number of other questions which could be settled sooner. It is important that the world should be warned when a satellite carrying a nuclear power source is to be launched. Moreover, adequate warning should be given by the launcher state, should such satellite malfunction. And finally, it is imperative that a UN conference on outer space is held as soon as possible. The conference should discuss the fundamental questions of how outer space should be used; this environment is at present being used largely for military purposes. The last such conference was held in 1968 in Vienna.

References

1. Long, F. A., 'Is there a future for peaceful nuclear explosions?' *Arms Control Today*, Vol. 5, No. 5, May 1975, p. 1.
2. Kruger, P., Depois, J., Sarda, J. P., and Nougarede, F., 'Peaceful nuclear explosions', *Transactions of the American Nuclear Society*, 21–25 April 1975, p. 792.
3. Ramanna, R., 'Peaceful nuclear explosions (PNEs)', in *Perspectives of India's Nuclear Policy*, T. T. Poulose, ed. (Young Asia Publications, New Delhi, 1978), p. 48.
4. 'New space nuclear power program planned', *Aviation Week and Space Technology*, Vol. 108, No. 5, 30 January 1978, p. 26.
5. 'Carter nuclear satellite ban could hurt research', *Science*, Vol. 199, No. 4330, 17 February 1978, pp. 752–53.

Chapter 11. Peaceful nuclear explosions

Paper 19. Peaceful applications of nuclear explosions

D. DAVIES

Square-bracketed numbers, thus [1], refer to the list of references on page 304.

I. Introduction

For more than 20 years there has been serious and extensive study of the possible use of nuclear explosions in non-military applications. Since the early 1960s, many ideas have been tried out, either in pilot projects on test-sites in the United States and the Soviet Union or actually in the field. In strictly engineering terms, a substantial proportion of these trials have been successful; in economic terms, results have been more mixed, and in the United States the 'Plowshare' programme ran into much opposition. Now the USA has abandoned its interest in Plowshare, although the Soviet Union still has a moderate programme (figure 19.1). But in the very serious discussions in progress on a comprehensive nuclear test ban, and with India detonating her 1974 nuclear device "for peaceful purposes" it is no longer possible to isolate such explosions and regard them as coming outside the scope of arms control and restraints on proliferation. This paper first briefly outlines the technical aspects of the national peaceful-use programmes. The way in which these programmes have been both stimulated and constrained at national and international level is then described. Finally, some of the future options, at international level, for preventing undesirable political side-effects from peaceful-use programmes are examined.

II.   What explosions can be used for

A nuclear explosion of one kiloton releases 4.2×10^{12} joules of energy almost instantaneously. Most of the interest expressed in peaceful applications has been for yields in the one to hundred kiloton range—a range in which many hundreds of military nuclear tests have been conducted. But there seems little doubt that nuclear devices can also be made with yields as low as tens of tons and as high as tens of megatons.

Non-military uses of underground explosions may be divided into engineering and scientific headings. In the former category we could place earth-moving, rock-breaking, oil or gas stimulation, underground-hole-making and sealing of wells. In the latter category would be nuclear and particle physics experiments and seismology. This list is by no means complete; a more comprehensive one is given in an IAEA report [1]. It does however include all known field experiments and operations. The point of this paper is not to dwell on detailed technical results, but it may be helpful briefly to summarize the successes and failures in the above categories.

Earth-moving

An underground explosion creates a cavity, several metres across, within seconds. If the overburden is a relatively soft rock it also forces this material upwards and outwards to produce a bulge on the surface. Over a period of minutes to hours, the cavity collapses and leaves a crater with a lip to it. Such craters are characteristic of many nuclear tests, peaceful or not. The linear dimensions of these craters, as of many nuclear-explosive phenomena, scale roughly as the cube root of yield and inversely as the depth of burial. It has been found that a 1-kiloton shot fired at the optimum depth can produce a crater about 100 metres across and 30 metres deep, whilst a 100-kiloton shot has left a crater 370 metres across and 100 metres deep. One such crater with a lip has been used as a circular reservoir in the Soviet Union; several appropriately spaced craters from simultaneous blasts could be used as the basis for a canal. There has been much discussion in the Soviet Union of such a canal to divert the waters of the Pechora River, flowing towards the Arctic Ocean, into the Volga—Kama river system and on to the Caspian Sea, the level of which is falling at present by 6 cm per year. A 65-km section of the canal could, it is estimated, be excavated by 250 nuclear explosions, in salvoes of 20 (with total yield per salvo about 3 megatons). Several preliminary experiments have been performed.

There are numerous other suggestions of earth-moving applications, such as in the creation of harbours, clearing of navigational obstructions, removal of overburden for mining and so on, but none of these has yet been attempted in the field.

Rock-breaking

If the device is in a hard-rock environment, a crater on the surface is unlikely to be formed, but there will be extensive fracturing of rock below ground, not just in a sphere surrounding the shot-point but also in a 'chimney' or cylinder of rubble above the shot-point. A 1-kiloton blast would probably crush rock out horizontally to 25 metres radius, and tensile-fracture out to 50 metres or more; the chimney would elongate this pattern vertically above the shot-point by at least a factor of two.

294

Applications are numerous [2a, 3]. Rock could be mined directly or ores could be leached out. Deep shots could be used to make geothermal heat recovery easier. Oil shales could be fractured for *in situ* retorting. Paper studies exist for many such applications, and the Soviet Union several years ago discussed proposals for ore-breaking [3], but it is unclear whether this experiment has been carried out. One form of rock-breaking operation, however, has been widely used and merits the following section to itself.

Oil and gas stimulation

It is generally considered that nuclear explosives would be used in oil and gas fields only when recovery by conventional methods was out of the question or was in decline. Fracturing either increases the permeability in hydrocarbon-bearing strata (sometimes making the effective well radius as large as the chimney radius itself) or breaks up impervious strata, allowing water at the reservoir pressure to drive oil out. There is still considerable uncertainty concerning precise mechanisms [3].

Oil stimulation experiments have been reported by the Soviet Union. In one of these, two 2.3-kiloton explosives 200 metres apart were fired close in time at a depth of 1 350 metres; a third 8-kiloton blast came three months later. These three increased production from the field by more than 30 per cent, it is estimated. In a second experiment, two 8-kiloton devices yielded production increases of 30−60 per cent.

There have been three gas-stimulation experiments in the United States —'Gasbuggy', a 29-kiloton explosion in New Mexico in 1967; 'Rulison', a 40-kiloton explosion in Colorado in 1970, and 'Rio Blanco', also in Colorado in 1973 but with three simultaneous detonations of 30-kiloton devices each vertically separated by more than 100 metres. Gasbuggy and Rulison were technically successful and undoubtedly stimulated gas flow, but there was general disappointment with Rio Blanco, partly because the three chimneys failed to interconnect, but largely because gas flow seems to be 6 to 10 times lower than predicted, probably because of inaccurate pre-shot estimates of reservoir capacity [4a].

The Soviet Union has also carried out gas stimulation experiments and plans more, but few details are available.

Underground holes

An explosion in salt will generally leave a roughly spherical cavity that does not collapse. Its radius for a 1-kiloton shot at a depth of 200 metres would be around 10 metres. The cavity could be used for liquid hydrocarbon or gas storage, or it could be used to dump radioactive wastes. The Soviet Union has detonated three devices, with yields 1, 15 and 25 kilotons, for hydrocarbon storage, and at least one cavity is in regular industrial use at present.

Sealing of wells

If an explosion is fired in a relatively soft stratum within a few tens of metres laterally of an oil-well, the material shifted sideways by the blast should block the well.

The Soviet Union has twice extinguished runaway wells in this fashion, with 30- and 40-kiloton devices, in 1966 and 1968.

Nuclear and particle physics

Nuclear explosions are rich sources of neutrons and indeed might yield valuable information to physicists, although not much enthusiasm has been detected among the academic community. Numerous detonations in the United States have been described as 'heavy-element experiments' or 'neutron-physics experiments', and presumably there have been many other military tests which have had scientific appendages.

Seismology

Every nuclear explosion, military or otherwise, is potentially valuable to seismologists in helping to refine models of the Earth's interior, both shallow and deep [2b]. The particular value is that the precise location, time and size of a simple source is known, whereas normally seismologists have to work with earthquakes, which are complex sources and whose location and time have to be inferred from the very data also being used to study the Earth's interior.

The United States almost always releases sufficient parameters of its nuclear tests for their seismological aspects to be capitalized on, and has occasionally given advance warning of a test, enabling special instrumentation to be deployed.

The Soviet Union does not reveal shot locations and times but does seem in the past few years to have been using several-kiloton blasts fairly regularly in conjunction with large seismic exploration projects studying the crust and upper mantle. These are included in figures 19.1 and 19.2.

III.   Encouragement and discouragement to peaceful-use programmes

The foregoing would suggest that peaceful-use nuclear explosions are an unmixed blessing. They certainly open up a wide variety of engineering and

296

Figure 19.1. An indication of the numbers of nuclear explosions used in testing and implementing peaceful-use concepts

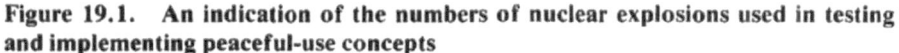

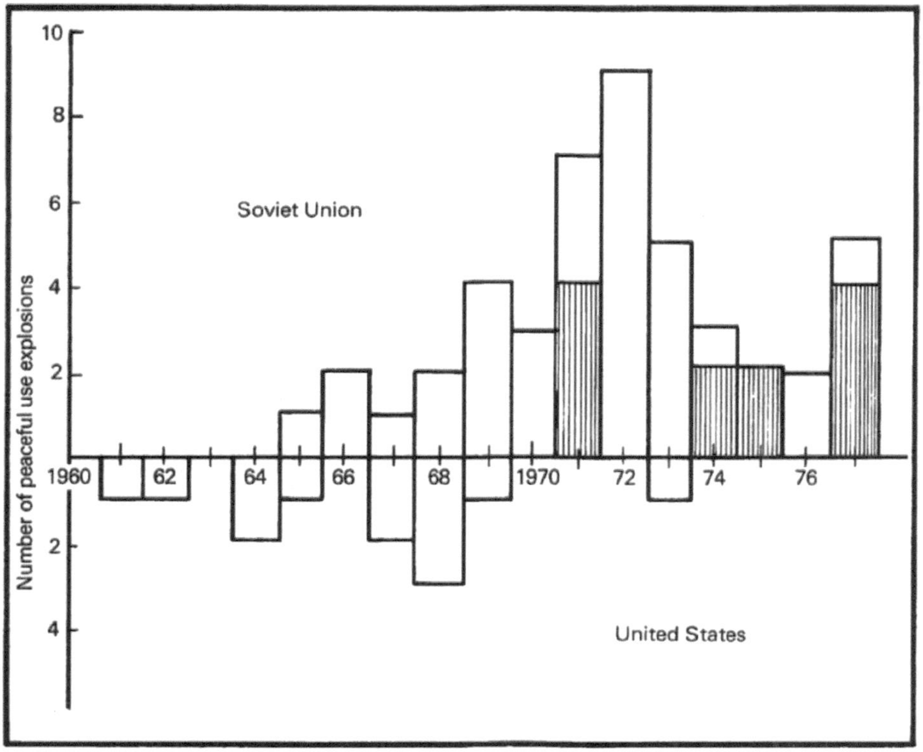

Note: Soviet numbers have been taken from lists of presumed nuclear explosions away from the two main weapons-testing sites in the Soviet Union. There is no way of being absolutely sure that all these events are nuclear, but their yields are mostly in the kiloton or tens of kiloton range. US numbers are based on announced peaceful-use explosions including those at the Nevada Test Site but excluding those described as 'heavy-element experiments', 'neutron-physics experiments' or 'device-development tests', as there is no way of making any comparison with any Soviet nuclear explosions of this nature. All but four of the US explosions were within the Nevada Test Site and purely experimental.

Within the Soviet total, a further cross-hatched subdivision marks those events believed to be for seismological purposes (see under Seismology). These events are often on long lines across Soviet territory (figure 19.2), and tend to be fired late in the day or at night, in contrast with almost all other explosions, military or non-military. Presumably this is because micro-seismic noise is lower at these times.

Sources: Official US announcements and seismic bulletins, largely collated in *Monitoring Underground Nuclear Explosions* [8].

scientific prospects and technically they must be deemed largely successful. Even the failures and disappointments doubtless contain lessons which could be used in future. But for all this, the enthusiasm has undoubtedly slackened; Project Plowshare is now terminated, and even the vigorous Soviet programme of the early 1970s has declined somewhat. Interest in peaceful-use explosions by other industrialized nations is very largely

297

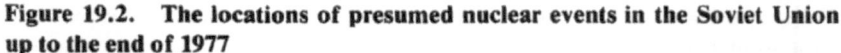

Figure 19.2. The locations of presumed nuclear events in the Soviet Union up to the end of 1977

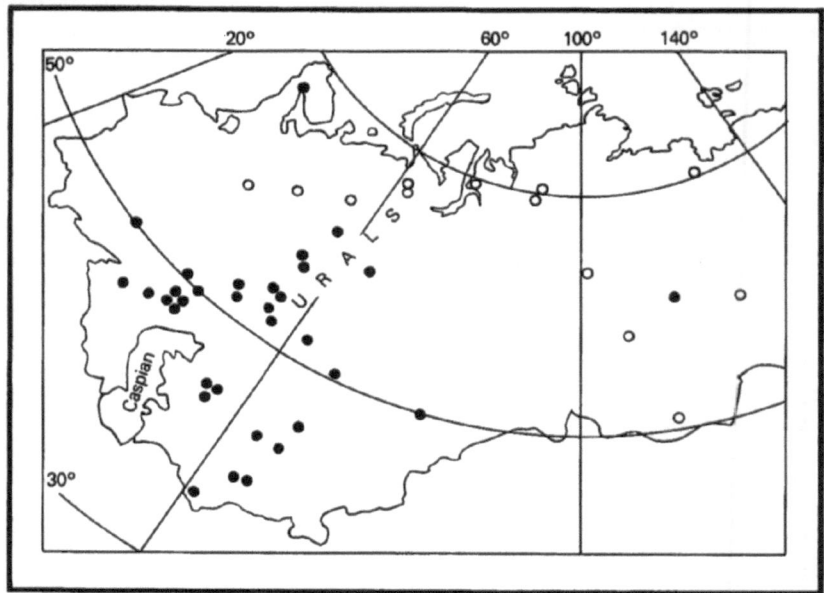

Note: The purpose of many explosions has been revealed in general terms without locations being given; it has usually been possible however to tie particular projects to particular locations. Explosions presumed to be for seismic investigation are shown as open circles. The purpose of more than ten of the events on this map is as yet unknown.

Source: Seismic bulletins largely collated in *Monitoring Underground Nuclear Explosions* [8].

limited to an attitude of wait-and-see, while the developing world, with the exception of India, has shown even less curiosity.

This global unconcern cannot be attributed to ignorance of the potential. The Non-Proliferation Treaty positively encouraged signatories to maximize benefits from peaceful explosions, saying that "Each Party to the Treaty undertakes to . . . ensure that . . . potential benefits . . . will be made available to non-nuclear-weapon states Party to the Treaty on a non-discriminatory basis and that the charge be as low as possible and exclude any charge for research and development". This is a clear green light to peaceful-use explosions and was followed by vigorous activity by the International Atomic Energy Agency (IAEA) to provide the sort of international framework that the Non-Proliferation Treaty called for ("Non-nuclear-weapon states . . . shall be able to obtain such benefits . . . through an appropriate international body"). Four major meetings on the subject have been held by the IAEA and their proceedings published [2, 4, 5, 6]. Other studies have also been conducted [1]. The meetings have seen the release of a substantial amount of data by the Soviet Union and the United States. And yet the IAEA has been approached very little by member states for assistance. Romania and Czechoslovakia have requested information on specific techniques in hydrocarbon recovery, Madagascar enquired about removing a rock in a harbour, and Egypt has made enquiries about an irrigation project in the Qattara depression. As far as is known, none of these enquiries has as yet led to anything definite.

States are not obliged to pass their requests through the IAEA, and it is known that Thailand made approaches to US consultant firms in 1972 for help in a pre-feasibility survey on a sea-level canal between the Gulf of Thailand and the Andaman Sea, that the US and Australia held some discussions in 1969 on a deep-water harbour in north-west Australia, and that a survey was made in the 1960s of the possibility of a new canal across the Central American isthmus [7]. Again none of this has come to anything.

There are several reasons why enthusiasm has waned. One is economics. In some circumstances the alternative of using more conventional techniques seems to make economic sense, although it must be said in defence of peaceful-use nuclear explosions that to expect new and complex technology to show a profit straight away would be unrealistic. In other circumstances, however, conventional techniques are plainly uncompetitive. This particularly applies in remoter parts of the Soviet Union where nuclear explosions are perhaps the only possible approach to some problems; this presumably explains in part the substantial difference in size between the Soviet and American programmes.

A second obstacle is changed public acceptance of nuclear activities. Although health and safety aspects of the work have been thoroughly pursued, and there are no grounds for believing peaceful-use explosions pose any special hazards, American experience has been that explosions outside the Nevada Test Site have met with fierce opposition even when placed on a remote island in the Aleutians. If other industrial countries were to embark now on domestic peaceful-use programmes, they could probably expect even fiercer opposition, as their population densities are higher than those of the United States.

A third obstacle to global acceptance of peaceful-use explosions, however, is growing awareness among politicians that it is impossible entirely to separate peaceful from military applications. The phrase 'peaceful-use explosion' is somewhat pedantically used here rather than the more common 'peaceful-purposes explosion' or just 'peaceful explosion', if only to serve as a reminder that it is most unlikely that the entire purpose of any nuclear explosion is non-military. Even if a 'peaceful' device is fired only as part of stockpile-testing of weapons, it is fulfilling a military function which cannot be monitored by an outsider. Peaceful-use explosions can doubtless fulfil other more complex military functions, too, without in any way looking suspicious. Further, there must be strong pressure for 'clean' devices to fulfil engineering applications—and for many purposes the device has to be as small and rugged as possible. Such pressures also exist on the military side, so development of peaceful-use devices is by no means a matter without relevance to weapons' programmes. With a fairly widespread desire at present among the nuclear powers to inhibit weapon testing, unbridled enthusiasm for peaceful uses is hardly compatible with attempts to put a lid on military nuclear activities.

Politicians in the developing world must have had another concern. The IAEA does not run its own nuclear-explosions enterprise, so any request for help from a developing country has eventually to be channelled to the nuclear powers. This would open up yet another way in which developing countries find themselves dependent on the two great powers. It would also help to justify the nuclear programmes of these two powers, by giving them a new altruistic aspect at a time when many non-nuclear weapon states are complaining bitterly that restraint on horizontal proliferation has had no response in restraint on vertical proliferation.

IV. The Indian explosion and horizontal proliferation

The 1974 Indian explosion has greatly helped to focus thinking on these matters. Until then, probably only a few politicians had given much thought to the dual purpose that peaceful-use devices could serve. But India's vigorous claims that the device was only peaceful must have opened many eyes. There is a certain irony in this, because India is probably one country that could profit considerably from nuclear explosives engineering; there now seems little prospect in the near future that she will be able to accumulate such benefits.

It seems unlikely that any other country will follow the Indian path, because world opinion was so unfavourable in 1974. If a country wanted to go the same way, it would probably first put its toe in the water, declaring that there were many interesting engineering projects it wished to pursue, and so it was beginning to accumulate the expertise for strictly non-military explosions; this would give other nations and international agencies a chance to react. It would seem more likely that any nation acquiring a military nuclear capability would proliferate vertically by a peaceful-use rationale, adopting an attitude of 'now we have a nuclear device to reassure us on security matters, we are turning our attention much more to bringing benefits to our people'. As already remarked, there is nothing to prevent any peaceful-use device having a parallel military purpose.

There is one other aspect of horizontal proliferation which is just worth a brief dismissive mention. There is little possibility of individual scientists and engineers in potential nuclear countries gaining access to valuable technological information either through an international agency such as the IAEA or by contact with the peaceful-use wing of a nuclear power. In the former case, no international agency is yet remotely in the business of making nuclear devices; in the latter we have to assume that it is overwhelmingly in the interests of nuclear powers not to let any foreign national have access to plans for nuclear devices, even if that foreign nation is host for a peaceful use experiment.

If horizontal proliferation were the only concern, this essay could stop here with a fairly bland assessment that peaceful-use explosions are still firmly in the hands of the United States and the Soviet Union, and that present non-nuclear nations are now unlikely to justify going nuclear for the civil benefits. But peaceful-use explosions do provide an excellent cover for military activities, so ways of inhibiting vertical proliferation must be examined in greater detail.

V. How international arms-control agreements affect peaceful-use explosions

We have already seen that the Non-Proliferation Treaty encouraged the peaceful benefits of nuclear explosions to be made more widely available. At the same time (late 1960s, early 1970s) many more non-nuclear weapon states were pressing regularly at the Conference of the Committee on Disarmament in Geneva for a comprehensive ban on testing of nuclear weapons to supplement the limited test ban of 1963. This could not come to much without cooperation from the United States and the Soviet Union and this only emerged slowly. Eventually, however, the United States and the Soviet Union agreed to certain limitations on underground testing. These were enshrined in a bilateral treaty, signed in 1974, putting a ceiling of 150 kilotons on any one test. The limit was not to apply until March 1976, and it was made clear that peaceful applications were excluded from the treaty and were to be the subject of a separate agreement.

This separate agreement emerged in May 1976 and was a document of immense complexity. The two states party to the treaty declared in the preamble that they desired "to assure that underground nuclear explosions for peaceful purposes shall not be used for purposes related to nuclear weapons" and desired "to develop appropriately cooperation in the field of underground nuclear explosions for peaceful purposes". They agreed that no individual explosion should exceed 150 kilotons, either fired on their own territory or provided for a third party. Further, any group explosion having an aggregate yield of more than 150 kilotons (which could include either well-separated devices such as would be used for canal-building or devices in very close proximity) had to be constituted of devices that separately had yields of no more than 150 kilotons, and facilities had to be provided for the non-firing party to the treaty to determine the yields of each separate device. The aggregate yield could not exceed 1.5 megatons. A detailed protocol laid down procedures for mutual notification and monitoring. The clear intention of the treaty was to allow the benefits of large peaceful-use explosions without permitting them to become the vehicle for large military tests forbidden by the earlier treaty. Nevertheless, there was a let-out clause that the questions of carrying out any individual explosion of yield greater than 150-kilotons "will be considered . . . at an appropriate time".

It should be noted that the treaty says nothing about using peaceful-use explosions, whatever the aggregate yield, for non-military purposes only, despite the phrase in the preamble giving that impression. The protocol of notification and modification concerns itself solely with precise determinations of yield to see that the 150-kiloton limit is not breached. Whether peaceful-use explosions do supplement military tests below this ceiling is not a concern of the treaty.

The 1974 and 1976 treaties have yet to be ratified by the US Senate, but certainly the 1974 treaty is observed in spirit, although it has come in for a lot of critical comment as an arms-control measure. Both could presumably be overtaken, however, by a comprehensive test-ban treaty in the near future.

This is no place to speculate on whether the present trilateral negotiations will end in a treaty and whether this treaty will gain wide adherence among other nations. But if a treaty does emerge, it presumably has to take note of peaceful-use explosions in one of the following ways:

1. by devising an exemption from the ban so that activity can continue as at present with nuclear powers free to fire and non-nuclear powers free to seek assistance,

2. by passing the responsibility for the provision of explosions on to an international agency, or

3. by banning all explosions, regardless of avowed purpose.

Option 1 would need detailed technical assurances that the device fulfilled no military function, and there is no way that this can be done. Even the information that a device had exploded and an estimate of its yield from standard seismological stations could be valuable. So this option is likely to be widely unacceptable.

Option 2 would be even more unacceptable if it opened up the prospect of an agency making its own nuclear devices, as this would inevitably expose a wider range of people to nuclear-explosives technology, as noted in section III. Nuclear powers would veto this on the grounds of horizontal proliferation, and although some non-nuclear powers might find attractions, there would be fierce opposition from others who would find access discriminatory. For this and a variety of other reasons, we must assume that an international agency cannot at present go into the explosives-manufacturing business.

There is one variant which cannot be so easily dismissed however: the agency could act as a stockpile for devices of a wide range of yields made by present nuclear powers. There would need to be many devices available at each yield, and they would have to be selected at random to fulfil requests, thereby masking the identity of the manufacturer of each individual device. This idea certainly gets around many objections, but it has a major stumbling block in that no nuclear power has yet, as far as is known, transferred control of a nuclear device to another state, let alone an international agency on which it had but one vote among many. Indeed, purely from the legal aspect, such a transfer would be in clear violation of Article 1 of the Non-Proliferation Treaty, although nuclear powers' objections would go well beyond legal considerations. There are probably also technical objections that devices off the shelf can fulfil only a restricted range of functions. We must take it that option 2 is a non-starter in any form, which leaves only option 3—no explosions of any sort.

It is commonly believed that it is on this basis that negotiations are proceeding, but as the Soviet Union does appear to value its peaceful-use programme, it is most likely that it will ask for a concession from other nuclear powers, that the treaty be of finite duration, probably three or five years, at the end of which negotiations are reopened.

Whilst it would be pleasing to regard three to five years as time enough for enthusiasm for testing to dissipate, this is no more than a pious hope. Between 1958 and 1961, a moratorium on testing of nearly three years ended in a vigorous bout of activity. So nuclear testing cannot be completely cleared from the arms-control agenda if a finite duration treaty is negotiated, and it must be expected that when the subject comes up again, the role of peaceful-use explosions, once a peripheral issue, will be central to discussions if only because the Soviet Union has been so active in them in recent years. It is exceedingly unlikely that any technical development in the next few years will enable us to monitor such explosions and assure that they have no military value whatsoever, so there are elements of confrontation ahead, with the Soviet Union demanding the right to indulge in non-military applications and the United States deeply suspicious of the uses to which explosions are put.

It should perhaps be noted in connection with this that although figure 19.1 shows that the Soviet peaceful-use programme is still active, a large part of it in recent years has been for presumed seismic investigations of the Earth's crust and mantle. Of course the information gained is valuable and possibly has economic benefit, but it is doubted that the information would be indispensable if the price were an arms-control treaty. Other peaceful use explosions seem to have tailed off.

VI. Conclusions

On the whole, peaceful-use explosions have been technically successful, even if they have made relatively little headway outside the Soviet Union.

And even in the Soviet Union it may be that enthusiasm has waned. What is clear, however, is that they offer a justification for horizontal proliferation and a cover for vertical proliferation. This is not primarily proliferation through the uncontrolled spread of fissionable materials—it is the result of attaching an acceptable label to a device which has less acceptable aspects to it.

What measures can be taken to dissuade nations from using peaceful-use explosions for military purposes? By far the greatest priority is that the United States and the Soviet Union set a good example themselves by avoiding ambiguity in these activities. There seems no way of doing this except by making the comprehensive test ban thoroughly comprehensive. They must then work hard to ensure that the ban is adhered to by as many nations as possible. Potentially nuclear nations could make it clear too, that they are not going to sign even a thoroughly comprehensive test ban treaty without some assurance that deadlock over peaceful-use aspects of the treaty will not force the treaty to be torn up in a few years time. The apparent decline in recent Soviet explosive engineering activities should be used to point the way to a more permanent phasing out of peaceful-use explosions.

Should anything be done in the Non-Proliferation Treaty environment? The review of the operation of the treaty in 1980 would seem a very suitable time to look at the peaceful-use explosion content. If by then a truly comprehensive test ban is in force, amendments will presumably have to be made to remove references to benefits of nuclear explosions being made available to all parties to the treaty. This would be a significant step forward.

VII. Tailpiece

Nothing has been said about China as a potential user of nuclear explosives in civil engineering since little is known about the subject. There is, however, scope for much speculation.

References

1. *'Nuclear Explosions for Peaceful Purposes'*, Report of an ad hoc advisory group (Chairman, A. R. W. Wilson) to the IAEA (IAEA, Vienna, 1977).
2. *Peaceful Nuclear Explosions II—Their Practical Application*, Proceedings of a panel on the peaceful uses of nuclear explosions, Panel Proceedings Series (IAEA, Vienna, 1971).
   (a) —, Nordyke, M. D., 'Other applications of PNE', pp. 223-32.
   (b) —, Thirlaway, H. I. S., 'Explosion seismology', pp. 275–81.

3. Nordyke, M. D., '*A Review of Soviet Data on the Peaceful Uses of Nuclear Explosions*', Lawrence Livermore Laboratory Report No. UCRL-51414 (Lawrence Livermore Laboratory, 1973).
4. *Peaceful Nuclear Explosions IV*, Proceedings of a technical committee on the peaceful uses of nuclear explosions, Panel Proceedings Series (IAEA, Vienna, 1975).

   (a) —, Toman, J., 'Project Rio Blanco. Pt II: Production test data and preliminary analysis of top chimney/cavity', pp. 117−40.
5. *Peaceful Nuclear Explosions—Phenomenology and Status Report 1970*, Proceedings of a panel on the peaceful uses of nuclear explosions, Panel Proceedings Series (IAEA, Vienna, 1970).
6. *Peaceful Nuclear Explosions III—Applications, Characteristics and Effects*, Proceedings of a panel organized by the IAEA, Panel Proceedings Series (IAEA, Vienna, 1974).
7. Bolt, B. A., *Nuclear Explosions and Earthquakes* (Freeman, San Francisco, 1976).
8. Dahlman, O. and Israelson, H., *Monitoring Underground Nuclear Explosions* (Elsevier, Amsterdam, 1977).

Acknowledgements

Technical information supplied by Georges Delcoigne (IAEA) and Peter Marshall (UK Ministry of Defence, Blacknest) is gratefully acknowledged. The views expressed are entirely those of the author.

Paper 20. Technical aspects of peaceful nuclear explosions relevant to their possible role in proliferation of weapon-usable nuclear materials

A. R. W. WILSON

Square-bracketed numbers, thus [1], refer to the list of references on page 317.

I. Introduction

The concept of peaceful nuclear explosions (PNE) has proved a complicating and disruptive factor in non-proliferation and arms control discussions and negotiations. One of the problems has been the tendency of states to base their arguments on different technical assessments. The willingness of states to cooperate in measures to limit the dangers which uncontrolled involvement in PNE activities would present for the international community, will depend in part upon their perception of the potential value of PNE and of the risks associated with their development. An Ad Hoc Advisory Group to the Board of Governors of the International Atomic Energy Agency (IAEA) has recently developed, and adopted by consensus, a report on the procedural, legal, health and safety economic aspects of PNE and other related issues [1]. The work of the Ad Hoc Advisory Group, in which 41 member states of the IAEA participated, has provided the first broadly based international assessment of PNE. This paper presents, in the context of a series of questions and answers, the author's[1] understanding of the Ad Hoc Advisory Group's views on the safety, technical feasibility and economic viability of PNE applications together with his appreciation of weapons-related technical aspects. It is intended to serve as a concise status report for those interested in the nuclear weapon proliferation implications of PNE.

Why are peaceful nuclear explosions of interest?

Nuclear explosives are relatively cheap sources of explosive energy. US figures [2] suggest that at low yields they offer a cost advantage of the order of 10 over the equivalent amount of the chemical explosive trinitrotoluene (TNT). At high yields the cost advantage is of the order of 1 000. Since energy released explosively can be used to break and move earth materials,

---

[1] The author was the chairman of the Ad Hoc Advisory Group.

nuclear explosives may provide a means of achieving significant cost savings in civil engineering projects.

The other particularly useful feature of nuclear explosives is their compactness. Because they can be designed to fit into boreholes constructed with conventional large-bore drilling equipment, it is possible to contemplate exploding them within, or close to, hydrocarbon reservoirs and mineral deposits. Resource-directed PNE applications seek to utilize the resultant displacement and fracturing of the adjoining strata to facilitate production from the reservoir or deposit.

II. PNE applications

What sort of PNE applications are envisaged?

All the applications considered to date only involve underground explosions.

Cratering explosions (that is, explosions at shallow depths which rupture the Earth's surface) have been considered as a potential means of excavating canals, harbours, water storages and railway cuttings and as a means of removing overburden in open-cut (strip) mining.

Contained explosions (that is, deeply buried explosions which produce rock-filled cavities) have been seen, *inter alia*, as offering potential for stimulating natural gas and oil flows, for facilitating *in-situ* leach mining of minerals and *in-situ* retorting of oil shales, and for fracturing deep granitic layers with a view to the recovery of geothermal heat. Contained explosions have also been used for scientific purposes.

What PNE research and development information is available?

Since PNE research and development involves the design, production and explosion of nuclear devices along with investigations into explosion effects, only the nuclear weapon powers have been able to engage in comprehensive PNE programmes. The USA and the USSR alone have chosen to do so.

The US PNE programme began in 1957 and government funding of the programme over the next two decades amounted to some $160 million. During the late 1960s and the early 1970s, industry contributed both funds and technical expertise. Both the US government and industry have now ceased all funding [3]. The US investigations into explosion effects and ways of utilizing them have been well reported in the scientific literature. Most explosive-related information has been withheld from publication.

The Soviet PNE programme has been at least comparable in magnitude to that of the USA and has encompassed a wider range of test projects. Information made available through IAEA technical panels [4] has provided a broad insight into the Soviet programme along with specific details of some of the projects. The programme appears to be continuing, although the current level of research and development activity has not been reported in the literature.

III.   The health and environmental risks of PNE

What are the public health safety hazards of PNE?

The main public safety considerations associated with cratering explosions are the radiation from fall-out, the possibility that fall-out might contaminate human food chains, ground shock and air blast. For contained explosions, they are the possibility that the residual radioactivity might contaminate aquifers, the possibility of product contamination and ground shock. As all applications considered to date involve underground explosions, thermal hazards are not an important concern.

Could cratering projects be undertaken within current guidelines for the radiation exposure of members of the public?

Most of the radioactivity ejected by a cratering explosion is trapped in the rubble and fall-back. The fraction which escapes as airborne particulate is deposited downwind as fall-out, the deposition concentration decreasing with distance from the explosion. For radiological safety purposes, the amount of fall-out at a particular location is characterized by the external radiation dose it would deliver to an individual spending a lifetime out of doors at the location and present from the time of deposition.

For PNE planning purposes, it has customarily been assumed that to be consistent with the most widely accepted international guidelines for radiation exposure, cratering projects must be planned on the basis that no member of the public receive a total radiation exposure in excess of 5 mSv (millisievert) from all explosions involved in the project. By using nuclear explosives and emplacement techniques designed to minimize the amount of radioactivity bcoming airborne, single cratering explosions of any realistic yield could now be conducted without causing fall-out levels in excess of 5 mSv beyond a downwind distance of 10 km. Since cratering projects are unlikely to be contemplated in other than sparsely populated areas, only very large excavation projects involving multiple and/or repeated cratering explosions are likely to require the evacuation of significant numbers of members of the public in order to hold their radiation exposures to the limits used for planning purposes to date.

The cost—benefit equation for a cratering project should take full account of the radiation doses which might be received by members of the public. The total population exposure should be seen as a cost even if the individual doses were fully consistent with international guidelines.

Are the environmental consequences of PNE applications predictable?

Particular radionuclides in the fall-out may be transferred to humans through various geochemical and biological chains. The likely extent and significance of any such transfer must be investigated on a project-to-project basis. Laboratory and field techniques for assessing the fate of the radionuclides in the fall-out are sufficiently developed to permit realistic predictions of the potential internal radiation dose to humans through such pathways.

The main environmental concern with contained explosions is the possibility that the residual radioactivity may contaminate aquifers. Because of the low flow rates in aquifers, only long-lived radionuclides are of concern. Of these, most of the biologically significant are held in the cavity in relatively non-leachable form; moreover, their transport is greatly slowed by ionic adsorption on clays. Tritium, which is readily available for transport and whose exchange with the rock matrix is negligible, is the only radionuclide likely to appear in significant concentrations far from the explosion point. Its significance requires careful evaluation, but the degree of tritium migration to be expected under average conditions is unlikely to be a threat to ground water supplies at any reasonable distance from the explosion site.

Product contamination is another environmental complication involved in many resource-recovery PNE applications. The demonstrated contamination of the produced gas with residual tritium from the explosion is a major consideration in assessing the case for using PNE to increase the productivity of gas wells. Various measures such as reduction of residual radioactivity by improvements in design of nuclear explosive devices, flushing and dilution can be adopted to reduce the level of product contamination. The acceptability to the consumer of a contaminated product and the incremental contribution of the product to both individual and population exposure constitute important factors in the cost—benefit equation.

The successful development of nuclear excavation would highlight another more general environmental concern. All large-scale civil engineering projects result in a number of long-term environmental effects. Such projects must be approached with considerable caution until more experience in assessing their environmental impact is available. This general

310

environmental reservation applies particularly to the scale of project likely to be facilitated by the successful development of nuclear excavation.

Any assessment of the potential environmental consequences of PNE must be project-specific. Moreover, because of the longer time-scale and the greater diversity of effects involved, the environmental consequences cannot be predicted with the same degree of confidence as can other hazards.

What limitations do the ground motion and air blast hazards place on the use of PNE?

The ground motion associated with an underground nuclear explosion can cause damage to buildings and structures up to distances of some kilometres for small explosions and some tens of kilometres for large explosions. Reliable methods have been developed for predicting both the intensity of the ground motion at locations of interest and the likely responses of structures to such ground motion. Since the safety of individuals can be assured, if necessary by evacuation of buildings at risk, the primary interest lies in assessing the damage cost which the project would have to carry. Ground motion usually imposes the main safety constraint on a contained explosion and is often more limiting than radioactivity dispersal or blast for cratering explosions.

Concern is sometimes expressed that the ground motion from a nuclear explosion may trigger a major earthquake. While this is a theoretical possibility, the chance of its happening is small. It can be discounted because the amplitude of the ground waves from even large nuclear explosions at seismically significant depths is small, compared with the wave amplitudes associated with frequently recurring minor earthquakes. Indeed, if the situation were otherwise, nuclear explosions would provide a useful means of releasing strain energy.

While the air blast resulting from contained explosions is trivial, direct propagation of the air blast from a cratering explosion can cause minor structural damage at distances out to some tens of kilometres. The intensity of the air blast at specific locations can be predicted with adequate accuracy to judge the acceptability of the proposed project and to implement any necessary precautions. When the meteorological conditions are unfavourable, the refraction and focusing of the blast waves in the upper atmosphere can cause minor damage at even greater distances. Damage by refracted waves can be minimized by timing explosions to avoid unfavourable meteorological conditions.

Ground motion and air blast can be regarded essentially as hazards to property, the cost of which must be reflected in the economic analysis of project viability.

IV. The technical and economic feasibility of PNE

Has the technical feasibility of any application been established?

Before it can be claimed that the technical feasibility of a new excavation method has been established, the method must have been widely employed for major projects and shown to allow the completion of civil works which perform satisfactorily over their planned operating lifetime. Similarly, to be considered technically feasible, a new mineral-recovery method must have been widely employed on an industrial scale and shown to allow satisfactory mineral recovery over the planned economic lifetime of the project.

On the basis of this criterion, no excavation or mineral-recovery PNE application can be said to have been taken anywhere near the stage of development where its technical feasibility, let alone its economic viability, has been established. Major technical reservations apply, even to those applications which have been the subject of most research and development.

Thus, while the USSR has demonstrated that nuclear explosions can be used to raise a rubble barrier across a watercourse to create a large reservoir, it has yet to be demonstrated that the barrier will remain stable and impermeable over the economic lifetime of the reservoir. The stability of crater walls, particularly under wave and wind action, is also one of the major technical uncertainties attached to the possible use of nuclear explosions for harbour and canal construction.

In the mineral-recovery area, gas stimulation has been the application which has received most attention to date. Whilst it has been shown that nuclear explosions can be used to stimulate gas wells, the technical feasibility of achieving the interconnection between the fracture systems created by individual explosions, which is needed to make the application economically relevant in the USA, has yet to be demonstrated.

The situation is as much a reflection of the relatively modest effort which has been directed to PNE research and development, as it is a consequence of the technical difficulties encountered. For most applications, the technical difficulties are not such as to preclude the possibility that, with further research and development, the relevant applications will be shown to be technically feasible. However, it must be remembered that even if proven technically feasible, PNE applications will be of no more than academic interest if they cannot be shown to be economically viable.

**What are the prospects that PNE applications will be found
to be economically viable?**

To be judged economically viable, PNE applications must pass a two-stage test—viz.: (a) the use of nuclear explosions in the particular circumstances of the project must be shown to offer significant cost savings over conventional techniques; and (*b*) the sum of the economic benefits of the PNE executed project must be shown to exceed the sum of the economic costs, taking appropriate account of factors such as community values which cannot necessarily be expressed in monetary terms.

Nuclear excavation is most cost-effective for very large civil works such as canal construction or the excavation of channels in hard rock. Cost analyses of two potential canal projects [5, 6] suggest that large savings might be achieved by the use of nuclear explosives. However, the possible remedial costs associated with the technical unknowns are comparable with, if not larger than, the possible cost savings. The economic viability of nuclear excavation, even in the favourable circumstances of very large projects, seemingly depends on the satisfactory resolution of technical problems. Even if the technical problems are satisfactorily resolved, it remains to be seen whether the community costs such as temporary evacuation, radiation exposure and environmental damage will be judged acceptable.

Technical uncertainties also cloud the picture so far as contained applications are concerned. Most attention has been directed to the possibility of using nuclear explosions to produce gas from 'tight' reservoir formations. An alternative technique for stimulating gas flow—massive hydraulic fracturing (MHF)—is also under development. MHF appears likely to prove more cost-effective than PNE. However, because neither method has yet been shown to be technically feasible, large uncertainties attach to the likely cost-effectiveness of both. The range of uncertainty is such that the apparent cost advantage of MHF may be illusory. For the present, neither method can be said to offer much prospect of producing gas at today's prices. Nevertheless, as readily recoverable gas resources are depleted, gas prices must rise, thus increasing the relevance of stimulation methods. In the end it may be that the critical element in the cost–benefit equation will prove to be the public acceptability of gas slightly contaminated with radioactivity and of the repeated ground shocks associated with the stimulation of a single gas field. Most other mineral recovery applications are even less technically developed and to that extent the cost uncertainties are even greater.

The contained application which can be approached with the greatest cost certainty is the use of PNE to create underground cavities for the storage of natural gas under pressure.

The USSR has operated successfully for gas storage a cavity created in a salt formation by a contained nuclear explosion. It has not been established that gas-tight cavities can be created in more widely occurring media by contained explosions, but the available evidence suggests that some low-permeability sedimentary deposits should prove suited to this application. The cost situation *vis-à-vis* alternative storage methods is strongly dependent on the site considered. Although detailed costing of the Soviet experience is not available, paper studies suggest that there are likely to be some sites, particularly off-shore, where PNE cavities would provide cheaper storage than alternative methods. However, the cost-saving projections are not such as to provide gas-supply operators with any great incentive to face the remaining technical uncertainties, the probable public reaction to nuclear explosions close to demand centres and the possibility of slight radioactive contamination of the stored gas.

What sort of numbers of nuclear explosion devices might be involved in the commercial industrial application of PNE?

Major excavation projects such as canal projects might each involve the explosion of a total of some hundreds of nuclear devices. However, the number of major projects where nuclear excavation might be used is likely to be small. Moreover, it seems unlikely that nuclear excavation will be widely used for the construction of more conventional works such as reservoirs and harbours. Accordingly, nuclear excavation could involve the explosion of some tens to perhaps two hundred nuclear devices annually.

Natural gas stimulation and *in-situ* oil shale retorting are the contained applications which, if successfully developed and widely applied, would probably involve the greatest number of explosions annually. To sustain a 5 per cent increase in US natural gas production through PNE stimulation would apparently require of the order of 1 000 explosions annually [7].

Their use on such a scale would seriously increase the risk of nuclear explosive devices being stolen or involved in accidents.

V. The relation between PNE and nuclear weapon proliferation

Are cratering explosions allowable under the Partial Test Ban Treaty (PTBT)?

A strictly technical interpretation of the language of the PTBT suggests that it was not the intention of the drafters to preclude cratering explosions, if these could be conducted within the restriction that they should not cause "radioactive debris to be present outside the territorial limits of the State under whose jurisdiction or control such explosion is conducted" [8]. The possibility that a cratering explosion can be conducted without any radioactivity, no matter how trivial, crossing territorial borders is remote, although it cannot be technically excluded. Large cratering projects such as canal excavations will almost certainly result in appreciable amounts of radioactivity crossing territorial borders, whereas small cratering explosions conducted deep inside a state's territory may involve only trivial amounts crossing borders. The USSR, which is a depositary state for the PTBT, appears to be continuing its cratering programme on the assumption that the small amounts of radioactive debris which might cross its territorial borders as a consequence of small cratering experiments conducted deep inside its territory, would not constitute a violation of the treaty.

Do the nuclear explosive devices used for PNE differ from those used as weapons?

The efficient use of a nuclear explosive device requires that its design characteristics be optimized to the critical requirements of its intended use. Many of these requirements differ significantly, not only as between weapon use and use for PNE applications in general, but also as between the various specific PNE applications.

The shape, size and weight of nuclear weapons will be optimized to the characteristics of the delivery vehicle, whereas for a PNE device, of these parameters only the diameter, which directly affects the cost of drilling the emplacement hole, is likely to be of critical importance.

Ruggedness requirements will also differ. An ability to withstand high accelerations and to remain reliable under the conditions of a service environment are two obvious requirements for some military uses. A PNE device on the other hand may have to be able to withstand the high temperatures and pressures associated with emplacement at depth.

The amount and nature of the radioactivity produced by the explosion is likely to be a much more critical consideration in the design of PNE devices than in the design of weapons. For gas and oil stimulation, it is important that the amount of residual tritium be as small as possible. For excavation, low total residual radioactivity is a critical requirement.

A highly predictable yield and low production cost are other potential design requirements for some PNE applications.

The design of PNE devices is thus both highly sophisticated and highly specific to their intended use.

Can PNE devices be used as weapons?

Despite their sophistication and specificity, PNE devices share the fundamental weapon characteristics of being both transportable and able to release amounts of explosive energy which can cause widespread destruction. Moreover, PNE devices necessarily include both fission and fusion explosives and the range of yields relevant to PNE applications spans that of interest for tactical and strategic nuclear weapons. Thus, any PNE device must be regarded as a potential weapon, although with major limitations on its deliverability and other important weapons-related characteristics. Nor is there any basis for believing that a PNE device incapable of use as a weapon can be developed.

Can a non-nuclear weapon state acquire a PNE development capability without upgrading its weapon development capability?

Even the first step along the PNE development road can contribute significantly to a state's weapon development capability, since the testing of even a primitive device can be used to confirm the underlying design concepts and calculations. It is to be expected that the process of improving nuclear explosive devices in the context of a development programme for PNE devices will correspondingly improve the state's ability to design and produce nuclear weapons.

Since PNE devices are highly sophisticated nuclear explosives, an ability to design and produce them certainly implies an ability to design and produce advanced nuclear weapons. The point has already been made that PNE devices, although not optimized to weapon use, can be employed as weapons.

Irrespective of the restrictions placed on a state's ability to acquire explosive-grade nuclear material for other than peaceful purposes, limitations on a state's weapon development capability obviously represent an important defence against nuclear weapon proliferation.

Would a PNE exclusion clause subvert the purpose of a comprehensive test ban treaty (CTBT)?

To the extent that it allowed nuclear weapon states to gain information relevant to upgrading or maintaining the effectiveness of their nuclear arsenals, an accommodation for PNE would be incompatible with the objectives of a CTBT. Whether controls can be developed which would allow nuclear weapon states to continue their involvement in the provision of nuclear-explosive-related services without infringing this limitation is an open technical question. Certainly any further development of specialized PNE devices would have to stop and means would have to be found to ensure that any nuclear explosive device used for PNE testing conformed to frozen designs. In particular, measures would be necessary to prevent the substitution of stockpiled weapons for PNE devices as a means of testing their continued reliability. Moreover, strict supervision of PNE tests would be essential to ensure that they were not used to obtain weapon-effects information which could reflect back on the design or use of existing nuclear weapon stockpiles. Even if these problems could be solved, a PNE accommodation would remain open to the criticism that it allowed the maintenance of skills and facilities which could be misused.

VI. Concluding observations

The technical facts clearly establish that unless adequately controlled, PNE can considerably increase the risks of horizontal nuclear weapon proliferation and compromise efforts to limit vertical proliferation. While the research and development carried out to date suggests that the safety problems are manageable, it has not gone far enough to establish whether any application will be technically feasible and economically viable or to discount the possibility that some applications may yet prove so. PNE research and development activity has declined over recent years, but diminishing resources and increasing population pressures may yet provide the momentum for further investigations. In the light of this situation, the sensible course would seem to be to promote open informed discussion of PNE issues with a view to developing an international arrangement which would avoid the proliferation risks posed by PNE, ahead of any resurgence of interest in PNE use.

References

1. *Nuclear Explosions for Peaceful Purposes*, Report No GOV/1854, IAEA Board of Governors (Restricted distribution), 1976.
2. Frank, W. J., *Characteristics of Nuclear Explosives; Engineering with Nuclear Explosives*, University of California, USAEC Report No. TID-7695 (USAEC Division of Technical Information Extension, Oak Ridge, Tennessee, 1964), pp. 8–9.
3. US Arms Control and Disarmament Agency, *The American Experience with Peaceful Nuclear Explosions*, Report No. 0-261-134 (US Government Printing Office, Washington, D.C., 1978).
4. *Peaceful Nuclear Explosions I, II, III, IV and V*, Panel Proceedings Series (IAEA, Vienna, 1970, 1971, 1974, 1975 and 1978).
5. Atlantic Pacific Interoceanic Canal Study Commission, *Interoceanic Canal Studies 1970*, Report No. 0-410-974 (US Government Printing Office, Washington, D.C., 1971), pp. 33–46.
6. Nordyke, M. D., *A Review of Soviet Data on the Peaceful Uses of Nuclear Explosions*, Lawrence Livermore Laboratory, Report No. UCRL-51414 (Lawrence Livermore Laboratory, 1973).
7. Rubin, B., Schwartz, L. and Montan, D., *An Analysis of Gas Stimulation using Nuclear Explosives*, Lawrence Livermore Laboratory Report No. UCRL-51226 (Lawrence Livermore Laboratory, 1972).
8. *Treaty Banning Nuclear Weapons Tests in the Atmosphere, in Outer space and Under Water*, United Nations Treaty Series, Vol. 480, New York, 1963.

Chapter 12. Reactors in satellites

Paper 21. Nuclear reactors in satellites

D. PAUL

Square-bracketed numbers, thus [1], *refer to the list of references on page 331.*

I. Introduction

At present the nuclear reactors in space are power generators designed to supply the electrical needs of instruments carried aboard the spacecraft. Although nuclear reactors could also be used as energy sources for propelling spacecraft, this second class of devices will not be discussed.

The following statement has been extracted from a UN document:

The importance of nuclear power sources in providing the energy for space programmes may be ascribed to their considerable potential advantages over other power sources. These include: self-contained operation and independence of the amount of illumination in a given region . . . , compact construction . . . simplifying the orientation system and greatly reducing the energy required to keep the vehicle in its orbit; high stability in the radiation belts of the Earth; improved weight and size characteristics, beginning with certain levels of power, as well as the possibility of improving the specific mass−energy characteristics with increased power and [assured] life [1].

There is little connection, if any, between the launching of satellites which bear nuclear reactors into space orbits and horizontal proliferation, that is, the increase in the number of states or groups which possess nuclear weapons. However, the capability of such satellites to assist in such tasks as military targetting could be used as an argument by the major powers for needing more nuclear weapons. In addition, such satellites contribute automatically to the qualitative arms race, although whether they do this in such a way as to increase or decrease the chance of nuclear war is debatable. Since the arms race is continuing both quantitatively and qualitatively in many directions, and there are alternatives to nuclear reactors for powering Earth satellites, this paper does not directly discuss the influence of such reactors on vertical proliferation, as this must be a small perturbation on the overall military situation.

Two nations, the Soviet Union and the United States, have been responsible for all of the nuclear fission reactors so far launched into space. The Soviet Union has launched many such reactors, while the United States has launched only one (Snap-10A), and has preferred since then to use a radioisotope as a nuclear heat source for the production of electric energy. However, the United States Department of Defense has not abandoned nuclear reactors as possibilities for future space missions requiring greatly increased electric power [2]. The differences in technologies employed by the two nations are presently overshadowed by the differences in the availability of information from them.

II. The development and use of Snap devices by the USA

A summary of nuclear power systems launched spaceward by the USA for the years 1961−77 is given in table 21.1. Of the large number of launchings, only Snapshot carried a nuclear fission reactor. The other Snap devices, Transit RTG and MHW, are radioisotope packages in which the heat from radioactive disintegration is used to produce electric power.

Snap-10A

The objectives of the Snap-10A programme were to develop, make, test and deliver to the US Air Force for flight test, a 500-watt (electrical) reactor−thermoelectric system having a one-year life. The flight test was intended to achieve public acceptance of nuclear reactors in space by launching, with all necessary safety approvals, a simple forerunner to larger, more complex nuclear systems. To do this, the reactor had to develop 50 kW(t) (thermal kilowatts). The reactor used was a zirconium hydride/uranium-235 type cooled by sodium−potassium liquid metal which transports the heat to the thermoelectric hot junction. The cold thermoelectric junction is cooled by radiation into space.

Snapshot was launched at 1324 hours PST (Pacific Standard Time), on 3 April 1965, being essentially non-radioactive at the time apart from the weak natural radioactivity of uranium-235. It achieved criticality at 2315 hours that evening and was operating at full power two and a half hours later. Six days later the reactor control system was deactivated by ground command, leaving the reactor in operation, subject to its own inherent control ('negative temperature coefficient'). The reactor ran for 43 days of power operation and was prematurely shut down because of a failure of the voltage regulator in the spacecraft.

On shutdown the system contained an inventory of radioactivity of 2×10^5 curies,[1] which will have decayed to 100 curies after 15 years (1980)

---

[1] One curie is a unit of rate of radioactive disintegration and is equal to 3.7×10^{10} disintegrations per second.

Table 21.1. Summary of space nuclear power systems launched by the USA, 1961–77

System (electrical power)	Mission	Launch date	Status
Snap-3A (2.7We)	Transit-4A	29 Jun 1961	Successfully achieved orbit
Snap-3A	Transit-4B	15 Nov 1961	Successfully achieved orbit
Snap-9A	Transit-5BN-1	28 Sep 1963	Successfully achieved orbit
Snap-9A	Transit-5BN-2	5 Dec 1963	Successfully achieved orbit
Snap-9A	Transit-5BN-3	21 Apr 1964	Mission aborted: burned-up on re-entry
Snap-10A reactor	Snapshot	3 Apr 1965	Successfully achieved orbit: reactor now shut down
Snap-19B2	Nimbus-B-1	18 May 1968	Mission aborted; heat source retrieved
Snap-19B3	Nimbus-111	14 Apr 1969	Successfully achieved orbit
Snap-27	Apollo-12	14 Nov 1969	Successfully placed on lunar surface
Snap-27	Apollo-13	11 Apr 1970	Mission aborted on way to Moon. Heat source returned to South Pacific Ocean
Snap-27	Apollo-14	31 Jan 1971	Successfully placed on lunar surface
Snap-27	Apollo-15	26 Jul 1971	Successfully placed on lunar surface
Snap-19	Pioneer-10	2 Mar 1972	Successfully operated to Jupiter
Snap-27	Apollo-16	16 Apr 1972	Successfully placed on lunar surface
Transit-RTG	Transit	2 Sep 1972	Successfully achieved orbit
Snap-27	Apollo-17	7 Dec 1972	Successfully placed on lunar surface
Snap-19	Pioneer-11	5 Apr 1973	Successfully operated to Jupiter; on to Saturn
Snap-19	Viking 1 & 2	20 Aug & 9 Sep 1975	Successfully launched; on Mars
MHW	LES 8/9	14 Mar 1976	Successfully achieved orbit
MHW	Voyager 11	20 Aug 1977	Successfully launched
MHW (475We)	Voyager 1	5 Sep 1977	Successfully launched

All nuclear devices listed are RTGs except Snap-10A.

Source: Reference [3].

and 0.1 curie after 100 years. Snapshot was placed in a 4 000-year orbit before reactor start-up, so that the level of radioactivity from fission fragments will be insignificant when the satellite re-enters the atmosphere.

Re-entry into the atmosphere—history and policy

One of the decisions which has to be made in planning satellite missions involving ultimate re-entry is whether the satellite should be designed to burn up in the stratosphere or whether it should re-enter intact. Early US experience with Snap-9A in the Transit-5BN-3 flight (which failed to achieve orbit) indicated complete burn-up in the stratosphere, as planned. However, the long-term problem of contaminating the stratosphere with successive re-entries led to a decision that RTGs (radioisotope thermo-electric generators) and reactors should re-enter intact. Subsequent re-entries, Nimbus B-1 and Apollo 13 (table 21.1), left the RTGs intact, as was verified by the recovery of the Nimbus RTG from the Santa Barbara channel and post-re-entry surveys of the Apollo-13 RTG which now rests in the four-mile deep Tonga ocean trench. No radioactivity was released to the atmosphere or the ocean.

Future trends in nuclear power for space vehicles

The low ratio of available electric power to the total rate of heat energy generation, in Snap-10A for example, reflects two factors: (*a*) the narrow relative temperature interval of the thermal cycle of the reactor, which results in low thermodynamic efficiency, and (*b*) the low efficiency of thermoelectric devices of that period. An obvious combination of routes towards higher electrical output without creating greater radioactive containment problems is to concentrate on the improvement of efficiencies and of the overall electric power output per unit mass of the generator (figure 21.1). The current US programmes on thermoelectric generation are directed towards more efficient conversion of radioisotope heat energy in RTGs.

Radioisotope thermoelectric generators

All except one of the nuclear power packs listed in table 21.1 fall into this category. As their name implies, they consist of a radioisotope suitably packed and coupled to a system of parallel thermoelectric generators. The isotope which has become established for the purpose is plutonium-238.[2] "This isotope has become the workhorse heat source for the majority of systems: past, present or future. Its 87.7-year half-life and its alpha emission decay scheme combine to enable design and development of space power supplies with light weight and long term operational reliability" [3]. Figure 21.2 shows the decay scheme of plutonium-238. Regarding the trends towards increased electric power outputs and efficiencies of RTGs, the current goal of 1−2 kW(e) for flight test in 1982−83 seems quite realizable and should obviate the need for further nuclear fission reactors in space in the near future.

Re-entry of RTGs

The intact re-entry policy adhered to by the USA renders the re-entry of RTGs and reactors about as dangerous as the re-entry of a similar sized vehicle not containing radioactive material. The safety tests to which RTGs have been subjected are very comprehensive and the procedures are outlined in *Use of Nuclear Power in Space by the United States of America* [3]. For

---

[2] The even-numbered plutonium isotopes are not weapon material.

Figure 21.1. Electrical power per kilogram of RTG mass

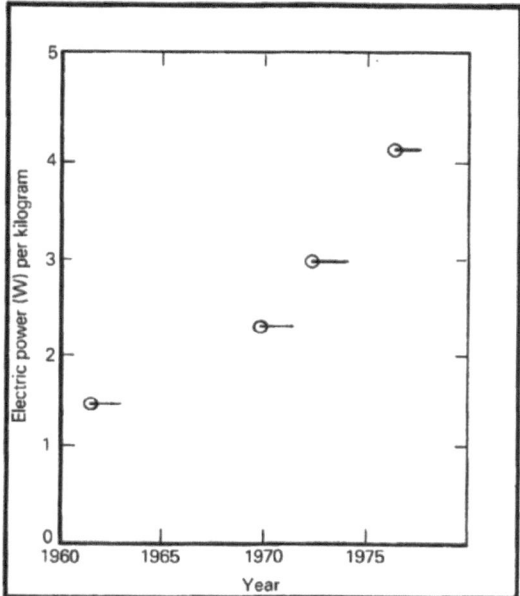

Note: The above figure shows the specific power in electrical watts output per kilogram of RTG mass for the US Snap and MHW devices. The dates are the first launchings into space [3].

arbitrary re-entry of a body of effective cross-sectional area A to a planetary surface of area A_p, in which the population of living beings L is N_L, the probability of death or injury, P, is given by

$$P = \frac{N_L A}{A_p} \lambda \tag{1}$$

for small angles λ, where λ is the angle of inclination of the orbit to the equator, and where it is to be understood that the area A includes allowance for the size of the struck being and possible perimetric effects at the point of impact. For an effective area A of 12 square metres and a world population of 4×10^9 people, the chance of injury is about one per 10 000 re-entries. In a future with a quadrupled world population and 1 000 such re-entries per year, the number of expected deaths would still be less than one per annum on average, that is, too low to be recorded in typical statistical records, but nevertheless non-zero. These estimates assume that people are spread over the Earth's surface, so that the probability of several people being struck by a single re-entering satellite is negligible. For an effective area 10 times larger, 120 square metres, the number of re-entries would have to be held to approximately 1 200 each year to keep accidents below 1 per annum out of a 16×10^9 world population.

The number of people killed by lightning in Canada in 1975 and 1976 was 12 people each year, or about one per million per two years. Clearly the expected re-entry casualties are several orders of magnitude lower than this.

Figure 21.2. The decay scheme of plutonium-238

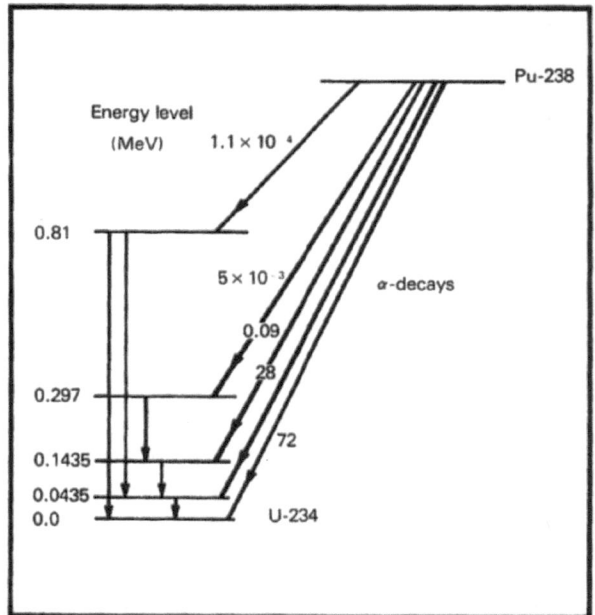

Note: The decay scheme of plutonium-238, showing the five strongest α-decay branches, and their percentages alongside the inclined arrows. All other α-decay branches are much weaker than 10^{-4} per cent and have been omitted for clarity. The α-particles will be entirely absorbed within the RTG, mainly within the plutonium itself, and thus give rise to the predominant source of heat. The gamma-ray decays which follow four of the five α branches are shown by vertical arrows. The lower energy gammas will be strongly absorbed in the plutonium, and only the radiation from the 0.81 MeV level of U-234 will have an appreciable probability of escape. Since this radiation occurs only once in about 10^6 decays, the radiation observed externally to a strong plutonium-238 source is relatively feeble [9].

The comparison with lightning is useful because lightning is not generally considered a major menace to life and limb. Too much attention to the intact re-entry problem could divert useful human energy from attention to much greater dangers which we face every day.

III. The Canadian experience with the Cosmos-954 satellite: operation 'Morning Light'

The story of the search for the fallen pieces of this satellite which re-entered the atmosphere on 24 January 1978 is well known from many newspaper accounts. A map of the areas which were searched during the winter is given in figure 21.3 and details of items recovered are listed in table 21.2. The search began without delay and continued until 10 April. A second search

Figure 21.3. Map of winter search areas

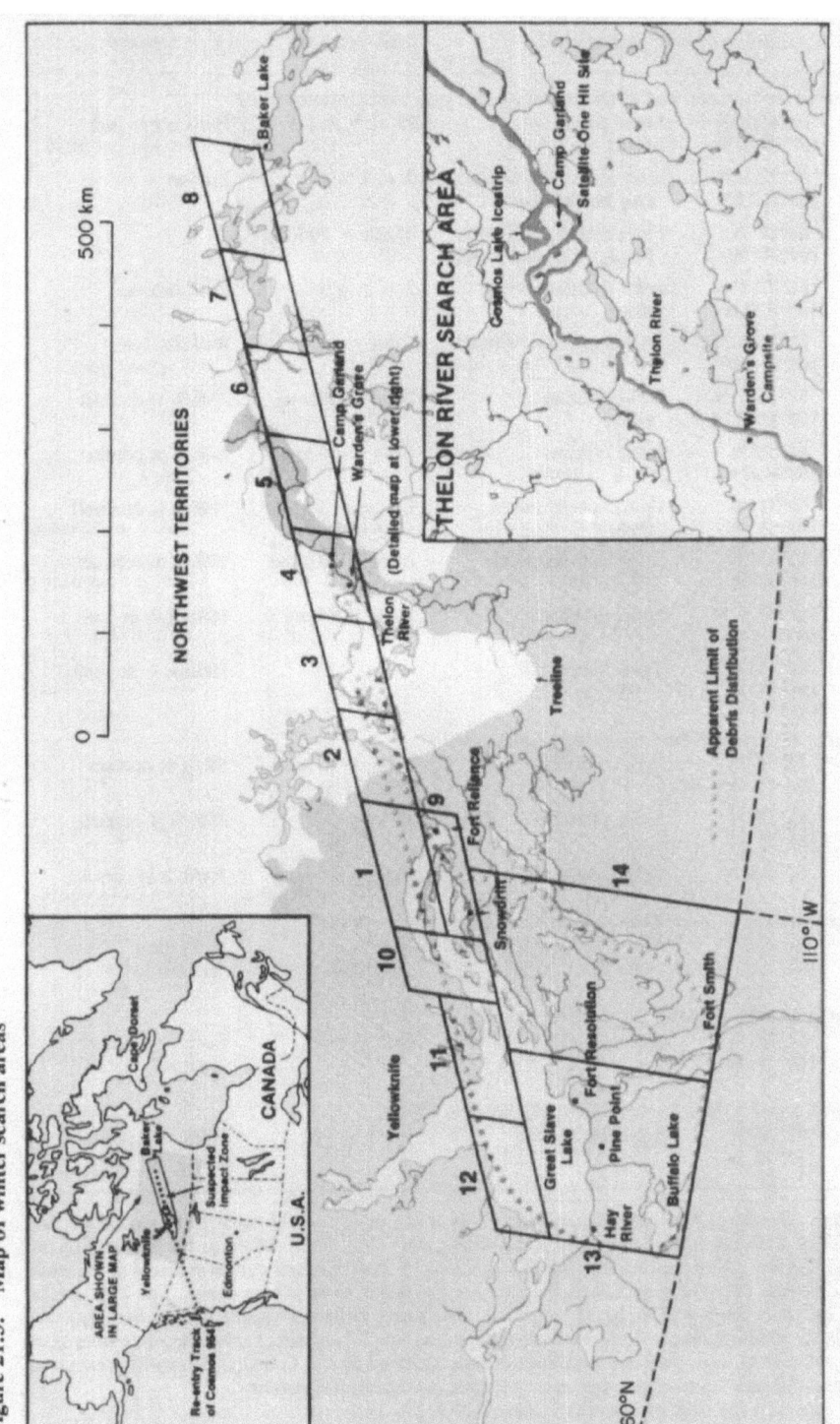

Source: Reference [10].

Table 21.2. Items recovered from the Cosmos-954 satellite

Part no.	Location	Description	Dimensions (cm)	Radioactivity
Search area one: east end of Great Slave Lake, near Fort Reliance NWT				
1	62°50.5'N 109°29.5'W	Metal piece	23 × 9 × 1.3	Fission product
2	62°52.1'N 109°32'W	Grey metal crushed ring black liner	2 × 2 × 8	Fission
3	62°52'N 109°40'W	Pipe, metals (Mn, Mg, Pb, Fe, Cu) and fibres	25 dia × 30 long	
4	62°57.3'N 109°8.5'W	Semi-circular metal plate	2 × 2 × 8	Radioactive
5	62°54.3'N 109°01'W	Thin metal, leaf-ribbed	7 dia	Radioactive
6	62°56.5'N 109°8'W	Solid cylinder metal	2 dia × 10 long	25R/h at contact
7	62°56'N 109°8.4'W	Solid cylinder	2 dia × 10 long	30R/h at contact
8	62°43'N 109°59'W	Oval, curved metal flake	Longest dimension 5	10R/h at contact
9	62°42'W 110°42'W	Solid cylinder metal	2 dia × 10 long	10R/h at contact
10	62°53.5'N 109°26'W	Solid cylinder	2 dia × 9 long	100mR/h at 1 m
11	62°33'W 109°33'W	Three flakes		100mR/h at 1 m
Search area two: vicinity of Artillery Lake, NWT				
1	62°58'N 108°57'W	Solid cylinder	2 dia × 10 long	5R/h at contact
2	62°59'N 108°55'W	Solid cylinder	10 long	25R/h at contact
3	63°0.5'N	Solid cylinder	2 dia × 10 long	100R/h at contact
Search area four: near Warden's Grove on Thelon River, NWT				
1	63°47.8'N 104°13.5'W	Metal struts attached to soft metal plate	3 long	Activation products
Search area nine: South-east of Great Slave Lake, NWT				
1	62°36'N 110°46'W	50 small pieces	each < 1 dia	Each 1−3R/h at contact
Search area ten: South-east of Great Slave Lake, NWT				
1	62°22'N 110°40'W	Solid black cube	1 × 1 × 1	500R/h at contact

Note: The list of parts found given in this table is incomplete: on 13 April the Canadian Department of External Affairs presented a diplomatic note (No. FLO—0840) to the embassy of the Soviet Union giving the total number of 10 cm long by 2 cm diameter cylinders found as 29, while six cylinders 25 cm long by 9.5 cm diameter are also stated to have been found. The 2 cm diameter cylinders are stated to be of beryllium. The larger cylinders appear also to be beryllium, possibly solid beryllium. A Soviet diplomatic note (No. 37) replying to the Canadian note gives the number of 2 cm diameter beryllium cylinders carried by the Cosmos-954 satellite reactor as ''several dozen'' and states that most of them have been discovered.
See also the map of the area in figure 21.3.

for small radioactive particles has since been contracted out to industry by the Atomic Energy Control Board, because it was known from the winter search that some particles had drifted, in the course of their descent, far south of the main search zones.

The Cosmos satellite was stated, through official channels of the Soviet Union, to contain an enriched uranium reactor. Some selected facts about the descent of the satellite and the search operation, Morning Light, are as follows:

1. No part of the satellite struck any person directly as it fell.

2. Nevertheless, two naturalists who were unaware that a satellite had fallen in their area found pieces of it sticking out of a crater in the Thelon River ice on 29 January. They are reported to have touched their find [4] not realizing that it might be radioactive. No radioactive contamination was found on either of the two men in subsequent testing [4].

3. Several of the satellite pieces which were recovered were highly radioactive.

4. One of the pieces was so highly radioactive that handling it without long tools could easily have been fatal. It required a special container to be made for its removal.

5. It is very unlikely that all the pieces, especially the tiny specks of uranium, have been found; but the fission-fragment activity will have died down to a safer level by mid-1978. However, this very feature of reduced activity also makes the search for these particles more difficult.

6. No particular human distress seems to have been caused by the Morning Light operation, but people unfamiliar with the general climate in the search area should bear in mind that temperatures of $-40°C$ were normal, sometimes with high winds. There was also a possibility of losing a Hercules aircraft and crew on a runway created on a frozen lake, the ice having been of minimum thickness to bear a large aircraft.

7. The total flying time for search aircraft of all types was 4 634 hours as of 10 April 1978.

IV. Discussion and legal questions

The re-entry of Cosmos-954 and subsequent finding of small pieces reveals first of all that this type of spacecraft falls between the two extremes of intact re-entry and complete burn-up in the stratosphere. This intermediate category of partial burn-up seems to be the most inconvenient. The effective area A for collision of a falling satellite with a person or people would generally be larger than it would be for intact re-entry, and most certainly the hazards of radioactive contamination at ground level are relatively large. The harmlessness of the Cosmos-954 re-entry is generally attributed to the extremely sparse population of the winter wilderness it fell into, yet two naturalists found part of the debris by chance, not knowing at the time what it was. A definite risk of serious radiation sickness occurred despite the very low population density in the district. The first questions which may very naturally be raised are those concerning satellites which are already in Earth orbits and for which it is therefore too late to plan a different technology. There are the questions of: (a) warning of an impending re-entry of a nuclear-powered satellite; (b) predicting and informing the state(s) where re-entry is likely to occur; (c) informing that re-entry has already occurred; (d) imparting of appropriate technical knowledge by the launching country to the states involved in post-re-entry search operations; and (e) supplying help for post-re-entry search operations. These first questions are all legal rather than technical, although some technical questions can emerge from them. None of these questions is necessarily adequately dealt with under the 1975 Convention on the Registration of Objects Launched into Outer Space, because of the special circumstances posed by the re-entry of radioactive materials. Accordingly a working paper [3] by 14 member states of the UN Committee on Peaceful Uses of Outer Space (UNCPUOS), dated 4 April 1978, called for a review by the Legal Subcommittee of UNCPUOS of these very questions, and also questions of notification before and at launching.

Questions regarding satellites which have not yet been launched can be both legal and highly technical. A change in planning or procedures for design and use of satellites containing radioactive material could require international agreement through the United Nations, although it is possible that a change might be made unilaterally by a nation which launches such satellites. For example, such a change was made by the United States when intact re-entry was decided upon. A change could also result from unprompted cooperation between nations.

It will also be useful to consider further developments at the United Nations and a range of questions about radioactive materials in space on which agreement should be reached.

On 13 February 1978 William H. Barton, Ambassador and Permanent Representative of Canada to the United Nations, issued a statement (Press Release No. 2) to the Scientific and Technical Subcommittee (STS) of UNCPUOS. The statement was devoted entirely to the Cosmos-954 re-entry and considerations following from it. Referring to the STS he said, "The overall objective of our efforts should be to develop a régime for the use of nuclear power sources in outer space which would ensure the highest standards of safety for mankind and protection for the environment. . . ." He proposed that a working group of technical and scientific experts be formed to carry out a detailed review and technical study for the STS, to prepare the ground for constructive action in the Legal Subcommittee of UNCPUOS, and in the UNCPUOS itself and the General Assembly. A decision to form such a working group was reached in the June–July session of UNCPUOS in New York.

Many of the questions which should be discussed in the working group have already been raised in a working paper [6]. Others, perhaps, may not yet have been considered. In what follows, a range of questions will be mentioned: (a) available power sources for satellites: their relative advantages and disadvantages; (b) radiation standards for re-entering satellites; (c) restrictions on the orbits of satellites which would supplement and reinforce (b); (d) design standards and tests which would ensure either intact re-entry on the one hand, or complete dispersal in the stratosphere on the other hand; (e) consideration of equipping intact re-entry satellites with rocket motors which could be activated years after satellite launching, so as to make it possible for re-entry to occur at a chosen location; and (f) legal inclusion of nuclear reactors and RTGs in satellites within the category of devices for which open exchange of information already exists. The first four questions (a) to (d), above, are already included in the proposed agenda [6] of the working group of the STS. Questions (e) and (f) appear to to have been formally proposed.

Question (e) refers only to intact re-entry and should be considered if intact re-entry is decided upon, especially the intact re-entry of reactors which could in principle go critical, for example, in the ocean. This suggestion poses problems which may be technically severe especially when the re-entry date is many years after launch. Nevertheless, this question should be investigated and could result in a strong preference for RTGs being launched in preference to nuclear reactors. The more restricted question of the control of the position of re-entry of an RTG is less sensitive because, as shown above, an intact re-entering RTG is unlikely to hurt anyone. The occurrence of uncontrolled RTG re-entry could therefore be felt to be one that could be classified legally as an 'inevitable accident'. Question (f) has interesting ramifications in nuclear cooperation, technological cooperation (for example, over thermoelectric generators), and even offers opportunities for international gestures of a tension-reducing nature. At one extreme one can even foresee the envisaged technological exchanges being coupled to moves in disarmament. Such exchanges may be relevant to the future design of surveillance satellites and would thus be in the interest of world security. For these and other reasons the technology of powering satellites should no longer be held in the realm of military secrets.

V.   Other power sources for satellites and the trend towards higher electrical power output

For space missions to remote parts of the solar system, far outside the radius of the Earth's orbit, there may be no suitable choice of power system other than reactors or RTGs. For Earth orbits, however, the use of solar cells is very well established. The practical efficiency of energy conversion using solar cells is furthermore fast approaching a value of 15 per cent [11, 12] which, combined with the solar constant of 1.38 kW m^{-2} at the position of the Earth, leads to the conclusion that solar panels in space may soon yield 100 W(e) m^{-2}. The limits to the power output would be dictated by the maximum mass which one could afford to launch, for example, as dictated by the maximum mass for a single shuttle.

The comparative limitations of various systems are illustrated in figure 21.4, which is intended to apply to missions contemplated by the US Department of Defense for geosynchronous orbits. The solar energy power boundary is seen to flatten at the level of 40 to 50 kW(e), presumably because some batteries must be carried aboard the power unit to take care of occultation. It is the mass of the batteries, even using the most favourable projections [12] from work on new types of battery, which sets a low ceiling on the power which can be expected to be brought into orbit in a single shuttle. The solar power capability could increase by more than a factor of three where batteries are not required.

Also the limitation of RTGs in figure 21.4 is not purely technical but represents the design limits of RTGs currently being developed for test and space missions in the early 1980s. Doubtless it also reflects to some extent the high cost of plutonium-238, which was $650 per thermal watt for the unencapsulated isotope (in 1973 prices). If only energy possibilities are considered, at 20 kW(e) the plutonium-238 mass could be held to roughly 200 kg, given thermoelectric converters of the improved efficiencies now being developed, so that the mass of the RTG could be kept within the rule of thumb value [13] of 955 kg for a single shuttle. Clearly, somewhere between 20 kW(e) and 200 kW(e) the RTG ceases to be practical by comparison with the nuclear reactor. The problems of cooling reactors and RTGs in space are similar. The assembly and launch problems with RTGs are likely to be worse than with reactors, since the latter are launched before start-up, whereas the plutonium-238 is always hot, thermally. Although the radioactivity is mainly self-contained, the rarer radioactive branch decays of the radioisotope (figure 21.2) will add to the hazards of launching very active RTGs.

Finally, the RTG offers the only sound prospect of a reliable power source having a lifetime considerably in excess of the seven to ten years indicated at the right in figure 21.4. Solar cells will deteriorate over periods of years due to radiation damage, and reactors would require refuelling.

VI. Concluding remarks

This paper has been based mainly on the knowledge which is available of

Figure 21.4. Régimes of possible space power applicability for DOD missions

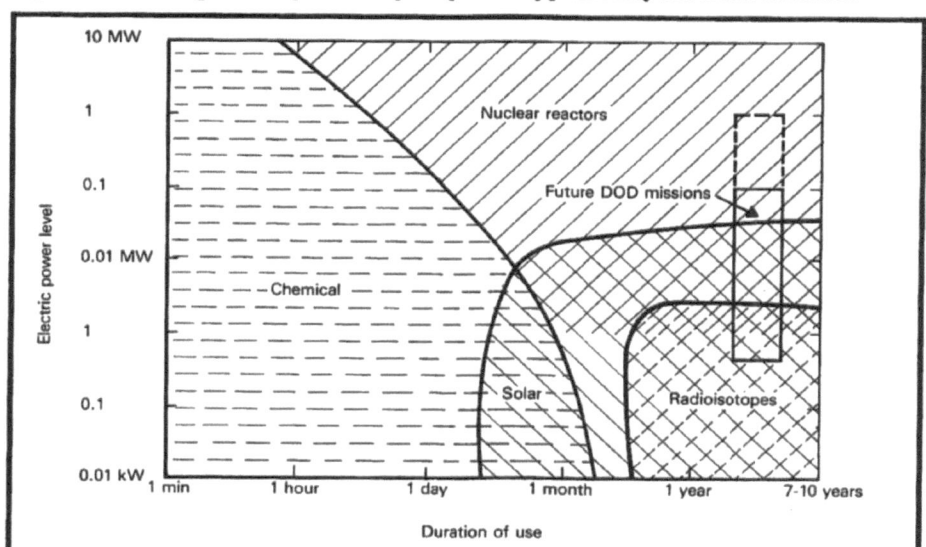

the current achievements in the production of electric power for space applications using nuclear sources, and on the foreseeable developments into the early 1980s. This could be too narrow a basis for an international agreement which is to be of reasonably long range. Therefore the more distant future prospects have been touched upon by making a few comments about figure 21.4, so as to begin to examine some of the questions which will be faced if much greater electric power is required in future space missions. To do justice to these very complex problems, it would be necessary to assemble the separate views of technical experts in each of the fields of reactor, RTG and solar power system design. It would not appear that reactors offer distinct advantages over RTGs up to power levels of at least 20 kW(e), apart from the high cost of the isotope plutonium-238. For satellites intended to have a powered lifetime much longer than seven years, reactors and solar cells may not be practical, and the RTG may be the only choice. Solar cell arrays for space use are moving towards a power-to-mass ratio of 200 W(e)/kg; already power outputs of over 150 kW(e) (single shuttle) seem realistic for a satellite which does not have to carry batteries. However, most systems to date have required electrical storage, so that power is maintained during occultation. Even projecting to the most advanced future batteries, this would reduce the solar power system output to about 40 kW(e) for the same mass.

References

1. UN document A/AC.105/PV.187.
2. Buden, D., *Nuclear Reactors for Space Electric Power*, Report No. LA-7290-SR, UC-33 and UC-80, Los Alamos, June 1978.

3. *Use of Nuclear Power in Space by the United States of America* (Department of Energy, Washington, D.C., 1978).
4. *Operation Morning Light*, General information fact sheet, Department of National Defence (Ottawa, April 1978).
5. UN document A/AC.105/C.2/L115.
6. UN document A/AC.105/C.1/L.103.
7. UN document A/AC.105/214.
8. UN document A/AC.105/217.
9. Lederer, C. M., Hollander, J. M. and Perlman, I., *Table of Isotopes* (J. Wiley, New York, 1968).
10. Aikman, Major B., 'Operation Morning Light', *Sentinel*, Vol. 14, No. 2, 1978.
11. Rittner, E. S., 'Recent advances in components of space power systems', *Proceedings of the 25th International Astronautical Congress* (Pergamon, Oxford, 1976), pp. 137–46.
12. Goldsmith, P. and Reppucci, G. M., 'Advanced photovoltaic synchronous-orbit spacecraft power systems', *Journal of Energy*, Vol. 2, No. 81 (1978).
13. *Reactors for Space Electrical Power*, ERDA Contract No. W7405-ENG. 36, Los Alamos Scientific Laboratory, January 1977.

Acknowledgements

I would like to thank the following people for help in finding information or for sending material on which this article is based: Dr R. J. Drachman and Dr Jerome Mullin of NASA; Mr B. J. Rock of the US Department of Energy; Dr D. Buden and Dr C. R. Emigh of the Los Alamos Scientific Laboratory; Dr E. S. Rittner of the Comsat Laboratories; Capt. S. B. Burton of Canada's Department of National Defence; Messrs Hugh Spence and G. B. Knight of the Atomic Energy Control Board; and lastly Messrs Erik Wang and P. McRae of the Department of External Affairs. Personal thanks are also due to Dr K. Jung and to A. L. who made useful comments on earlier versions of this paper.

Part V

Chapter 13. Implementation of the Non-Proliferation Treaty

J. GOLDBLAT

Article VIII of the Treaty on the Non-Proliferation of Nuclear Weapons (NPT) provides for periodic conferences of the parties to review the operation of the treaty with a view to assuring that its purposes and provisions are being realized. (For the text of the NPT, see Appendix A.) The first review conference was held on 5–30 May 1975.

Although the NPT is generally considered to be the most important multilateral arms-control treaty concluded so far, attendance at that conference was poor: only some 60 per cent of the 96 states which were party to the NPT in May 1975 participated. Seven states which had signed but not ratified the NPT were present at the conference without the right to take part in its decisions, while a few states which had neither signed nor ratified the treaty applied for and were accorded observer status.

The review conference concluded its work with the adoption, by consensus (that is, without a vote being taken), of a final declaration. (For the text of the declaration, see Appendix B.) However, in spite of the formal acceptance of the declaration, a number of delegations expressed dissatisfaction about the outcome of the conference, made interpretative statements contradicting the consensus, or objected outright to various formulations. Proposals for additional protocols to the NPT, as well as resolutions dealing with various matters related to the implementation of the NPT, were submitted by several participants but did not obtain sufficient support. On the insistence of the sponsors, they were included in the final document of the conference for subsequent consideration by governments.

The conference reaffirmed the role of the NPT in international efforts to avert further proliferation of nuclear weapons, to achieve the cessation of the nuclear arms race and to undertake effective measures in the direction of nuclear disarmament, as well as to promote cooperation in the peaceful uses of nuclear energy under adequate safeguards. But the results of the conference can be properly evaluated only in the light of how its recommendations have been implemented.

I. Non-transfer and non-acquisition of nuclear weapons

The first two articles of the NPT contain the essence of the non-proliferation undertakings: the nuclear weapon states are committed not to transfer to any recipient, while the non-nuclear weapon states are under the obligation not to receive, manufacture or otherwise acquire, nuclear weapons or other nuclear explosive devices, or control over them.

These provisions seem to have been complied with. No complaints have been made about the transfer of nuclear weapons or other nuclear explosive devices, or of control over them, by the nuclear weapon powers; neither has any non-nuclear weapon party to the NPT been formally accused of manufacturing these weapons or devices or acquiring them by other means. In spite of this, it would be wrong to conclude that the very purpose of the NPT has been achieved.

As long as nuclear weapons remain deployed on the territories of non-nuclear weapon states, there will be a danger of sudden change in the control over these weapons in times of severe international crisis. Furthermore, since the NPT has not been universally subscribed to, its observance by the parties alone cannot guarantee that other states will abstain from acquiring nuclear weapons. In fact, the number of states known to have come into possession of nuclear weapons or other nuclear explosive devices, which the treaty was intended to restrict to five, increased in 1974, when India carried out a nuclear explosion. Moreover, Israel has been reported to possess several untested nuclear bombs, while South Africa was rumoured to have been on the brink of testing a nuclear device. But these countries cannot be charged with a breach of the NPT which they never signed. If anyone is to account for the further proliferation that occurred, it is, in the first place, the parties to the treaty themselves. For it seems unlikely that India, or other countries, would have been in a position to manufacture a nuclear explosive device, certainly not that soon, if the undertaking under Article I not "in any way" to assist any non-nuclear weapon state to manufacture nuclear weapons or other nuclear explosive devices, had been fully respected. Indeed, by providing nuclear material, equipment and know-how to countries which refuse to adhere to the NPT and to renounce thereby, formally, the nuclear weapon option, the suppliers party to the NPT (including France which, while not being party to the treaty, stated that it would behave as if it were one) have perforce contributed to the building of new nuclear weapon capabilities. Thus, the fissile material for the Indian explosive device was obtained in a Canadian-supplied reactor, with the use of US-supplied heavy water.[1]

Nevertheless, a year after the Indian nuclear explosion, and just a few weeks after the conclusion of the first NPT review conference, an

---

[1] In the pre-NPT period, it was the transfer of nuclear equipment and technology by the USSR to China, in the late 1950s, that enabled the latter power to manufacture its own bomb and test it as early as 1964.

agreement on nuclear supplies was signed between the Federal Republic of Germany, a party to the NPT, and Brazil, a non-party. Under the terms of this agreement Brazil is buying a complete nuclear fuel cycle from the FRG. The cycle will cover prospecting, mining and processing uranium ores in Brazil, as well as production of uranium compounds; uranium enrichment; construction of nuclear power stations; manufacture of fuel elements; and reprocessing of irradiated fuels. The cooperation includes exchanges of technological information, and several joint enterprises are envisaged. No such comprehensive nuclear deal has ever before been concluded.

There is special concern about the sale to Brazil of a uranium-enrichment facility—a novel item on the nuclear shopping list. The technology needed to enrich natural uranium to nuclear fuel levels is more difficult than the further step of reaching weapon-grade levels, and highly enriched uranium can be used in a nuclear bomb or as a trigger for a thermonuclear bomb. Moreover, the acquisition of plutonium-reprocessing technology (without any evident commercial need) might in itself be enough for Brazil to secure a nuclear military potential. Assurances that the plants for enrichment and reprocessing will be used exclusively to make reactor fuel, and the envisaged IAEA safeguards to prevent diversion, though unaffected by the termination of the cooperation agreement, apply only to the equipment, installations and materials supplied by the Federal Republic of Germany. Brazil is under no legal obligation that would prevent it from constructing an unsafe-guarded fuel cycle.

Brazil undertook not to use the technological information received, including that on plutonium-reprocessing, for the manufacture of nuclear weapons, and appropriate international safeguards have been devised to ensure compliance with this undertaking. But restrictions on the use of transferred technology may be difficult to enforce, and since they are also limited in time,[2] replication of facilities will eventually become a possibility. Once Brazil achieves nuclear self-sufficiency and starts operating its own, indigenously built plants, it will be able to manufacture nuclear weapons. This may not be imminent, in view of the size of the required investments. Nonetheless, the German supplies of sensitive elements of the fuel cycle which will enable Brazil to keep the nuclear weapon option open, are at variance with the 'non-assistance' clause of the NPT.

It is true that there is no express prohibition for a non-nuclear weapon state, party to the NPT, to provide assistance, encouragement or inducement to manufacture nuclear explosive devices to another non-nuclear weapon state, which is not party to the NPT. But, as early as 1968, in response to a proposal to close this apparent loophole in the NPT, the

---

[2] Any nuclear facility or specified equipment designed, constructed or operated "within a period of twenty years" will be deemed to be designed, constructed or operated on the basis of or by the use of transferred relevant technological information if its design, construction or operation is based on the "same or essentially the same" physical and chemical processes as those specified and communicated to the IAEA by the transferor of the relevant technological information which does not include information available to the public.

Soviet Union made it clear that "if a non-nuclear-weapon State Party to the Treaty were to assist another non-nuclear State to manufacture and acquire nuclear weapons, such a case would be regarded as a violation of the Treaty". (This understanding was reiterated during the 1975 NPT review conference.) The USA then argued that "it seems clear that a non-nuclear-weapon State which accepts the Treaty's restrictions on itself would have no reason to assist another country not accepting the same restrictions to gain advantage from this fact in the field of nuclear weapon development". But it also stated that "if a non-nuclear-weapon Party did nevertheless attempt to provide such assistance in the territory of a non-party, the presumption would immediately arise that these acts had the purpose of developing nuclear weapons for itself, in violation of the Treaty". This interpretation given by the powers responsible for the formulation of the relevant provisions of the NPT has not been contested by any state. Consequently, in supplying Brazil with nuclear equipment, the FRG also runs the risk of weakening, in the long run, its own non-proliferation 'credentials'.

Because of the link between the peaceful and military aspects of nuclear energy, it can be argued that any supplies destined for nuclear power programmes facilitate the acquisition of nuclear weapon capabilities. This may be so, but the risk is considerably greater when the recipient countries are unhampered by a legally binding commitment not to manufacture nuclear weapons and, especially, when they claim the right to conduct nuclear explosions, as in the case of Brazil or Argentina and, until recently, India.[3] The reactor power cycle may be a more complicated way towards a nuclear weapon capability, both technically and economically, than constructing specialized plants for the production of nuclear explosives, but it is a convenient way in the sense that the intention to produce a bomb is not visible in a civilian nuclear programme, and untoward international repercussions can be avoided while the nuclear weapon potential is being built. There may, of course, be cases when even parties to the NPT plan clandestinely to acquire nuclear weapons. But it must be assumed that nations which adhere to the treaty do so in good faith, unless and until firm evidence to the contrary has been provided.

It can also be argued that certain non-nuclear weapon states, not party to the NPT, might acquire a nuclear weapon capability independently, using their own domestic resources, or in cooperation with other non-parties. This is possible,[4] and, as yet, there exists no international mechanism which could prevent this from happening; a similar situation might arise with other arms control agreements. But if the parties to the NPT speed up proliferation by making it easier and less costly for non-parties to traverse the route towards a nuclear bomb, they undermine the very foundation of the treaty they have themselves constructed.

---

[3] In 1978, the Indian Prime Minister stated that India would not conduct nuclear explosions even for peaceful purposes, but it still refused to sign the NPT.

[4] It will be noted that components of uranium enrichment installations, as well as materials and equipment for the separation of plutonium are now commercially available.

The most immediate threat to the non-proliferation régime is posed by the spread of reactor-grade plutonium, which is readily convertible into an explosive device, the manufacture of the device itself being no longer a very difficult task. Some non-nuclear weapon states have probably designed nuclear weapons and perhaps even developed their non-nuclear components, or may do so in the future, since there is nothing in the NPT or in agreements on nuclear transfers to forbid this kind of activity. For such states, access to plutonium is all that is needed to cross the nuclear weapon threshold at any time, just a few kilograms of plutonium being enough to make a bomb. To minimize the availability of this material, states should refrain from supplying plutonium-separation equipment, or such separation services as would result in the plutonium being stored under national control of non-nuclear weapon states.

The most radical solution would be to renounce reprocessing of spent fuels and to avoid, thereby, the separation of plutonium. This solution may imply giving up the energy potential that fast breeders needing large quantities of plutonium would offer, if breeders ever became an economically attractive proposition for stretching uranium resources or a way to secure independent fuel supplies. But this is a question of priorities, and, in signing the NPT, states have already accepted significant restrictions on their national sovereign rights, and have implicitly agreed that the interest of the international community in halting the spread of nuclear weapons must have precedence over other considerations.

In a 'plutonium economy', it might be well-nigh impossible to preserve a firebreak between nuclear power technology and nuclear weapon capability (not to mention environmental risks as well as the increased danger of plutonium being stolen by sub-national groups and used for terrorist purposes), irrespective of IAEA safeguards. These safeguards cannot be made fool-proof; moreover, their function is not to avert abuses, but merely to detect diversion of a significant quantity of weapon-grade material in time for some kind of response. In the case of plutonium, it may require no more than a few days for diverted material to be transformed into an explosive, which is too short a period for effective international action to be mounted. Furthermore, safeguards agreements can be abrogated at short notice, and even withdrawal from the NPT by a country claiming that its "supreme interests" have been jeopardized (Article X) should not be ruled out, notwithstanding the political risks involved in such a drastic step. Emergence of yet another nuclear weapon state would probably be considered reprehensible by most nations, but a stage has not yet been reached where such an event would be regarded as a threat to peace, as defined by the UN Charter, requiring coercive measures to be taken by the UN Security Council against the new proliferator. The defaulting state could, of course, be penalized by a denial of further nuclear material and equipment deliveries by the suppliers, as foreseen in the IAEA Statute. But such a belated sanction will probably not be sufficient to prevent a state, which already possesses the wherewithal, from 'going nuclear'.

The feasibility of so-called proliferation-resistant nuclear fuel cycles has been considered at the meetings of the International Nuclear Fuel Cycle Evaluation (INFCE) set up in 1977. However, it would not be realistic to expect that the possibility of national nuclear power programmes being utilized for military ends could be entirely removed and, in any event, general acceptance of alternative fuel cycles seems unlikely. In this situation, internationalization of the sensitive parts of the nuclear fuel cycle, combined with an internationally guaranteed supply of fuel, could make nuclear energy safer from the point of view of weapon proliferation.

II. Nuclear safeguards

Under Article III of the NPT, the non-nuclear weapon states undertook to conclude safeguards agreements with the IAEA covering all their peaceful nuclear activities, within the prescribed time limits, of 24 months for the original parties, and 18 months for states acceding later. The stated purpose of these safeguards is to verify the fulfilment of the treaty obligations with a view to preventing diversion of nuclear energy from peaceful uses to nuclear weapons or other nuclear explosive devices.[5]

Although the control clauses constitute an inseparable part of the NPT commitments, by March 1979, only 62 out of 103 non-nuclear weapon parties had concluded the required agreements. Many of the defaulting parties are states which as yet have no significant nuclear activities, and there may be nothing to safeguard on their territories. Nevertheless, from the point of view of observance of the treaty provisions, this is an unsatisfactory state of affairs. The first review conference's recommendation, that states party to the NPT that have not yet done so should conclude safeguards agreements with the IAEA "as soon as possible", has still not been implemented.

The 1975 conference declaration attached considerable importance to the continued application of safeguards to the nuclear activities of the non-nuclear weapon parties to the NPT, "on a non-discriminatory basis". However, the problem of discriminatory treatment of the parties to the NPT as compared with non-parties has not been settled. The latter, as distinct from the former, are still not subject to safeguards comprehensively covering their nuclear activities: safeguards applied in their territories continue to be facility-oriented, which means that they may place nuclear material under IAEA safeguards only in certain facilities, and retain unsafeguarded all or part of a nuclear fuel cycle.

---

[5] Military uses of nuclear energy for non-explosive purposes, for example, for the propulsion of warships or submarines, are not to be covered by controls performed in accordance with the NPT.

The conference further recommended that "more attention and fuller support" should be given to the improvement of safeguards techniques. As a matter of fact, during the past few years IAEA safeguards techniques have developed considerably, reaching a high degree of sophistication in surveillance and containment. The number of IAEA inspectors has increased parallel with the increased number of nuclear plants placed under safeguards. However, all these measures remain of limited significance, since they apply only to states which have already forsaken the nuclear weapon option by becoming party to the NPT, while significant and sensitive parts of the nuclear programme of certain states not parties to the NPT remain outside international safeguards. The latter states have little incentive to join the NPT and accept safeguards on all their nuclear activities as long as they are assured of continued nuclear supplies; they may have nothing tangible to gain from abandoning their freedom of action. In other words, for non-proliferation purposes, improving safeguards which apply merely to the nuclear material supplied is inadequate.

In 1977, realizing that commercial competition hampers the pursuance of non-proliferation objectives, a group of fifteen supplier states, members of the so-called London Club, adopted guidelines for nuclear transfers, streamlining the terms for transfer of nuclear items and technology. In particular, they drew up a so-called trigger list, that is, a list of goods which, when exported, 'trigger' the application of safeguards in the recipient countries. But no agreement has been reached on the question of full-scope safeguards to be required as a condition for nuclear supplies, and exports of highly sensitive nuclear facilities have not been prohibited. Since then, certain supplier countries have unilaterally adopted more restrictive export policies than those required by the London guidelines, with or without special national legislative acts, but supplies of nuclear materials, plants or know-how, without safeguards on the full fuel cycle have not been brought to a halt. Thus, after a period of hesitation following the adoption in 1978 of the Nuclear Non-Proliferation Act, which specifically requires IAEA safeguards for all nuclear materials in all peaceful activities of the recipient states, the USA resumed its shipments of enriched uranium to India, in spite of the latter country's reiterated categorical rejection of full-scope safeguards. In addition, the USSR shipped a large amount of heavy water to India, in spite of its advocacy of full-scope safeguards in non-nuclear weapon states.

NPT safeguards are applied in many countries without hampering their economic, scientific or technological development. A concerted denial of nuclear material deliveries to states unwilling to accept NPT safeguards would not, therefore, be a measure promoting particular political or commercial interests, as was the case with the oil embargo. At any rate, nuclear items are not ordinary items of international trade; they demand special policies, even if such policies are seen as discriminatory. Isolation from international cooperation in the peaceful uses of nuclear energy can provide considerable leverage: a few countries have already been pressured into acceding to the NPT in order to qualify as nuclear material importers.[6] If, in addition to these restrictions, the parties to the NPT undertook to import nuclear material or equipment only from other parties, pressure would be put on the exporting countries to observe the non-proliferation rule, with the result that the quantities of nuclear material entering the world market outside the framework of the NPT would be further reduced. The suggested restrictions would apply only to the provision of elements of the nuclear fuel cycle, bearing in mind that for the majority of nations nuclear power is still a rather distant prospect. For them, the use of nuclear science and nuclear techniques in food preservation and production, in agricultural research, in medicine, in water resources development and in geological and industrial applications, is of more immediate concern. International cooperation in these fields should continue under all circumstances.

Ever since the signing of the NPT, a number of countries, especially in the industrialized world, have insisted that all nuclear weapon powers party to the treaty should agree to apply IAEA safeguards to their peaceful nuclear activities, even though they are not obliged to do so under the treaty. Since the first review conference, the United Kingdom has voluntarily submitted its non-military nuclear installations to safeguards under IAEA supervision, and the United States has negotiated a similar agreement. Also France, a non-party to the NPT, signed an agreement under which part of its nuclear facilities will be placed under IAEA safeguards. However, the right of these states to withdraw nuclear material from civilian activities and to use it for military purposes has remained unaffected.

Wider openness of nuclear weapon states to verification may somewhat reduce the sense of discrimination of non-nuclear weapon states and satisfy the commercial interests of the nuclear industry. But, as far as non-proliferation of nuclear weapons is concerned, safeguarding peaceful activities in countries unrestricted in their military nuclear programmes seems pointless: it amounts to verifying the fulfilment of non-existing obligations. On the other hand, it would appear useful to safeguard nuclear items imported by nuclear weapon powers, in order to ensure that these items do not contribute to a further build-up of nuclear weapon arsenals.

III. Physical protection of nuclear materials

The first review conference recognized the need for physical protection of nuclear materials in storage, use or transit. It called upon all states engaging

---

[6] Recently, the USA has gone even further in this respect by deciding to wind down its economic assistance to Pakistan, which refused to place under international safeguards the uranium enrichment plant it was building.

in peaceful nuclear activities to enter into such international agreements and arrangements as may be necessary to ensure this protection.

In 1977, the IAEA published recommendations for the physical protection of nuclear materials (a modified and extended version of recommendations issued first in 1972, and revised in 1975), which were accepted by the 'London Club' of supplier countries as a basis for guiding recipient countries in designing a system of physical protection measures and procedures. At the same time, preparations started for an international convention on the physical protection of nuclear materials for peaceful purposes during international transport, and a meeting of governmental representatives was convened to draft such a convention.

The aim of the convention would be to prohibit export, import or transit of nuclear material in the absence of assurances that such material will be protected during international transport at levels to be described in the convention. In the case of theft of nuclear material, or threat thereof, the parties would cooperate and assist one another in the protection and recovery of such material. The committing of acts prohibited by the convention would be considered a punishable (and possibly also extraditable) offence by each state. The implementation and the adequacy of the convention would be subject to periodic review.

Strictly speaking, the matter lies outside the framework of the NPT. Nevertheless, unlawful seizure of nuclear material might have serious repercussions for the durability of the NPT, and for the security of nations, in general. To reduce further the risk of such occurrence, there should also be internationally binding rules for the protection of nuclear material in domestic use, storage and transport. Even without formal treaties, observance of the minimum standards, as have already been agreed to, should be a condition for supplying nuclear material and equipment. It is, of course, essential that the nuclear weapon powers take all the necessary measures to protect their stockpiles of nuclear weapons against theft or other abuse.

IV. Peaceful uses of nuclear energy

Article IV of the NPT deals with the contribution by states in a position to do so, to the development of the applications of nuclear energy for peaceful purposes, "especially in the territories of non-nuclear-weapon States Party to the Treaty, with due consideration for the needs of the developing areas of the world".

The above provision has made little impact on international nuclear collaboration, and its implementation was seriously questioned by many participants at the review conference when statistics showed that non-parties to the NPT had benefited considerably more from international exchanges in the field of peaceful uses of nuclear energy than had the parties to the treaty. This anomaly was recognized in the final declaration of the conference which recommended, *inter alia*, that in reaching decisions on the provision of equipment, materials, services and scientific and technological information for the peaceful uses of nuclear energy, on concessional and other appropriate financial arrangements and on the furnishing of technical assistance in the nuclear field, including cooperation related to the continuous operation of peaceful nuclear facilities, states party to the treaty "should give weight" to adherence to the NPT by recipient states.

During the past years, a few states have voluntarily contributed funds, or grants in kind, to the IAEA, especially earmarked for technical assistance to the NPT parties.[7] As far as supply of nuclear material and equipment on a bilateral basis is concerned, it is difficult to assess whether the recommendation to accord preferential treatment to parties has actually been carried out. In any event, the NPT is primarily an arms control treaty, its provision for international cooperation in the application of nuclear energy for peaceful purposes being subordinated to the non-proliferation obligations.

V. Nuclear fuel cycle centres

Admitting that wide availability of nuclear technology and fissionable material aggravates the risk of nuclear weapon proliferation, the first review conference considered a proposal for setting up regional or multinational nuclear fuel cycle centres. It recommended that the IAEA should study the subject and, among other things, identify the practical and organizational difficulties, which would need to be dealt with in connection with such projects.

In 1977, the IAEA issued a report of its study on regional nuclear fuel cycle centres, which covers the back-end of the cycle, from the discharge of irradiated fuel from power reactors through storage, reprocessing, fabrication of new mixed oxide fuel elements and radioactive waste management. (It does not address the front-end of the cycle, such as enrichment.) The study has shown that the multinational concept[8] offers a number of advantages in meeting non-proliferation objectives, when compared to the alternative of a further expansion of national capabilities.

The most important of these advantages is that states are offered an

---

[7] Under its Statute, the IAEA is to give due consideration to the needs of the underdeveloped areas of the world, but it is not authorized, on its own, to differentiate between parties and non-parties to other international treaties, such as the NPT.

[8] It is assumed that multinational groupings of participants would be formed on the basis of common needs and interests, and would not be limited by geographical considerations.

incentive to engage in multinational alternatives to national reprocessing and thereby to reduce the number of national facilities constructed.[9] In addition to the possible attractive economic, waste disposal and environmental aspects, intergovernmental agreements envisaged for the fuel cycle centres would enhance controls on the transfer and use of nuclear materials and relevant technologies, and provide for physical protection requirements for the facilities. The centres would be established with the application of full IAEA safeguards to their activities.

In spite of the encouraging results of the IAEA study, no multinational fuel cycle centre has been set up, or even seriously considered, since the NPT review conference (apart from the joint Western European enrichment facilities which had been in existence for some time) and there are now doubts as to whether this approach is at all promising. There is, however, increased interest in international management and storage of plutonium and spent fuel, with specific reference to the IAEA rights under its Statute to require deposit with it of 'excess' stocks of plutonium.[10]

VI.   Peaceful nuclear explosions

Article V of the NPT contains an obligation to ensure that "potential" benefits from any peaceful applications of nuclear explosions should be made available to non-nuclear weapon states party to the treaty, under "appropriate" international observation and through "appropriate" international procedures, and that the charge for the explosive devices used should be as low as possible and exclude any charge for research and development. A basic agreement defining the functions of an appropriate international body, through which the benefits from peaceful nuclear explosions could be obtained, was envisaged, and the possibility of concluding bilateral agreements was left open. In 1968, the USA and the USSR promised to start the necessary consultations promptly and to consider the matter even before the entry into force of the NPT. This did not happen, but the two powers discussed certain technical aspects of the use of peaceful nuclear explosions. In 1976, they concluded a treaty (not yet in force) regulating explosions for peaceful purposes (the Peaceful Nuclear Explosions Treaty—PNET). Although the PNET deals mainly with peaceful applications of nuclear explosions within areas under the jurisdiction or

---

[9] It is envisaged that existing or planned national installations could serve as the initial core of the centres.
[10] According to its new non-proliferation policy, the USA has been trying to induce countries to store spent uranium fuel instead of reprocessing it to extract plutonium. It has recently proposed using a Pacific island as a regional storage facility for spent fuel from Japan, South Korea and Taiwan.

control of the USA and the USSR, it has some relevance to the implementation of the NPT provisions in so far as the bilaterally agreed yield restrictions and verification procedures would also apply to possible US and Soviet services in this field in the territories of non-nuclear weapon states.

The NPT is not explicit as to whether or not non-parties may benefit from peaceful nuclear explosion services. It would seem that such services, if provided, would not be concordant with the policy of creating incentives for states to join the NPT. On the other hand, it can be argued that non-parties might be dissuaded from, or deprived of an excuse for, developing an independent nuclear explosive capability if they were assured of foreign assistance. The latter point of view prevailed and the review conference decided that any potential benefits from peaceful nuclear explosions could be made available also to non-nuclear weapon states not party to the treaty. It is assumed, however, that parties to the NPT would still enjoy preferential treatment as regards supply of relevant services, including preferential charges.

A point which has remained unanswered is the status of non-nuclear weapon states, non-parties to the NPT, which have already started an autonomous nuclear explosion programme for peaceful purposes. According to the letter of the NPT, these states are to be considered non-nuclear weapon states, in spite of the explosions, because only states which had carried out a nuclear explosion before *1 January 1967* are, for the purposes of the treaty, nuclear weapon states (Article IX.3). The countries in question could, therefore, claim the same rights as other non-nuclear weapon countries, not party to the NPT. But such a formalistic approach would contradict the spirit of the treaty. The inclusion of non-parties engaged in nuclear explosions (whatever the declared aim of the explosions) in the category of possible beneficiaries under Article V of the NPT could be construed as a 'premium' for undermining the international non-proliferation régime.

The review conference declaration noted that the technology of nuclear explosions for peaceful purposes was still "at the stage of development and study" and that there were a number of interrelated aspects of such explosions which still needed to be investigated. Nevertheless, considering that the IAEA was an appropriate body through which the task of providing nuclear explosion services could be performed, it urged the Agency "to expedite work on identifying and examining the important legal issues involved in, and to commence consideration of, the structure and content of the special international agreement or agreements contemplated in Article V of the Treaty", taking into account the view of the Conference of the Committee on Disarmament (CCD) and the UN General Assembly. Accordingly, in 1975, the IAEA established an *ad hoc* advisory group which in a report submitted in 1977 proposed the following alternative international legal instruments for providing nuclear explosions for peaceful purposes:

1. An international multilateral 'umbrella' agreement establishing principles which should form the legal framework for the detailed provisions to be negotiated by the parties to individual project agreements. A model project agreement could be annexed to the umbrella agreement.

2. A comprehensive list of general principles which would not be embodied in an agreement binding under international law, but would form the basis of project agreements. It would be submitted for approval by the IAEA Board of Governors, and contained in an IAEA document. A model project agreement could also be annexed to the general principles.

3. Bilateral master agreements concluded between the IAEA and individual supplier nuclear weapon states, by which these states would be bound in all projects in which they participate. Separate agreements for projects would have to take account of guidelines or codes of practice developed by the IAEA.

4. A memorandum of understanding embodying the principles which should govern project services, and operating on a good-faith basis. The memorandum would contemplate the conclusion of individual project agreements between supplier, recipient and consultant states, and would be complemented by codes of practice developed by the IAEA.

Variations on above-mentioned alternatives were also envisaged.

However, the scepticism as to the technical feasibility and economic viability of nuclear explosions for peaceful purposes has remained.[11] Apart from health, safety and environmental problems, which in many countries reduce the chances for public acceptance of such explosions, a question has arisen about the implications for existing and possible future arms-control agreements. The prevailing opinion in this respect can be summarized as follows.

1. It is probably impossible to develop nuclear explosive devices which would be capable only of peaceful application. 'Peaceful' devices could also be used as weapons: they are transportable and the amount of energy they are able to release could cause mass destruciton. Thus, any state conducting peaceful nuclear explosions has a nuclear weapon capability.

2. As distinct from contained underground nuclear explosions, the use of cratering explosions, especially on a large scale, to dig canals, harbours, or reservoirs, would release radioactive products into the atmosphere. Should the debris cross the territorial limits of the state under whose jurisdiction or control the explosion was conducted (and this may prove inevitable), the state in question may be brought to task for violating the 1963 Partial Test Ban Treaty which prohibits such events.

3. Any agreed restraints on underground nuclear weapon testing by the nuclear powers must be accompanied by corresponding limitations on

---

[11] This scepticism may not be shared by the USSR. In 1978, it carried out a number of explosions (seven out of 27), which because of their location outside the usual weapon testing sites are presumed to be for non-military purposes. (The USA stopped conducting its peaceful nuclear explosion programme several years ago.)

peaceful nuclear explosions, because peaceful programmes could be used to obtain weapon-related information.

4. With a treaty banning all nuclear weapon tests, for all time, the incentive for seeking military benefits from peaceful nuclear explosions would be even greater. No verification system could deny such benefits to nuclear weapon states conducting peaceful nuclear explosions on their own territory, or elsewhere under Article V of the NPT.

In the prevailing conditions, elaboration of an agreement regulating the question of nuclear explosive services under the NPT seems redundant from the practical point of view. From the point of view of arms control, such an agreement could be counterproductive in hampering efforts to reach a treaty for a comprehensive prohibition of nuclear testing. Consequently, the implementation of the relevant NPT provision would better be kept in abeyance.

VII. Disarmament obligations

Article VI of the NPT contains a commitment to pursue negotiations "in good faith" on effective measures relating to the cessation of the nuclear arms race and to nuclear disarmament, as well as on a treaty on general and complete disarmament under strict and effective international control.

Although, formally, all parties undertook the above obligation, and the depositary states are usually keen to stress this point, it is clear that nuclear disarmament, which is of paramount importance in a treaty dealing with nuclear proliferation, can be effected only by the nuclear powers. It is therefore these powers, parties to the NPT, that were subjected to criticism at the first review conference for not fulfilling the relevant undertakings. The non-nuclear weapon participants at the review conference, in particular representatives of the non-aligned countries, drew attention to and showed concern about the continuing nuclear weapon test programmes and the steady increase of nuclear arsenals in spite of the negotiations on their limitation. In response to the Soviet contention that the basic problems of nuclear disarmament can be solved only with the participation of all nuclear weapon powers, two of which had not adhered to the NPT, opinion was expressed that the USA and the USSR, being by far the most powerful nations, should take the lead in the disarmament process, thereby encouraging other states to join. Various proposals were put forward with a view to speeding up the conclusion of arms-control agreements which would substantially reduce the levels of nuclear armaments and halt their qualitative development. All these proposals proved unacceptable to the nuclear weapon states. They refused to discuss any timetable for nuclear arms-control measures, even though according to the NPT such measures should be carried out "at an early date". They contended that the review conference was not competent to deal with a matter which was their exclusive concern, and that it was up to the SALT negotiators to determine the pace of progress in nuclear arms limitation. And yet the review conference recognized that it is essential to maintain in the implementation of the NPT an acceptable balance of mutual responsibilities and obligations of all the parties to the treaty. The proposals presented at the conference were intended precisely to redress the balance by matching the cessation of 'horizontal' proliferation with a halt to 'vertical' proliferation.

Doubts are sometimes expressed as to whether there exists a relationship between the two types of proliferation. Indeed, if at this stage any new country acquires nuclear weapons, it will do so presumably in order to intimidate or impress its immediate neighbours, or to enhance its international standing and gain more political prestige, influence and consideration in world councils, rather than to compete militarily with the present nuclear weapon powers, especially the USA and the USSR. Whether or not nuclear weapons will spread any further will also depend on the resolution of the most acute regional conflicts. Be that as it may, a treaty denying a powerful weapon to most nations in order to preserve a firebreak between the 'haves' and 'have-nots' is not likely to withstand the pressures of a continued arms race. Since nuclear weapons appear to have political and military usefulness for the nuclear powers, the non-nuclear weapon countries may feel that they too must obtain these advantages. A dynamic process of nuclear disarmament is therefore necessary to de-emphasize the role and utility of nuclear weaponry in world diplomacy and military strategy and to generate political and moral inhibitions dampening the nuclear ambitions of certain non-nuclear weapon states.

Consequently, the first review conference appealed to the nuclear weapon parties to the NPT to make every effort to reach agreement on the conclusion of an effective comprehensive test ban. It also called upon the USA and the USSR meanwhile to limit the number of their underground nuclear weapon tests to a minimum. Furthermore, the conference appealed to the two major powers to endeavour to conclude at the earliest possible date an agreement on the limitation of strategic arms outlined by their leaders in November 1974, and stated that it was looking forward to the commencement of follow-up negotiations on "further limitations of, and significant reductions in, their nuclear weapons systems" as soon as possible following the conclusion of such an agreement. Also the CCD was urged to increase its efforts to achieve effective disarmament agreements on all subjects on its agenda. These recommendations and appeals have not been fulfilled.

The treaty on the cessation of all nuclear weapon tests has not materialized and the rate of testing has not decreased. In fact, the number of nuclear explosions conducted by the USA and the USSR in 1978 was 20 per cent higher than that in 1975. Moreover, there has been a significant increase in the size of the nuclear arsenals. Since 1975, the total number of nuclear warheads on US and Soviet strategic bombers and missiles has grown by 30 per cent.

The second round of the US–Soviet strategic arms limitation talks (SALT II) will introduce new rules in the nuclear competition between the two powers. It will result mainly in a modest reduction of obsolete nuclear delivery vehicles, over the next few years, down to agreed levels, in some restructuring of the strategic forces, and in a few temporary restraints on the qualitative improvement of the nuclear weapon systems. But certain important missile deployments will still be carried out, as planned, thus further increasing the destructive power of the US and Soviet strategic arsenals, while non-strategic nuclear weapons are subject to no restrictions whatsoever.

The UN Special Session on Disarmament, held in 1978, called for negotiations on the cessation of the production of fissionable material for weapon purposes. Considering that such a cut-off measure would contribute towards the efforts to promote non-proliferation, limit the production of nuclear weapons and facilitate nuclear disarmament, the 33rd UN. General Assembly decided to transmit the matter to the Disarmament Committee. However, there are no prospects for an early agreement halting the production of weapon-grade fissionable material. In any case, the arms-control effect of such an agreement would not be significant: the amounts of weapon-grade material already accumulated by the nuclear weapon states make it possible for them to continue the manufacture of arms for the foreseeable future. Nevertheless, the introduction of international safeguards on all the relevant activities of the nuclear weapon states, which an effective cut-off treaty would require, might rectify one of the unbalanced provisions of the NPT: that which imposes control only on non-nuclear weapon states.

The final declaration of the review conference contains a reference to Article VII of the NPT, which re-affirms the right of any group of states to conclude regional treaties in order to assure the absence of nuclear weapons in their respective territories. The declaration expresses the conviction that the establishment of nuclear-weapon-free zones is an effective means of curbing the spread of nuclear weapons. At present, there exists only one such zone in the populated part of the world, namely in Latin America. Since the 1975 NPT review conference, the number of parties to the Treaty of Tlatelolco, which set up the zone, has increased, but its principal goal has not yet been achieved: Argentina and Brazil, the only countries in the area with any nuclear potential and aspirations, are still not bound by its provisions. Proposals for creating nuclear-weapon-free zones in Africa, the Middle East or South Asia were amply discussed, but no steps have been taken towards their realization. Equally, no progress has been made in the work of the Disarmament Committee. Since 1975, this Committee has been unable to work out any new agreement, with the exception of the convention prohibiting the use of environmental modification techniques for hostile purposes, which is of doubtful arms control value. (For an analysis of the continuing arms races and the hitherto unsuccessful attempts to stop them, see *SIPRI Yearbook 1979*.)

Nevertheless, a big-power nuclear rivalry and lack of progress in disarmament negotiations should not be used as justification for other nations to acquire or seek to acquire nuclear weapons. The NPT serves the interests of all, and the emergence of more new nuclear weapon powers would jeopardize international security in general.

VIII. The security of non-nuclear weapon states

In a UN Security Council resolution, adopted in 1968, the states renouncing the acquisition of nuclear weapons under the NPT have received an assurance of immediate assistance, in accordance with the UN Charter, in the event that they become "a victim of an act or an object of a threat of aggression in which nuclear weapons are used". But the significance of this document has been repeatedly questioned on the following grounds:

1. The resolution, and the declarations by the UK, the USA and the USSR associated with it, merely reaffirm the existing UN Charter obligation to provide or support assistance to a country attacked, irrespective of the type of weapon employed.
2. As long as all the nuclear weapon powers, that is, powers capable of using nuclear weapons, are also permanent members of the Security Council, any decision concerning military or non-military measures against the delinquent states would require their approval, and it is inconceivable that an aggressor nation would consent to a collective action being taken against itself.
3. Immediate active intervention, as envisaged by the resolution, is deemed unacceptable by some non-aligned and neutral states, unless assistance has been specifically requested by the victim.
4. The resolution in question relates to a possible action by the Security Council only when a threat of nuclear attack has been made or the attack has actually occurred; it does not offer assurance for the prevention of the use or threat of use of nuclear weapons.

These deficiencies were pointed out by many delegations at the review conference. At the same time, doubts were expressed as to whether it was at all possible in the present world situation to devise such 'positive' security guarantees which would be both credible and effective as well as acceptable to all countries, aligned and non-aligned. There was, therefore, wide support for additional, no-threat and no-use assurances in the form of 'negative' security guarantees. However, all requests for such guarantees put forward by the non-nuclear weapon states, especially those not covered by the protective 'umbrella' of the great powers, were rejected.

Now, after many years of refusal, the nuclear weapon powers seem to be yielding to these demands. By January 1979, China, France, the UK, the USA and the USSR had ratified Protocol II of the Treaty of Tlatelolco and are now all legally bound to respect the denuclearized status of Latin America. In addition, at the Special Session of the UN General Assembly devoted to disarmament, the USSR declared that it would never use nuclear weapons against those states which "renounce the production and acquisition of such weapons and do not have them on their territories". The USA announced that it would not use nuclear weapons against any non-nuclear weapon state which is party to the NPT or "any comparable internationally binding agreement not to acquire nuclear explosive devices". The UK issued a similar statement.

However, the above declarations do not go far enough in providing adequate guarantees. The Soviet Union has proposed that "appropriate" bilateral security agreements be concluded between the nuclear and non-nuclear weapon countries—something that is obviously unacceptable to many neutral and non-aligned states, not to speak of the members of the military alliances. The United States, on the other hand (similarly to the United Kingdom), has explicitly excluded from its non-use commitment non-nuclear weapon states allied to a nuclear weapon power or "associated" with such a power in carrying out an attack on the USA or its allies. As a matter of fact, none of these states is prepared to commit itself not to be the first to use nuclear weapons "under any circumstances", as China has done. (France is willing to give non-use guarantees only to nuclear-weapon-free zones.)

The question of security assurances is essentially a multilateral proposition. Unilateral statements should, therefore, be converted into internationally agreed, legally binding instruments. The matter has been placed on the agenda of the Committee on Disarmament meeting in Geneva.

IX. The second review conference

By March 1979, the number of parties to the NPT had reached 106. This number, which includes three nuclear weapon powers—the UK, the USA and the USSR—as well as many highly developed countries not possessing nuclear weapons (for the list, see Appendix C), may be taken as evidence that the non-proliferation idea has been accepted by a substantial portion of the international community. However, the non-proliferation régime will be in constant danger as long as the NPT has not been subscribed to by all states having significant nuclear activities, and there are now about a dozen states belonging to this category which remain outside the treaty. Only such universal adherence to the NPT could reinforce the legal barrier against further nuclear weapon dissemination. The second review conference, to be held in 1980, provides an opportunity to promote this goal through concrete measures directed at both parties and non-parties to the NPT:

1. Pressure should be brought to bear upon non-parties by denying them supplies of nuclear material and equipment, while an outright defiance of the treaty should be met with even more stringent measures.

2. The parties should undertake to import nuclear material only from other parties, and greater assistance should be granted to them in the use of nuclear energy for peaceful purposes, as well as in the development of alternative sources of energy.

3. The sensitive parts of the nuclear fuel cycle should be internationalized for the benefit of the parties.

4. Safeguards procedures should be improved, and the authority of the IAEA strengthened, in order to enable quick detection of any diversion of fissionable material for weapon purposes, as well as quick action following detection.

5. More severe sanctions than those limited to nuclear material supplies should be envisaged in the case of breaches of the safeguards agreements.

6. Adherence to the internationally agreed standards for the physical protection of nuclear material should be made a condition for supplying such material.

7. Participation in the treaty should be made more attractive through the provision of internationally agreed, legally binding security assurances to the non-nuclear weapon parties.

8. The nuclear weapon powers should clearly commit themselves to reversing the nuclear arms race; they could start fulfilling their part of the NPT bargain by halting nuclear weapon tests, as well as undertaking to reduce their nuclear armaments, both strategic and tactical, by significant amounts and by a specified date, and to stop, or at least slow down, the qualitative improvement of these armaments.

9. The obligation not to assist others in the manufacture of nuclear weapons should apply not only to non-nuclear weapon states, but to all states without exception; consequently, all exports of nuclear material and equipment to nuclear weapon powers should be subject to IAEA safeguards with a view to preventing their use for weapon purposes.

Of the measures suggested above, those dealing with political aspects of the problem of non-proliferation are of primary importance, because the problem itself is basically political. But they ought to be accompanied by technical measures of control to assure a clear distinction between nuclear power and nuclear weapon. All this can be achieved through agreed statements of understanding of the NPT provisions and/or international instruments complementary to the treaty. The NPT is the main tool in stemming the dangerous proliferation drift, and no efforts must be spared to avert its collapse. It is, however, essential for the next review conference formally to recognize that the NPT is not an end in itself, but merely a transitional stage in the process of nuclear disarmament.

Appendix A

Treaty on the Non-Proliferation of Nuclear Weapons

Signed at London, Moscow and Washington on 1 July 1968.
Entered into force on 5 March 1970.
Depositaries: UK, US and Soviet governments.

The States concluding this Treaty, hereinafter referred to as the "Parties to the Treaty",

Considering the devastation that would be visited upon all mankind by a nuclear war and the consequent need to make every effort to avert the danger of such a war and to take measures to safeguard the security of peoples,

Believing that the proliferation of nuclear weapons would seriously enhance the danger of nuclear war,

In conformity with resolutions of the United Nations General Assembly calling for the conclusion of an agreement on the prevention of wider dissemination of nuclear weapons,

Undertaking to co-operate in facilitating the application of International Atomic Energy safeguards on peaceful nuclear activities,

Expressing their support for research, development and other efforts to further the application, within the framework of the International Atomic Energy Agency safeguards system, of the principle of safeguarding effectively the flow of source and special fissionable materials by use of instruments and other techniques at certain strategic points,

Affirming the principle that the benefits of peaceful applications of nuclear technology, including any technological by-products which may be derived by nuclear-weapon States from the development of nuclear explosive devices, should be available for peaceful purposes to all Parties to the Treaty, whether nuclear-weapon or non-nuclear-weapon States,

Convinced that, in furtherance of this principle, all Parties to the Treaty are entitled to participate in the fullest possible exchange of scientific information for, and to contribute alone or in co-operation with other States to, the further development of the applications of atomic energy for peaceful purposes,

Declaring their intention to achieve at the earliest possible date the cessation of the nuclear arms race and to undertake effective measures in the direction of nuclear disarmament,

Urging the co-operation of all States in the attainment of this objective,

Recalling the determination expressed by the Parties to the 1963 Treaty banning nuclear weapon tests in the atmosphere, in outer space and under water in its Preamble to seek to achieve the discontinuance of all test explosions of nuclear weapons for all time and to continue negotiations to this end,

Desiring to further the easing of international tension and the strengthening of trust between States in order to facilitate the cessation of the manufacture of nuclear weapons, the liquidation of all their existing stockpiles, and the elimination from national arsenals of nuclear weapons and the means of their delivery pursuant to a Treaty on general and complete disarmament under strict and effective international control,

Recalling that, in accordance with the Charter of the United Nations, States must refrain in their international relations from the threat or use of force against the territorial integrity or political independence of any State, or in any other manner inconsistent with the Purposes of the United Nations, and that the establishment and maintenance of international peace and security are to be promoted with the least diversion for armaments of the world's human and economic resources,

Have agreed as follows:

Article I

Each nuclear-weapon State Party to the Treaty undertakes not to transfer to any recipient whatsoever nuclear weapons or other nuclear explosive devices or control over such weapons or explosive devices directly, or indirectly; and not in any way to assist, encourage, or induce any non-nuclear-weapon State to manufacture or otherwise acquire nuclear weapons or other nuclear explosive devices, or control over such weapons or explosive devices.

Article II

Each non-nuclear-weapon State Party to the Treaty undertakes not to receive the transfer from any transferor whatsoever of nuclear weapons or other nuclear explosive devices or of control over such weapons or explosive devices directly, or indirectly; not to manufacture or otherwise acquire nuclear weapons or other nuclear explosive devices; and not to seek or receive any assistance in the manufacture of nuclear weapons or other nuclear explosive devices.

Article III

1. Each non-nuclear-weapon State Party to the Treaty undertakes to accept safeguards, as set forth in an agreement to be negotiated and concluded with the International Atomic Energy Agency in accordance with the Statute of the International Atomic Energy Agency and the Agency's safeguards system, for the exclusive purpose of verification of the fulfilment of its obligations assumed under this Treaty with a view to preventing diversion of nuclear energy from peaceful uses to nuclear weapons or other nuclear explosive devices. Procedures for the safeguards required by this Article shall be followed with respect to source or special fissionable material whether it is being produced, processed or used in any principal nuclear facility or is outside any such facility. The safeguards required by this Article shall be

applied on all source or special fissionable material in all peaceful nuclear activities within the territory of such State, under its jurisdiction, or carried out under its control anywhere.

2. Each State Party to the Treaty undertakes not to provide: (*a*) source or special fissionable material, or (*b*) equipment or material especially designed or prepared for the processing, use or production of special fissionable material, to any non-nuclear-weapon State for peaceful purposes, unless the source or special fissionable material shall be subject to the safeguards required by this Article.

3. The safeguards required by this Article shall be implemented in a manner designed to comply with Article IV of this Treaty, and to avoid hampering the economic or technological development of the Parties or international co-operation in the field of peaceful nuclear activities, including the international exchange of nuclear material and equipment for the processing, use or production of nuclear material for peaceful purposes in accordance with the provisions of this Article and the principle of safeguarding set forth in the Preamble of the Treaty.

4. Non-nuclear-weapon States Party to the Treaty shall conclude agreements with the International Atomic Energy Agency to meet the requirements of this Article either individually or together with other States in accordance with the Statute of the International Atomic Energy Agency. Negotiation of such agreements shall commence within 180 days from the original entry into force of this Treaty. For States depositing their instruments of ratification or accession after the 180-day period, negotiation of such agreements shall commence not later than the date of such deposit. Such agreements shall enter into force not later than eighteen months after the date of initiation of negotiations.

Article IV

1. Nothing in this Treaty shall be interpreted as affecting the inalienable right of all the Parties to the Treaty to develop research, production and use of nuclear energy for peaceful purposes without discrimination and in conformity with Articles I and II of this Treaty.

2. All the Parties to the Treaty undertake to facilitate, and have the right to participate in, the fullest possible exchange of equipment, materials and scientific and technological information for the peaceful uses of nuclear energy. Parties to the Treaty in a position to do so shall also co-operate in contributing alone or together with other States or international organizations to the further development of the applications of nuclear energy for peaceful purposes, especially in the territories of non-nuclear-weapon States Party to the Treaty, with due consideration for the needs of the developing areas of the world.

Article V

Each Party to the Treaty undertakes to take appropriate measures to ensure that, in accordance with this Treaty, under appropriate international observation and through appropriate international procedures, potential benefits from any peaceful applications of nuclear explosions will be made available to non-nuclear-weapon States Party to the Treaty on a non-discriminatory basis and that the charge to such Parties for the explosive devices used will be as low as possible and exclude any charge for research and development. Non-nuclear-weapon States Party to the Treaty shall be able to obtain such benefits, pursuant to a special international agreement or agreements, through an appropriate international body with adequate representation of non-nuclear-weapon States. Negotiations on this subject shall commence as soon as possible after the Treaty enters into force. Non-nuclear-weapon States Party to the Treaty so desiring may also obtain such benefits pursuant to bilateral agreements.

Article VI

Each of the Parties to the Treaty undertakes to pursue negotiations in good faith on effective measures relating to cessation of the nuclear arms race at an early date and to nuclear disarmament, and on a treaty on general and complete disarmament under strict and effective international control.

Article VII

Nothing in this Treaty affects the right of any group of States to conclude regional treaties in order to assure the total absence of nuclear weapons in their respective territories.

Article VIII

1. Any Party to the Treaty may propose amendments to this Treaty. The text of any proposed amendment shall be submitted to the Depositary Governments which shall circulate it to all Parties to the Treaty. Thereupon, if requested to do so by one-third or more of the Parties to the Treaty, the Depositary Governments shall convene a conference, to which they shall invite all the Parties to the Treaty, to consider such an amendment.

2. Any amendment to this Treaty must be approved by a majority of the votes of all the Parties to the Treaty, including the votes of all nuclear-weapon States Party to the Treaty and all other Parties which, on the date the amendment is circulated, are members of the Board of Governors of the International Atomic Energy Agency. The amendment shall enter into force for each Party that deposits its instrument of ratification of the amendment upon the deposit of such instruments of ratification by a majority of all the Parties, including the instruments of ratification of all nuclear-weapon States Party to the Treaty and all other Parties which, on the date the amendment is circulated, are members of the Board of Governors of the International Atomic Energy Agency. Thereafter, it shall enter into force for any other Party upon the deposit of its instrument of ratification of the amendment.

3. Five years after the entry into force of this Treaty, a conference of Parties to the Treaty shall be held in Geneva, Switzerland, in order to review the operation of this Treaty with a view to assuring that the purposes of the Preamble and the provisions of the Treaty are being realised. At intervals of five years thereafter, a majority of the Parties to the Treaty may obtain, by submitting a proposal to this effect to the Depositary Governments, the convening of further conferences with the same objective of reviewing the operation of the Treaty.

Article IX

1. This Treaty shall be open to all States for signature. Any State which does not sign the Treaty before its entry into force in accordance with paragraph 3 of this Article may accede to it at any time.

2. This Treaty shall be subject to ratification by signatory States. Instruments of ratification and instruments of accession shall be deposited with the Governments

of the United Kingdom of Great Britain and Northern Ireland, the Union of Soviet Socialist Republics and the United States of America, which are hereby designated the Depositary Governments.

3. This Treaty shall enter into force after its ratification by the States, the Governments of which are designated Depositaries of the Treaty, and forty other States signatory to this Treaty and the deposit of their instruments of ratification. For the purposes of this Treaty, a nuclear-weapon State is one which has manufactured and exploded a nuclear weapon or other nuclear explosive device prior to 1 January, 1967.

4. For States whose instruments of ratification or accession are deposited subsequent to the entry into force of this Treaty, it shall enter into force on the date of the deposit of their instruments of ratification or accession.

5. The Depositary Governments shall promptly inform all signatory and acceding States of the date of each signature, the date of deposit of each instrument of ratification or of accession, the date of the entry into force of this Treaty, and the date of receipt of any requests for convening a conference or other notices.

6. This Treaty shall be registered by the Depositary Governments pursuant to Article 102 of the Charter of the United Nations.

Article X

1. Each Party shall in exercising its national sovereignty have the right to withdraw from the Treaty if it decides that extraordinary events, related to the subject matter of this Treaty, have jeopardized the supreme interests of its country. It shall give notice of such withdrawal to all other Parties to the Treaty and to the United Nations Security Council three months in advance. Such notice shall include a statement of the extraordinary events it regards as having jeopardized its supreme interests.

2. Twenty-five years after the entry into force of the Treaty, a conference shall be convened to decide whether the Treaty shall continue in force indefinitely, or shall be extended for an additional fixed period or periods. This decision shall be taken by a majority of the Parties to the Treaty.

Article XI

This Treaty, the English, Russian, French, Spanish and Chinese texts of which are equally authentic, shall be deposited in the archives of the Depositary Governments. Duly certified copies of this Treaty shall be transmitted by the Depositary Governments to the Governments of the signatory and acceding States.

IN WITNESS WHEREOF the undersigned, duly authorised, have signed this Treaty.

DONE in triplicate, at the cities of London, Moscow and Washington, the first day of July, one thousand nine hundred and sixty-eight.

Appendix B

Final declaration of the Review Conference of the Parties to the Treaty on the Non-Proliferation of Nuclear Weapons, 30 May 1975

Preamble

The States Party to the Treaty on the Non-Proliferation of Nuclear Weapons which met in Geneva in May 1975, in accordance with the Treaty, to review the operation of the Treaty with a view to assuring that the purposes of the Preamble and the provisions of the Treaty are being realized,

Recognizing the continuing importance of the objectives of the Treaty,

Affirming the belief that universal adherence to the Treaty would greatly strengthen international peace and enhance the security of all States,

Firmly convinced that, in order to achieve this aim, it is essential to maintain, in the implementation of the Treaty, an acceptable balance of mutual responsibilities and obligations of all States Party to the Treaty, nuclear-weapon and non-nuclear-weapon States,

Recognizing that the danger of nuclear warfare remains a grave threat to the survival of mankind,

Convinced that the prevention of any further proliferation of nuclear weapons or nuclear explosive devices remains a vital element in efforts to avert nuclear warfare, and that the promotion of this objective will be furthered by more rapid progress towards the cessation of the nuclear arms race and the limitation and reduction of existing nuclear weapons, with a view to the eventual elimination from national arsenals of nuclear weapons, pursuant to a Treaty on general and complete disarmament under strict and effective international control,

Recalling the determination expressed by the Parties to seek to achieve the discontinuance of all test explosions of nuclear weapons for all time,

Considering that the trend towards détente in relations between States provides a favourable climate within which more significant progress should be possible towards the cessation of the nuclear arms race,

Noting the important role which nuclear energy can, particularly in changing economic circumstances, play in power production and in contributing to the progressive elimination of the economic and technological gap between developing and developed states,

Recognizing that the accelerated spread and development of peaceful applications of nuclear energy will, in the absence of effective safeguards, contribute to further proliferation of nuclear explosive capability,

Recognizing the continuing necessity of full co-operation in the application and improvement of International Atomic Energy Agency (IAEA) safeguards on peaceful nuclear activities,

Recalling that all Parties to the Treaty are entitled to participate in the fullest possible exchange of scientific information for, and to contribute alone or in co-operation with other States to, the further development of the applications of atomic energy for peaceful purposes,

Reaffirming the principle that the benefits of peaceful applications of nuclear technology, including any technological by-products which may be derived by nuclear-weapon States from the development of nuclear explosive devices, should be available for peaceful purposes to all Parties to the Treaty, and

Recognizing that all States Parties have a duty to strive for the adoption of tangible and effective measures to attain the objectives of the Treaty,

Declares as follows:

Purposes

The States Party to the Treaty reaffirm their strong common interest in averting the further proliferation of nuclear weapons. They reaffirm their strong support for the Treaty, their continued dedication to its principles and objectives, and their commitment to implement fully and more effectively its provisions.

They reaffirm the vital role of the Treaty in international efforts

— to avert further proliferation of nuclear weapons

— to achieve the cessation of the nuclear arms race and to undertake effective measures in the direction of nuclear disarmament, and

— to promote co-operation in the peaceful uses of nuclear energy under adequate safeguards.

Review of Articles I and II

The review undertaken by the Conference confirms that the obligations undertaken under Articles I and II of the Treaty have been faithfully observed by all Parties. The Conference is convinced that the continued strict observance of these Articles remains central to the shared objective of averting the further proliferation of nuclear weapons.

Review of Article III

The Conference notes that the verification activities of the IAEA under Article III, of the Treaty respect the sovereign rights of States and do not hamper the economic, scientific or technological development of the Parties to the Treaty or international co-operation in peaceful nuclear activities. It urges that this situation be maintained. The Conference attaches considerable importance to the continued application of safeguards under Article III, 1, on a non-discriminatory basis, for the equal benefit of all States Party to the Treaty.

The Conference notes the importance of systems of accounting for and control of nuclear material, from the standpoints both of the responsibilities of States Party to the Treaty and of co-operation with the IAEA in order to facilitate the implementation of the safeguards provided for in Article III, 1. The Conference expresses the hope that all States having peaceful nuclear activities will establish and maintain effective accounting and control systems and welcomes the readiness of the IAEA to assist states in so doing.

The Conference expresses its strong support for effective IAEA safeguards. In this context it recommends that intensified efforts be made towards the standardization and the universality of application of IAEA safeguards, while ensuring that safeguards agreements with non-nuclear-weapon States not Party to the Treaty are of adequate duration, preclude diversion to any nuclear explosive devices and contain appropriate provisions for the continuance of the application of safeguards upon re-export.

The Conference recommends that more attention and fuller support be given to the improvement of safeguards techniques, instrumentation, data-handling and implementation in order, among other things, to ensure optimum cost-effectiveness. It notes with satisfaction the establishment by the Director General of the IAEA of a standing advisory group on safeguards implementation.

The Conference emphasizes the necessity for the States Party to the Treaty that have not yet done so to conclude as soon as possible safeguards agreements with the IAEA.

With regard to the implementation of Article III, 2 of the Treaty, the Conference notes that a number of States suppliers of nuclear material or equipment have adopted certain minimum, standard requirements for IAEA safeguards in connexion with their exports of certain such items to non-nuclear-weapon States not Party to the Treaty (IAEA document INFCIRC/209 and Addenda). The Conference attaches particular importance to the condition, established by those States, of an undertaking of non-diversion to nuclear weapons or other nuclear explosive devices, as included in the said requirements.

The Conference urges that:

(a) in all achievable ways, common export requirements relating to safeguards be strengthened, in particular by extending the application of safeguards to all peaceful nuclear activities in importing States not Party to the Treaty;

(b) such common requirements be accorded the widest possible measure of acceptance among all suppliers and recipients;

(c) all Parties to the Treaty should actively pursue their efforts to these ends.

The Conference takes note of:

(a) the considered view of many Parties to the Treaty that the safeguards required under Article III, 2 should extend to all peaceful nuclear activities in importing States;

(b) (i) the suggestion that it is desirable to arrange for common safeguards requirements in respect of nuclear material processed, used or produced by the use of scientific and technological information transferred in tangible form to non-nuclear-weapon States not Party to the Treaty;

(ii) the hope that this aspect of safeguards could be further examined.

The Conference recommends that, during the review of the arrangements relating to the financing of safeguards in the IAEA which is to be undertaken by its Board of Governors at an appropriate time after 1975, the less favourable financial situation of the developing countries be fully taken into account. It recommends further that on that occasion, the Parties to the Treaty concerned seek measures that would restrict within appropriate limits the respective shares of developing countries in safeguards costs.

The Conference attaches considerable importance, so far as safeguards inspectors are concerned, to adherence by the IAEA to Article VII.D of its Statute, prescribing, among other things, that "due regard shall be paid . . . to the importance of recruiting the staff on as wide a geographical basis as possible"; it also recommends that safeguards training be made available to personnel from all geographic regions.

The Conference, convinced that nuclear materials should be effectively protected at all times, urges that action be pursued to elaborate further, within the IAEA, concrete recommendations for the physical protection of nuclear material in use, storage and transit, including principles relating to the responsibility of States, with a view to ensuring a uniform, minimum level of effective protection for such material.

It calls upon all States engaging in peaceful nuclear activities (i) to enter into such international agreements and arrangements as may be necessary to ensure such protection; and (ii) in the framework of their respective physical protection systems, to give the earliest possible effective application to the IAEA's recommendations.

Review of Article IV

The Conference reaffirms, in the framework of Article IV, 1, that nothing in the Treaty shall be interpreted as affecting, and notes with satisfaction that nothing in the Treaty has been identified as affecting, the inalienable right of all the Parties to the Treaty to develop research, production and use of nuclear energy for peaceful purposes without discrimination and in conformity with Articles I and II of the Treaty.

The Conference reaffirms, in the framework of Article IV, 2, the undertaking by all Parties to the Treaty to facilitate the fullest possible exchange of equipment, materials and scientific and technological information for the peaceful uses of nuclear energy and the right of all Parties to the Treaty to participate in such exchange and welcomes the efforts made towards that end. Noting that the Treaty constitutes a favourable framework for broadening international co-operation in the peaceful uses of nuclear energy, the Conference is convinced that on this basis, and in conformity with the Treaty, further efforts should be made to ensure that the benefits of peaceful applications of nuclear technology should be available to all Parties to the Treaty.

The Conference recognizes that there continues to be a need for the fullest possible exchange of nuclear materials, equipment and technology, including up-to-date developments, consistent with the objectives and safeguards requirements of the Treaty. The Conference reaffirms the undertaking of the Parties to the Treaty in a position to do so to co-operate in contributing, alone or together with other States or international organizations, to the further development of the applications of nuclear energy for peaceful purposes, especially in the territories of non-nuclear-weapon States Party to the Treaty, with due consideration for the needs of the developing areas of the world. Recognizing, in the context of Article IV, 2, those growing needs of developing States the Conference considers it necessary to continue and increase assistance to them in this field bilaterally and through such multilateral channels as the IAEA and the United Nations Development Programme.

The Conference is of the view that, in order to implement as fully as possible Article IV of the Treaty, developed States Party to the Treaty should consider taking measures, making contributions and establishing programmes, as soon as possible, for the provision of special assistance in the peaceful uses of nuclear energy for developing States Party to the Treaty.

The Conference recommends that, in reaching decisions on the provision of equipment, materials, services and scientific and technological information for the peaceful uses of nuclear energy, on concessional and other appropriate financial arrangements and on the furnishing of technical assistance in the nuclear field, including co-operation related to the continuous operation of peaceful nuclear facilities, States Party to the Treaty should give weight to adherence to the Treaty by recipient States. The Conference recommends, in this connexion, that any special measures of co-operation to meet the growing needs of developing States Party to the Treaty might include increased and supplemental voluntary aid provided bilaterally or through multilateral channels such as the IAEA's facilities for administering funds-in-trust and gifts-in-kind.

The Conference further recommends that States Party to the Treaty in a position to do so, meet, to the fullest extent possible, "technically sound" requests for technical assistance, submitted to the IAEA by developing States Party to the Treaty, which the IAEA is unable to finance from its own resources, as well as such "technically sound" requests as may be made by developing States Party to the Treaty which are not Members of the IAEA.

The Conference recognizes that regional or multinational nuclear fuel cycle centres may be an advantageous way to satisfy, safely and economically, the needs of many States in the course of initiating or expanding nuclear power programmes, while at the same time facilitating physical protection and the application of IAEA safeguards, and contributing to the goals of the Treaty.

The Conference welcomes the IAEA's studies in this area, and recommends that they be continued as expeditiously as possible. It considers that such studies should include, among other aspects, identification of the complex practical and organizational difficulties which will need to be dealt with in connexion with such projects.

The Conference urges all Parties to the Treaty in a position to do so to co-operate in these studies, particularly by providing to the IAEA where possible economic data concerning construction and operation of facilities such as chemical reprocessing plants, plutonium fuel fabrication plants, waste management installations, and longer-term spent fuel storage, and by assistance to the IAEA to enable it to undertake feasibility studies concerning the establishment of regional nuclear fuel cycle centres in specific geographic regions.

The Conference hopes that, if these studies lead to positive findings, and if the establishment of regional or multinational nuclear fuel cycle centres is undertaken, Parties to the Treaty in a position to do so, will co-operate in, and provide assistance for, the elaboration and realization of such projects.

Review of Article V

The Conference reaffirms the obligation of Parties to the Treaty to take appropriate measures to ensure that potential benefits from any peaceful applications of nuclear explosions are made available to non-nuclear-weapon States Party to the Treaty in full accordance with the provisions of Article V and other applicable international obligations. In this connexion, the Conference also reaffirms that such services should be provided to non-nuclear-weapon States Party to the Treaty on a non-discriminatory basis and that the charge to such Parties for the explosive devices used should be as low as possible and exclude any charge for research and development.

The Conference notes that any potential benefits should be made available to non-nuclear-weapon States not Party to the Treaty by way of nuclear explosion services provided by nuclear-weapon States, as defined by the Treaty, and conducted under the appropriate international observation and international procedures called for in Article V and in accordance with other applicable international obligations. The Conference considers it imperative that access to potential benefits of nuclear explosions for peaceful purposes not lead to any proliferation of nuclear explosive capability.

The Conference considers the IAEA to be the appropriate international body, referred to in Article V of the Treaty, through which potential benefits from peaceful applications of nuclear explosions could be made available to any non-nuclear-weapon State. Accordingly, the Conference urges the IAEA to expedite work on identifying and examining the important legal issues involved in, and to commence consideration of, the structure and content of the special international agreement or agreements contemplated in Article V of the Treaty, taking into account the views of the Conference of the Committee on Disarmament (CCD) and the United Nations General Assembly and enabling States Party to the Treaty but not Members of the IAEA which would wish to do so to participate in such work.

The Conference notes that the technology of nuclear explosions for peaceful purposes is still at the stage of development and study and that there are a number of interrelated international legal and other aspects of such explosions which still need to be investigated.

The Conference commends the work in this field that has been carried out within the IAEA and looks forward to the continuance of such work pursuant to United Nations General Assembly resolution 3261 D (XXIX). It emphasizes that the IAEA should play the central role in matters relating to the provision of services for the application of nuclear explosions for peaceful purposes. It believes that the IAEA should broaden its consideration of this subject to encompass, within its area of competence, all aspects and implications of the practical applications of nuclear explosions for peaceful purposes. To this end it urges the IAEA to set up appropriate machinery within which intergovernmental discussion can take place and through which advice can be given on the Agency's work in this field.

The Conference attaches considerable importance to the consideration by the CCD, pursuant to United Nations General Assembly resolution 3261 D (XXIX) and taking due account of the views of the IAEA, of the arms control implications of nuclear explosions for peaceful purposes.

The Conference notes that the thirtieth session of the United Nations General Assembly will receive reports pursuant to United Nations General Assembly resolution 3261 D (XXIX) and will provide an opportunity for States to discuss questions related to the application of nuclear explosions for peaceful purposes. The Conference further notes that the results of discussion in the United Nations General Assembly at its thirtieth session will be available to be taken into account by the IAEA and the CCD for their further consideration.

Review of Article VI

The Conference recalls the provisions of Article VI of the Treaty under which all Parties undertook to pursue negotiations in good faith on effective measures relating

— to the cessation of the nuclear arms race at an early date and

— to nuclear disarmament and

— to a treaty on general and complete disarmament under strict and effective international control.

While welcoming the various agreements on arms limitation and disarmament elaborated and concluded over the last few years as steps contributing to the implementation of Article VI of the Treaty, the Conference expresses its serious concern that the arms race, in particular the nuclear arms race, is continuing unabated.

The Conference therefore urges constant and resolute efforts by each of the Parties to the Treaty, in particular by the nuclear-weapon States, to achieve an early and effective implementation of Article VI of the Treaty.

The Conference affirms the determination expressed in the preamble to the 1963 Partial Test Ban Treaty and reiterated in the preamble to the Non-Proliferation Treaty to achieve the discontinuance of all test explosions of nuclear weapons for all time. The Conference expresses the view that the conclusion of a treaty banning all nuclear weapons tests is one of the most important measures to halt the nuclear arms race. It expresses the hope that the nuclear-weapon States Party to the Treaty will take the lead in reaching an early solution of the technical and political difficulties on this issue. It appeals to these States to make every effort to reach agreement on the conclusion of an effective comprehensive test ban. To this end, the desire was expressed by a considerable number of delegations at the Conference that the nuclear-weapon States Party to the Treaty should as soon as possible enter into an agreement, open to all States and containing appropriate provisions to ensure its effectiveness, to halt all nuclear weapons tests of adhering States for a specified time, whereupon the terms of such an agreement would be reviewed in the light of the opportunity, at that time, to achieve a universal and permanent cessation of all nuclear weapons tests. The Conference calls upon the nuclear-weapon States signatories of the Treaty on the Limitation of Underground Nuclear Weapons tests, meanwhile, to limit the number of their underground nuclear weapons tests to a minimum. The Conference believes that such steps would constitute an incentive of particular value to negotiations for the conclusion of a treaty banning all nuclear weapons test explosions for all time.

The Conference appeals to the nuclear-weapon States Parties to the negotiations on the limitation of strategic arms to endeavour to conclude at the earliest possible date the new agreement that was outlined by their leaders in November 1974. The Conference looks forward to the commencement of follow-on negotiations on further limitations of, and significant reduction in, their nuclear weapons systems as soon as possible following the conclusion of such an agreement.

The Conference notes that, notwithstanding earlier progress, the CCD has recently been unable to reach agreement on new substantive measures to advance the objectives of Article VI of the Treaty. It urges, therefore, all members of the CCD Party to the Treaty, in particular the nuclear-weapon States Party, to increase their efforts to achieve effective disarmament agreements on all subjects on the agenda of the CCD.

The Conference expresses the hope that all States Party to the Treaty, through the United Nations and the CCD and other negotiations in which they participate, will work with determination towards the conclusion of arms limitation and disarmament agreements which will contribute to the goal of general and complete disarmament under strict and effective international control.

The Conference expresses the view that, disarmament being a matter of general concern, the provision of information to all governments and peoples on the situation in the field of the arms race and disarmament is of great importance for the attainment of the aims of Article VI. The Conference therefore invites the United Nations to consider ways and means of improving its existing facilities for collection, compilation and dissemination of information on disarmament issues, in order to keep all governments as well as world public opinion properly informed on progress achieved in the realization of the provisions of Article VI of the Treaty.

Review of Article VII and the security of non-nuclear weapon states

Recognizing that all States have need to ensure their independence, territorial integrity and sovereignty, the Conference emphasizes the particular importance of assuring and strengthening the security of non-nuclear-weapon States Parties which have renounced the acquisition of nuclear weapons. It acknowledges that States Parties find themselves in different security situations and therefore that various appropriate means are necessary to meet the security concerns of States Parties.

The Conference underlines the importance of adherence to the Treaty by non-nuclear-weapon States as the best means of reassuring one another of their renunciation of nuclear weapons as one of the effective means of strengthening their mutual security.

The Conference takes note of the continued determination of the Depositary States to honour their statements, which were welcomed by the United Nations Security Council in resolution 255 (1968), that, to ensure the security of the non-nuclear-weapon States Party to the Treaty, they will provide or support immediate assistance, in accordance with the Charter, to any non-nuclear weapon State Party to the Treaty which is a victim of an act or an object of a threat of aggression in which nuclear weapons are used.

The Conference, bearing in mind Article VII of the Treaty, considers that the establishment of internationally recognized nuclear-weapon-free zones on the initiative and with the agreement of the directly concerned States of the zone, represents an effective means of curbing the spread of nuclear weapons, and could contribute significantly to the security of those States. It welcomes the steps which have been taken toward the establishment of such zones.

The Conference recognizes that for the maximum effectiveness of any Treaty arrangements for establishing a nuclear-weapon-free zone the co-operation of the nuclear-weapon States is necessary. At the Conference it was urged by a considerable number of delegations that nuclear-weapon States should provide, in an appropriate manner, binding security assurances to those States which become fully bound by the provisions of such regional arrangements.

At the Conference it was also urged that determined efforts must be made especially by the nuclear weapon States Party to the Treaty, to ensure the security of all non-nuclear-weapon States Parties. To this end the Conference urges all States, both nuclear-weapon States and non-nuclear-weapon States to refrain, in accordance

with the Charter of the United Nations, from the threat or the use of force in relations between States, involving either nuclear or non-nuclear-weapons. Additionally,
it stresses the responsibility of all Parties to the Treaty and especially the nuclear-
weapon States, to take effective steps to strengthen the security of non-nuclear-
weapon States and to promote in all appropriate fora the consideration of all
practical means to this end, taking into account the views expressed at this Conference.

Review of Article VIII

The Conference invites States Party to the Treaty which are Members of the United
Nations to request the Secretary-General of the United Nations to include the
following item in the provisional agenda of the thirty-first session of the General
Assembly: "Implementation of the conclusions of the first Review Conference of the
Parties to the Treaty on the Non-Proliferation of Nuclear Weapons".

The States Party to the Treaty participating in the Conference propose to the
Depositary Governments that a second Conference to review the operation of the
Treaty be convened in 1980.

The Conference accordingly invites States Party to the Treaty which are
Members of the United Nations to request the Secretary-General of the United
Nations to include the following item in the provisional agenda of the thirty-third
session of the General Assembly: "Implementation of the conclusions of the first
Review Conference of the Parties to the Treaty on the Non-Proliferation of Nuclear
Weapons and establishment of a preparatory committee for the second Confer-
ence."

Review of Article IX

The five years that have passed since the entry into force of the Treaty have demonstrated its wide international acceptance. The Conference welcomes the recent
progress towards achieving wider adherence. At the same time, the Conference notes
with concern that the Treaty has not as yet achieved universal adherence. Therefore,
the Conference expresses the hope that the States that have not already joined the
Treaty should do so at the earliest possible date.

Appendix C

List of states which have signed, ratified, acceded or succeeded to the Treaty on the Non-Proliferation of Nuclear Weapons (NPT) as of March 1979

Total number of parties: 106.
NPT safeguards agreements are in force with 62 non-nuclear weapon states.

Note:

Abbreviations used in the table:

S = signature
R = deposit of instruments of ratification, accession or succession

Place of signature and/or deposit of the instrument of ratification, accession or succession:

L = London
M = Moscow
W = Washington

SA = safeguards agreement in force with the International Atomic Energy Agency (IAEA)

Afghanistan	S:	1 Jul	1968 LMW	Bahamas	R:[2]	11 Aug	1976 L
	R:	4 Feb	1970 W			13 Aug	1976 W
		5 Feb	1970 M			30 Aug	1976 M
		5 Mar	1970 L				
	S.A.:	20 Feb	1978	Barbados	S:	1 Jul	1968 W
Australia	S:[1]	27 Feb	1970 LMW	Belgium	S:	20 Aug	1968 LMW
	R:	23 Jan	1973 LMW		R:	2 May	1975 LW
	S.A.:	10 Jul	1974			4 May	1975 M
					S.A.:	21 Feb	1977
Austria	S:	1 Jul	1968 LMW				
	R:	27 Jun	1969 LMW	Benin	S:	1 Jul	1968 W
	S.A.:	23 Jul	1972		R:	31 Oct	1972 W

Country		Date		
Bolivia	S	1 Jul	1968	W
	R:	26 May	1970	W
Botswana	S:	1 Jul	1968	W
	R:	28 Apr	1969	L
Bulgaria	S:	1 Jul	1968	LMW
	R:	5 Sep	1969	W
		18 Sep	1969	M
		3 Nov	1969	L
	S.A.:	29 Feb	1972	
Burundi	R:	19 Mar	1971	M
Cameroon: see United Republic of Cameroon				
Canada	S:	23 Jul	1968	L W
		29 Jul	1968	M
	R:	8 Jan	1969	LMW
	S.A.	21 Feb	1972	
Central African Empire	R:	25 Oct	1970	W
Chad	S:	1 Jul	1968	M
	R:	10 Mar	1971	W
		11 Mar	1971	M
		23 Mar	1971	L
Colombia	S:	1 Jul	1968	W
Congo	R:	23 Oct	1978	W
Costa Rica	S:	1 Jul	1968	W
	R:	3 Mar	1970	W
Cyprus	S:	1 Jul	1968	LMW
	R:	10 Feb	1970	M
		16 Feb	1970	W
		5 Mar	1970	L
	S.A.:	26 Jan	1973	
Czecho-slovakia	S:	1 Jul	1968	LMW
	R:	22 Jul	1969	LMW
	S.A.:	3 Mar	1972	
Dahomey: see Benin				
Democratic Kampuchea	R:	2 Jun	1972	W
Democratic Yemen*	S:	14 Nov	1968	M
Denmark	S:	1 Jul	1968	LMW
	R:	3 Jan	1969	LMW
	S.A.:	21 Feb	1977	
Dominican Republic	S:	1 Jul	1968	W
	R:	24 Jul	1971	W
	S.A.:	11 Oct	1973	
Ecuador	S:	9 Jul	1968	W
	R:	7 Mar	1969	W
	S.A.:	10 Mar	1975	
Egypt	S:	1 Jul	1968	LM
El Salvador	S:	1 Jul	1968	W
	R:	11 Jul	1972	W
	S.A.:	22 Apr	1975	
Ethiopia	S:	5 Sep	1968	LMW
	R:	5 Feb	1970	M
		5 Mar	1970	L W
	S.A.:	2 Dec	1977	
Fiji	R:[2]	21 Jul	1972	W
		14 Aug	1972	L
		29 Aug	1972	M
	S.A.:	22 Mar	1973	
Finland	S:	1 Jul	1968	LMW
	R:	5 Feb	1969	LMW
	S.A.:	9 Feb	1972	
Gabon	R:	19 Feb	1974	W
Gambia	S:	4 Sep	1968	L
		20 Sep	1968	W
		24 Sep	1968	M
	R:	12 May	1975	W
	S.A.:	8 Aug	1978	
German Democratic Republic	S:	1 Jul	1968	M
	R:[3]	31 Oct	1969	M
	S.A.:	7 Mar	1972	
Germany, Federal Republic of	S:	28 Nov	1969	LMW
	R:[4]	2 May	1975	L W
	S.A.:	21 Feb	1977	
Ghana	S:	1 Jul	1968	MW
		24 Jul	1968	L
	R:	4 May	1970	L
		5 May	1970	W
		11 May	1970	M
	S.A.:	17 Feb	1975	
Greece	S:	1 Jul	1968	MW
	R:	11 Mar	1970	W
	S.A.:	1 Mar	1972	

* Yemen refers to the Yemen Arab Republic (Northern Yemen). Democratic Yemen refers to the People's Democratic Republic of Yemen (Southern Yemen).

Grenada	R:[2]	2 Sep	1975	L
		3 Dec	1975	W
Guatemala	S:	26 Jul	1968	W
	R:	22 Sep	1970	W
Guinea-Bissau	R:	20 Aug	1976	M
Haiti	S:	1 Jul	1968	W
	R:	2 Jun	1970	W
Holy See	R:[5]	25 Feb	1971	LMW
(Vatican City)	S.A.:	1 Aug	1972	
Honduras	S:	1 Jul	1968	W
	R:	16 May	1973	W
	S.A.:	18 Apr	1975	
Hungary	S:	1 Jul	1968	LMW
	R:	27 May	1969	LMW
	S.A.:	30 Mar	1972	
Iceland	S:	1 Jul	1968	MW
	R:	18 Jul	1969	LMW
	S.A.:	16 Oct	1974	
Indonesia	S:[6]	2 Mar	1970	LMW
Iran	S:	1 Jul	1968	LMW
	R:	2 Feb	1970	W
		10 Feb	1970	M
		5 Mar	1970	L
	S.A.:	15 May	1974	
Iraq	S:	1 Jul	1968	M
	R:	29 Oct	1969	M
	S.A.:	29 Feb	1972	
Ireland	S:	1 Jul	1968	MW
		4 Jul	1968	L
	R:	1 Jul	1968	W
		2 Jul	1968	M
		4 Jul	1968	L
	S.A.:	21 Feb	1977	
Italy	S:	28 Jan	1969	LMW
	R:[7]	2 May	1975	LW
		4 May	1975	M
	S.A.:	21 Feb	1977	
Ivory Coast	S:	1 Jul	1968	W
	R:	6 Mar	1973	W
Jamaica	S:	14 Apr	1969	LMW
	R:	5 Mar	1970	LMW
	S.A.:	6 Nov	1978	
Japan	S:	3 Feb	1970	LMW
	R:[8]	8 Jun	1976	LMW
	S.A.:	2 Dec	1977	
Jordan	S:	10 Jul	1968	W
	R:	11 Feb	1970	W
	S.A.:	21 Feb	1978	
Kampuchea: see Democratic Kampuchea				
Kenya	S:	1 Jul	1968	W
	R:	11 Jun	1970	M
Korea, South	S:[9]	1 Jul	1968	W
	R:[10]	23 Apr	1975	W
	S.A.:	14 Nov	1975	
Kuwait	S:	15 Aug	1968	MW
		22 Aug	1968	L
Lao People's	S:	1 Jul	1968	LMW
Democratic	R:	20 Feb	1970	M
Republic		5 Mar	1970	LW
Lebanon	S:	1 Jul	1968	LMW
	R:	15 Jul	1970	LM
		20 Nov	1970	W
	S.A.:	5 Mar	1973	
Lesotho	S:	9 Jul	1968	W
	R:	20 May	1970	W
	S.A.:	12 Jun	1973	
Liberia	S:	1 Jul	1968	W
	R:	5 Mar	1970	W
Libya	S:	18 Jul	1968	L
		19 Jul	1968	W
		23 Jul	1968	M
	R:	26 May	1975	LMW
Liechtenstein	R:[12]	20 Apr	1978	LMW
Luxembourg	S:	14 Aug	1968	LMW
	R:	2 May	1975	LW
		4 May	1975	M
	S.A.:	21 Feb	1977	
Madagascar	S:	22 Aug	1968	W
	R:	8 Oct	1970	W
	S.A.:	14 Jun	1973	
Malaysia	S:	1 Jul	1968	LMW
	R:	5 Mar	1970	LMW
	S.A.:	29 Feb	1972	
Maldives	S:	11 Sep	1968	W
	R:	7 Apr	1970	W
	S.A.:	2 Oct	1977	

Mali	S:	14 Jul	1969	W		Peru	S:	1 Jul	1968 W
		15 Jul	1969	M			R:	3 Mar	1970 W
	R:	10 Feb	1970	M					
		5 Mar	1970	W		Philippines	S:	1 Jul	1968 W
								18 Jul	1968 M
Malta	S:	17 Apr	1969	W			R:	5 Oct	1972 W
	R:	6 Feb	1970	W				16 Oct	1972 L
								20 Oct	1972 M
Mauritius	S:	1 Jul	1968	W			S.A.:	16 Oct	1974
	R:	8 Apr	1969	W					
		14 Apr	1969	L		Poland	S:	1 Jul	1968 LMW
		25 Apr	1969	M			R:	12 Jun	1969 LMW
	S.A.:	31 Jan	1973				S.A.:	11 Oct	1972
Mexico	S:[11]	26 Jul	1968	LMW		Portugal	R:	15 Dec	1977 LMW
	R:	21 Jan	1969	LMW					
	S.A.:	14 Sep	1973			Romania	S:	1 Jul	1968 LMW
							R:	4 Feb	1970 LMW
Mongolia	S:	1 Jul	1968	M			S.A.:	27 Oct	1972
	R:	14 May	1969	M					
	S.A.:	5 Sep	1972			Rwanda	R:	20 May	1975 LMW
Morocco	S:	1 Jul	1968	LMW		Samoa	R:	17 Mar	1975 M
	R:	27 Nov	1970	M				18 Mar	1975 W
		30 Nov	1970	L				26 Mar	1975 L
		16 Dec	1970	W			S.A.:	22 Jan	1979
	S.A.:	18 Feb	1975						
						San Marino	S:[9]	1 Jul	1968 W
Nepal	S:	1 Jul	1968	LMW				29 Jul	1968 L
	R:	5 Jan	1970	W				21 Nov	1968 M
		9 Jan	1970	M			R:	10 Aug	1970 L
		3 Feb	1970	L				20 Aug	1970 M
	S.A.:	22 Jun	1972					31 Aug	1970 W
Netherlands	S:	20 Aug	1968	LMW		Senegal	S:	1 Jul	1968 MW
	R:	2 May	1975	LMW				26 Jul	1968 L
	S.A.:	21 Feb	1977				R:	17 Dec	1970 M
								22 Dec	1970 W
New Zealand	S:	1 Jul	1968	LMW				15 Jan	1971 L
	R:	10 Sep	1969	LMW					
	S.A.:	29 Feb	1972			Sierra Leone	R:	26 Feb	1975 LMW
Nicaragua	S:	1 Jul	1968	LW		Singapore	S:	5 Feb	1970 LMW
	R:	6 Mar	1973	W			R:	10 Mar	1976 LMW
	S.A.:	29 Dec	1976				S.A.:	18 Oct	1977
Nigeria	S:	1 Jul	1968	LMW		Somalia	S:	1 Jul	1968 LMW
	R:	27 Sep	1968	L			R:	5 Mar	1970 L
		7 Oct	1968	W				12 Nov	1970 W
		14 Oct	1968	M					
Norway	S:	1 Jul	1968	LMW		Sri Lanka	S:	1 Jul	1968 LMW
	R:	5 Feb	1969	LMW			R:	5 Mar	1979 W
	S.A.:	1 Mar	1972						
						Sudan	S:	24 Dec	1968 M
Panama	S:	1 Jul	1968	W			R:	31 Oct	1973 W
	R:	13 Jan	1977	W				22 Nov	1973 M
								10 Dec	1973 L
Paraguay	S:	1 Jul	1968	W			S.A.:	7 Jan	1977
	R:	4 Feb	1970	W					
		5 Mar	1970	L		Suriname	R:[2]	30 Jun	1976 W
	S.A.:	20 Mar	1979				S.A.:	2 Feb	1979

Swaziland	S:	24 Jun	1969	L	United	S:	1 Jul	1968	LMW
	R:	11 Dec	1969	L	Kingdom	R:[13]	27 Nov	1968	LW
		16 Dec	1969	W			29 Nov	1968	M
		12 Jan	1970	M		S.A.:[14]	14 Aug	1978	
	S.A.:	28 Jul	1975						
Sweden	S:	19 Aug	1968	LMW	United	S:	17 Jul	1968	W
	R:	9 Jan	1970	LMW	Republic of		18 Jul	1968	M
	S.A.:	14 Apr	1975		Cameroon	R:	8 Jan	1969	W
Switzerland	S:	27 Nov	1969	LMW	United States	S:	1 Jul	1968	LMW
	R:[12]	9 Mar	1977	LMW		R:	5 Mar	1970	LMW
	S.A.:	6 Sep	1978			S.A.:[15]			
Syria	S:	1 Jul	1968	M					
	R:[9]	24 Sep	1969	M	Upper Volta	S:	25 Nov	1968	W
							11 Aug	1969	M
Taiwan	S:	1 Jul	1968	W		R:	3 Mar	1970	W
	R:	27 Jan	1970	W					
					Uruguay	S	1 Jul	1968	W
Thailand	R:	7 Dec	1972	L		R:	31 Aug	1970	W
	S.A.:	16 May	1974			S.A.:	17 Sep	1976	
Togo	S:	1 Jul	1968	W	Venezuela	S:	1 Jul	1968	W
	R:	26 Feb	1970	W		R:	25 Sep	1975	L
							26 Sep	1975	W
Tonga	R:[2]	7 Jul	1971	L			3 Oct	1975	M
		15 Jul	1971	W					
		24 Aug	1971	M	Yemen*	S:	23 Sep	1968	M
Trinidad and	S:	20 Aug	1968	W	Yugoslavia	S:	10 Jul	1968	LMW
Tobago		22 Aug	1968	L		R:[16]	4 Mar	1970	W
							5 Mar	1970	LM
Tunisia	S:	1 Jul	1968	LMW		S.A.:	28 Dec	1973	
	R:	26 Feb	1970	LMW					
Turkey	S:	28 Jan	1969	LMW	Zaire	S:	22 Jul	1968	W
							26 Jul	1968	M
Union of	S:	1 Jul	1968	LMW			17 Sep	1968	L
Soviet Social-	R:	5 Mar	1970	LMW		R:	4 Aug	1970	W
ist Republics						S.A.:	9 Nov	1972	

---

*Yemen refers to the Yemen Arab Republic (Northern Yemen). Democratic Yemen refers to the People's Democratic Republic of Yemen (Southern Yemen).

Notes:

[1] On signing the Treaty, Australia stated, *inter alia*, that it regarded it as essential that the Treaty should not affect security commitments under existing treaties of mutual security.
[2] Notification of succession.
[3] On 25 November 1969, the United States notified its non-acceptance of notification of signature and ratification by the German Democratic Republic which it then did not recognize as a state. On 4 September 1974, the two countries established diplomatic relations with each other.
[4] On depositing the instrument of ratification, the Federal Republic of Germany reiterated the declaration made at the time of signing: it reaffirmed its expectation that the nuclear weapon states would intensify their efforts in accordance with the undertakings under Article VI of the Treaty, as well as its understanding that the security of FR Germany continued to be ensured by NATO; it stated that no provision of the Treaty may be interpreted in such a way as to hamper further development of European unification; that research, development and use of nuclear energy for peaceful purposes, as well as international and multinational co-operation in this field, must not be prejudiced by the Treaty; that the application of the Treaty, including

the implementation of safeguards, must not lead to discrimination of the nuclear industry of FR Germany in international competition; and that it attached vital importance to the undertaking given by the United States and the United Kingdom concerning the application of safeguards to their peaceful nuclear facilities, hoping that other nuclear weapon states would assume similar obligations.

In a separate note, FR Germany declared that the Treaty will also apply to Berlin (West) without affecting Allied rights and responsibilities, including those relating to demilitarization. In notes of 24 July, 19 August, and 25 November 1975, respectively, addressed to the US Department of State, Czechoslovakia, the USSR and the German Democratic Republic stated that this declaration by FR Germany had no legal effect.

[5] On acceding to the Treaty, the Holy See stated, *inter alia*, that the Treaty will attain in full the objectives of security and peace and justify the limitations to which the states party to the Treaty submit, only if it is fully executed in every clause and with all its implications. This concerns not only the obligations to be applied immediately but also those which envisage a process of ulterior commitments. Among the latter, the Holy See considers it suitable to point out the following:

(*a*) The adoption of appropriate measures to ensure, on a basis of equality, that all nonnuclear weapon states party to the Treaty will have available to them the benefits deriving from peaceful applications of nuclear technology.

(*b*) The pursuit of negotiations in good faith on effective measures relating to cessation of the nuclear arms race at an early date and to nuclear disarmament, and on a treaty on general and complete disarmament under strict and effective international control.

[6] On signing the Treaty, Indonesia stated, *inter alia*, that the government of Indonesia attaches great importance to the declarations of the USA, the UK and the USSR affirming their intention to provide immediate assistance to any non-nuclear weapon state party to the Treaty that is a victim of an act of aggression in which nuclear weapons are used.

Of utmost importance, however, is not the action *after* a nuclear attack has been committed but the guarantees to prevent such an attack. The Indonesian government trusts that the nuclear weapon states will study further the question of effective measures to ensure the security of the non-nuclear weapon states. Its decision to sign the Treaty is not to be taken in any way as a decision to ratify the Treaty. The ratification will be considered after matters of national security, which are of deep concern to the government and people of Indonesia, have been clarified to their satisfaction.

[7] Italy stated that in its belief nothing in the treaty was an obstacle to the unification of the countries of Western Europe; it noted full compatibility of the Treaty with the existing security agreements; it noted further that when technological progress would allow the development of peaceful explosive devices different from nuclear weapons, the prohibition relating to their manufacture and use shall no longer apply; it interpreted the provisions of Article IX, paragraph 3 of the Treaty, concerning the definition of a military nuclear state, in the sense that it referred exclusively to the five countries which had manufactured and exploded a nuclear weapon or other nuclear explosive device prior to 1 January 1967, and stressed that under no circumstance would a claim of pertaining to such category be recognized by the Italian government to any other state.

[8] On depositing the instrument of ratification, Japan expressed the hope that France and China would accede to the Treaty; it urged a reduction of nuclear armaments and a comprehensive ban on nuclear testing; appealed to all states to refrain from the threat or use of force involving either nuclear or non-nuclear weapons; expressed the view that peaceful nuclear activities in non-nuclear weapon states party to the Treaty should not be hampered and that Japan should not be discriminated against in favour of other parties in any aspect of such activities. It also urged all nuclear weapon states to accept IAEA safeguards on their peaceful nuclear activities.

[9] A statement was made containing a disclaimer regarding the recognition of states party to the Treaty.

[10] On depositing the instrument of ratification, the Republic of Korea took note of the fact that the depositary governments of three nuclear weapon states had made declarations in June 1968 to take immediate and effective measures to safeguard any non-nuclear weapon state which is a victim of an act or an object of a threat of aggression in which nuclear weapons are used. It recalled that the UN Security Council adopted a resolution to the same effect on 19 June 1968.

[11] On signing the Treaty, Mexico stated, *inter alia*, that none of the provisions of the Treaty shall be interpreted as affecting in any way whatsoever the rights and obligations of Mexico as a state party to the Treaty for the Prohibition of Nuclear Weapons in Latin America (Treaty of Tlatelolco).

It is the understanding of Mexico that at the present time any nuclear explosive device is capable of being used as a nuclear weapon and that there is no indication that in the near future it will be possible to manufacture nuclear explosive devices that are not potentially nuclear weapons. However, if technological advances modify this situation, it will be necessary to amend the relevant provisions of the Treaty in accordance with the procedure established therein.

[12] On depositing the instruments of ratification and accession, Switzerland and Liechtenstein stated that activities not prohibited under Articles I and II of the Treaty include, in particular, the whole field of energy production and related operations, research and technology concerning future generations of nuclear reactors based on fission or fusion, as well as production of isotopes. Switzerland and Liechtenstein define the term "source or special fissionable material" in Article III of the Treaty as being in accordance with Article XX of the IAEA Statute, and a modification of their interpretation requires their formal consent; they will accept only such interpretations and definitions of the terms "equipment or material especially designed or prepared for the processing, use or production of special fissionable material", as mentioned in Article III of the Treaty, that they will expressly approve; and they understand that the application of the Treaty, especially of the control measures, will not lead to discrimination of their industry in international competition.

[13] The United Kingdom recalled its view that if a régime is not recognized as the government of a state, neither signature nor the deposit of any instrument by it, nor notification of any of those acts, will bring about recognition of that régime by any other state. The provisions of the Treaty shall not apply with regard to Southern Rhodesia unless and until the government of the United Kingdom informs the other depositary governments that it is in a position to ensure that the obligations imposed by the Treaty in respect of that territory can be fully implemented. Cameroon stated that it was unable to accept the reservation concerning Southern Rhodesia. Also Mongolia stated that the obligations assumed by the United Kingdom under the Non-Proliferation Treaty should apply equally to Southern Rhodesia. In a note addressed to the British Embassy in Moscow, the Soviet government expressed the view that the United Kingdom carries the entire responsibility for Southern Rhodesia until the people of that territory acquire genuine independence, and that this fully applies to the Non-Proliferation Treaty.

[14] This agreement, signed between the United Kingdom, Euratom and the IAEA, provides for the submission of British non-military nuclear installations to safeguards under IAEA supervision.

[15] This agreement, under which US civilian nuclear facilities will be placed under IAEA safeguards, was approved by the IAEA Board but was not in force by 31 December 1978.

[16] In connection with the ratification of the Treaty, Yugoslavia stated, *inter alia*, that it considered a ban on the development, manufacture and use of nuclear weapons and the destruction of all stockpiles of these weapons to be indispensable for the maintenance of a stable peace and international security; it held the view that the chief responsibility for progress in this direction rested with the nuclear weapon powers, and expected these powers to undertake not to use nuclear weapons against the countries which have renounced them as well as against non-nuclear weapon states in general, and to refrain from the threat to use them. It also emphasized the significance it attached to the universality of the efforts relating to the realization of the NPT.

Chapter 14. Nuclear energy and nuclear weapon proliferation

J. ROTBLAT

Square-bracketed numbers, thus [1], *refer to the list of references on page 433.*

I. Introduction

The peaceful and military aspects of nuclear energy are intrinsically linked and it is impossible to separate them. The links are psychological, historical and factual. Psychologically, the first time the public at large learned about the release of nuclear energy on a practical scale was the announcement in August 1945 of the destruction of Hiroshima by an atomic bomb. Historically, the first nuclear reactors, in the USA, the USSR and the UK, were built not to generate electricity but to manufacture plutonium for nuclear weapons. Factually, to this day it is impossible to generate electricity in a peaceful nuclear reactor without at the same time using or manufacturing materials which could be used for nuclear weapons.

Any nation which acquires nuclear reactors for peaceful purposes will have personnel trained in nuclear reactor technology, from which it is only a short step to the acquisition of nuclear weapon technology. If the materials for making nuclear weapons were also available, such a nation would become a potential nuclear weapon state. Thus, the widespread use of nuclear energy for peaceful purposes is likely to lead to 'horizontal proliferation', that is, an increase in the number of nuclear weapon states. Such proliferation constitutes a grave threat to the security of mankind, as it greatly increases the probability of the outbreak of a nuclear war.

This chapter surveys the development of nuclear energy for peaceful purposes with particular emphasis on the links with nuclear weapons. It then discusses the measures, both technological and institutional, that can be taken to reduce the danger of horizontal proliferation.

II. The current nuclear power situation

Nuclear fuels

The method of generating electricity in a nuclear power plant is the same as in a coal-fired station: heat is used to produce steam which in turn drives a turbo-alternator. The source of heat is different, however. Instead of coming from burning coal, it comes from the splitting of atomic nuclei. When atomic nuclei of some heavy elements are exposed to neutrons, they divide into two in a process called fission. In this process a large amount of energy is released which is soon converted into heat. At the same time several neutrons are released and these can be used to produce further fission and thus initiate a self-sustained chain reaction. Compared with chemical reactions, the release of energy in nuclear reactions is enormous; the complete burning of 1 kg of nuclear fuel can produce as much heat as the burning of 2 500 tonnes of coal.

The only naturally occurring element that can be used directly as nuclear fuel is uranium; actually only a tiny fraction of it can be used. The atoms of natural uranium consist mainly of two types of nuclide, uranium-235 and uranium-238, which occur in nature in proportions 0.7 per cent and 99.3 per cent. Only the former can be used directly as a nuclear fuel, although under some conditions uranium-238 also undergoes fission. However, the bulk of uranium can be utilized indirectly. When it is exposed to neutrons, U-238 is transformed into other nuclides and eventually into plutonium-239, which has similar properties to uranium-235 as a nuclear fuel. Another chemical element, thorium, can also be converted into a nuclear fuel in a similar fashion. When exposed to neutrons, thorium-232 is transformed into uranium-233, which is a nuclear fuel. To produce Pu-239 or U-233 on a practical scale, extremely high neutron intensities are needed to bombard the uranium or thorium for a long time. So far it has been possible to create these conditions only in nuclear reactors. Thus, to begin with, only U-235 provides the nuclear fuel.

Due to the very low concentration of U-235, a very large amount of uranium is needed for a nuclear reactor. Even so, it would have been impossible to sustain a chain reaction with natural uranium without making the neutrons more efficient. The neutrons emitted in the fission process are very fast, and as such have a low probability of causing further fission in a given mass of uranium. (Similarly, the faster a golf ball moves, the smaller is the probability of its falling into the hole.) This probability is greatly increased by slowing the neutrons down to so-called thermal velocities, at which atoms of gases move about at normal temperatures. Fast neutrons can be slowed down by making them collide with nuclei of light elements, such as hydrogen or carbon. A material used to slow down the neutrons is called a moderator. Thus, in order to maintain a fission chain reaction one needs a certain amount of uranium mixed with, or surrounded by, a moderator. A system in which a fission chain reaction is sustained by thermal neutrons is known as a thermal reactor.

Table 1. Characteristics of thermal reactors

Reactor	Fuel	Moderator	Coolant	Number	Power output (MW(e)) (March 1979)
Magnox	Natural uranium metal	Graphite	Carbon dioxide	26	6610
AGR	Uranium oxide 2.3%	Graphite	Carbon dioxide	4	2464
LWGR	Uranium metal 1–2%	Graphite	Light water	6	4970
CANDU	Natural uranium oxide	Heavy water	Heavy water[a]	12	6015
BWR	Uranium oxide 2.2%	Light water	Light water	52	33912
PWR	Uranium oxide 3%	Light water	Light water	86	58476

[a] One of these reactors is cooled with light water.

Thermal reactors

Characteristics of thermal reactors

The low concentration of U-235 not only makes it necessary to have a very large amount of uranium in the reactor, but also restricts the choice of moderating material to two—heavy water and pure graphite. Ordinary water is not suitable because the hydrogen in it absorbs neutrons excessively. However, this difficulty can be overcome if the concentration of U-235 in the uranium is significantly increased, say from 0.7 per cent to 3 per cent. Uranium with a higher than natural content of U-235 is called enriched uranium; it can be made in special enrichment plants (see under *Enrichment plants*, page 391). With enriched fuel, the total amount of uranium in the reactor can be greatly reduced. The great majority of thermal reactors in existence or being built use enriched uranium.

There are four essential parts of a thermal reactor: the fuel elements, the moderator, the coolant, and the control rods. The coolant is the material used to remove the heat from the fuel elements and to transfer it to the heat exchanger. The control rods are usually made of boron, a material which avidly absorbs thermal neutrons; by inserting these rods to a greater or lesser extent, the neutron intensity is regulated so that a steady power output is achieved. Different types of reactor do not differ significantly in the design of the control rods, but they differ in the other characteristics.

Figure 1 shows a diagram of a thermal reactor with the essential parts indicated. Table 1 lists the main types of thermal reactor which have been developed for commercial purposes.

The fuel is usually in the form of uranium oxide (UO_2), which can stand much higher temperatures than uranium metal; the latter is used in reactors based on natural uranium (Magnox and LWGR). The uranium oxide is manufactured into small ceramic pellets, a large number of which are assembled in rods, or 'pins', that is, cylinders, which are typically about 1 cm in diameter and about 4 metres long. A bundle of such rods forms a fuel element. A typical modern reactor may contain about 200 fuel elements, each consisting of about 200 rods.

Figure 1.   Diagram of a thermal reactor

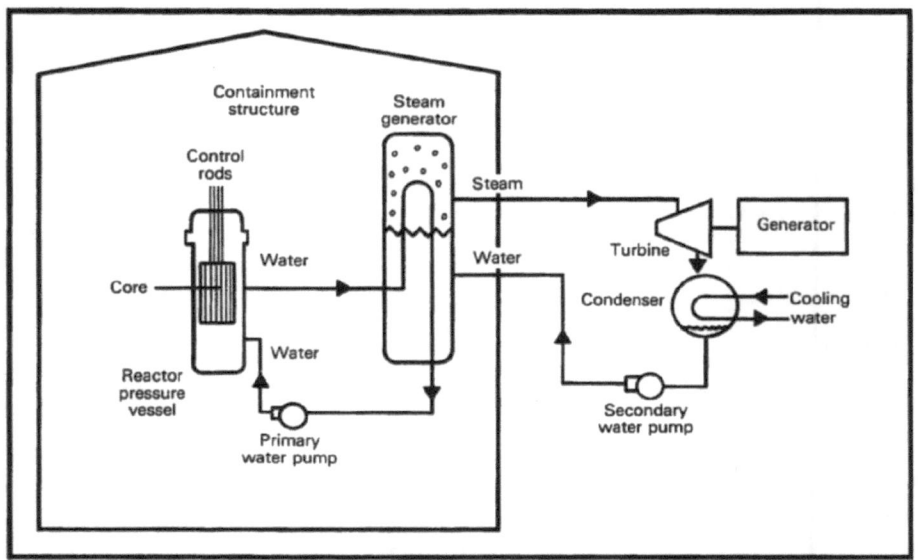

In order to prevent the escape of the radioactive elements which are formed as a result of fission, each rod has to be contained in a vacuum-tight sheath. In most reactors, the material of the sheath, or cladding, is zirconium alloy. Some earlier reactors, mainly in the UK, were sheathed in a magnesium alloy (hence the name Magnox for these reactors), and some corrosion problems have been encountered with them (see under *Reprocessing,* p. 393). An improved version of the Magnox is the Advanced Gas Cooled Reactor (AGR) which also uses graphite as a moderator and carbon dioxide as a coolant, but the fuel is enriched uranium and the cladding of the elements is stainless steel.

The CANDU reactor, built in Canada, uses natural uranium oxide as fuel, and heavy water both for moderating and for cooling.

The most popular type of reactor uses ordinary water both as moderator and coolant. There are two types of these light water reactors (LWRs): boiling water reactors (BWRs), in which the water is brought to the boil and the steam passed to the turbines; and pressurized water reactors (PWRs), in which the coolant water, prevented from boiling by high pressure, is passed through a heat exchanger to make steam. As seen from table 1, the LWRs account for 74 per cent of the number and 82 per cent of the power output of reactors in operation at the present time; of the two, the PWRs are gaining further ground in new orders.

Apart from the characteristics indicated in table 1, there are many other differences between reactors, relating to their structure and mode of operation. An important parameter in the operation of reactors is the burn-up. As the nuclear fuel is being burnt in the reactor, the fuel rods gradually accumulate the so-called waste products; these are mostly the radioactive elements which result from the splitting, or the transformation, of the uranium or plutonium nuclei. Some of these products absorb neutrons significantly, and as time goes on they may accumulate to such an extent that not enough neutrons would be left to sustain the chain reaction even with the control rods completely withdrawn. For this reason, the fuel elements have to be removed from the reactor after a certain time and replaced by fresh elements, even though some nuclear fuel in the form of U-235, as well as Pu-239, is still left in them. In a few reactors (for example CANDU), reloading can be done without interrupting their operation, but for the majority of reactors, they have to be shut down. How long the fuel rods can spend in the reactor depends on the running conditions and type of reactor, but usually it is between two and four years. For most LWRs the reactors are shut down once a year, when about one-third of their fuel load is removed and replaced by fresh fuel.

Table 2. Content of spent fuel elements (yearly discharge from 1GW(e) PWR)

	Nuclide	Weight (kg)
Uranium	U-235*	215
	U-236	114
	U-238	25 683
Plutonium	Pu-238	5.9
	Pu-239*	142.4
	Pu-240	58.7
	Pu-241*	27.5
	Pu-242	9.5
Actinides	Neptunium	22.2
	Americium	3.3
	Curium	1.2
Fission products		951

*Fissile nuclides

It is of course of great economic advantage to make the fullest use of the fuel. This usage is expressed in terms of 'burn-up', which gives the amount of energy (in megawatt days) obtained per kilogram of uranium in the elements. The early reactors had a burn-up of about 5 MW-days per kg, but current LWRs have a burn-up of about 30 MW-days per kg.

The main purpose of the reactor is the generation of electricity. From this point of view three parameters are of importance.

1. Thermal efficiency: According to the laws of thermodynamics, in any heat engine only a fraction of the energy released can be converted into other forms, for example electricity, this fraction depending primarily on the temperature of the steam reaching the turbine. In LWRs the temperature is in the range of 280–330°C, giving a thermal efficiency of about 33

Table 3. Thermal reactors in operation (March 1979) (net output > 150MW(e))

Country	Number of reactors	Total power output (MW(e))
Argentina	1	345
Belgium	3	1 665
Bulgaria	2	837
Canada	10	5 465
Finland	2	1 080
France	11	6 860
FRG	11	8 076
GDR	2	816
India	3	602
Italy	4	1 382
Japan	20	13 239
Korea	1	564
Netherlands	1	447
Spain	3	1 073
Sweden	6	3 700
Switzerland	3	1 006
Taiwan	2	1 208
UK	22	6 230
USA	67	51 183
USSR	12	6 669
Total	**186**	**112 447**

per cent; this means that only one-third of the nuclear energy is utilized as electricity, while the remaining two-thirds is wasted as heat which has to be dissipated, usually in seas or rivers.

2. Power output: From the practical point of view this is the most important parameter of the reactor. It is the amount of power in the form of generated electricity which the reactor can develop. It is expressed in MW(e) (megawatt electric) or GW(e) (gigawatt electric) and should be distinguished from the thermal output (MW(th)). The electrical output is the thermal output multiplied by the thermal efficiency. Note that burn-up refers to thermal and not electrical output.

The electrical power output is sometimes given as gross and sometimes as net. The difference between them corresponds to the electrical energy consumed in running the reactor itself; on average it amounts to about 5 per cent. The figures in tables 1, 3, 4 and 5 are the net output.

The designated power output of reactors has been gradually increasing over the years. Due to the high capital cost, it is much more economical to build large reactors. For example, the basic cost of a reactor operating at 600 MW(e) is about 33 per cent higher than that of a 1 000 MW(e) reactor per unit of electricity generated [1]. The average power output of the reactors listed in tables 1 and 3 is 605 MW(e) but this average is a reflection of the lower power rating of the older reactors. More modern reactors have ratings in the range 1 000−1 300 MW(e).

3. Load factor: as far as energy production is concerned, an important quantity is the load factor, which gives the fraction of the time

Table 4. **Thermal reactors in operation, under construction, or planned** (net output > 150 MW(e))

Country	Number of reactors	Total power ·output (MW(e))
Argentina	3	1 505
Austria	1	692
Belgium	8	6 483
Brazil	3	3 116
Bulgaria	4	1 677
Canada	24	15 217
Cuba	2	880
Czechoslovakia	11	4 541
Finland	5	3 160
France	34	29 495
FRG	34	35 916
GDR	12	4 896
Hungary	4	1 632
India	8	1 689
Iran	8	8 982
Israel	1	600
Italy	8	5 242
Japan	32	23 019
Korea	5	3 598
Mexico	2	1 308
Netherlands	1	447
Pakistan	1	600
Philippines	1	621
Poland	2	816
Romania	1	440
South Africa	2	1 843
Spain	19	15 991
Sweden	12	9 442
Switzerland	10	7 833
Taiwan	6	4 923
Thailand	1	600
Turkey	1	620
UK	32	12 408
USA	202	200 931
USSR	31	23 269
Yugoslavia	1	632
Total	**532**	**435 064**

the reactor is actually operating. Because of the need to shut down the reactor for reloading and routine maintenance, quite apart from unforeseen breakdowns, the load factor is considerably less than 100 per cent. In the earlier days it was about 50 per cent, but with the accumulation of operating experience the load factor has been gradually increasing, and the average for all reactors listed in tables 1 and 3 is now about 64 per cent.

Reactor inventories

The reactor inventory refers to the amount and type of reactor fuel needed

Table 5. Types of thermal reactor in operation, under construction, or planned (net output > 150MW(e))

	Number	Power output (MW(e))
Magnox	26	6 610
AGR	14	8 642
LWGR	14	13 970
CANDU	34	18 644
BWR	119	103 627
PWR	315	275 151
Unspecified	10	8 420

Figure 2. Flow sheet of material in a PWR

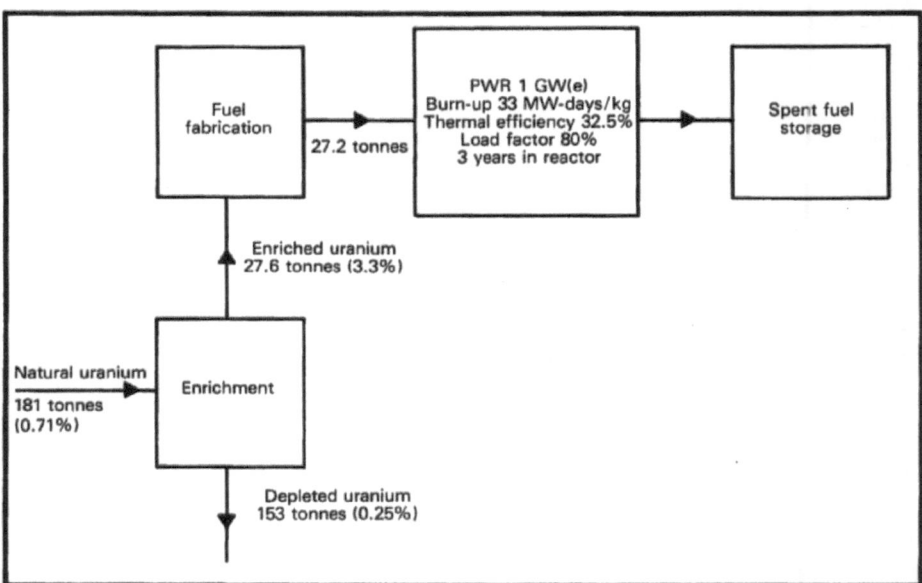

to start the reactor, and to the quantities and composition of the fuel elements which come out of the reactor and the fresh fuel going in every year.

Calculations of the inventory were made [2] for a PWR with a power output of 1 GW(e), thermal efficiency 32.5 per cent, load factor 80 per cent (even though this has not yet been achieved), and burn-up 33 MW-days per kg. To start with, such a reactor would require about 82 tonnes of uranium enriched to 3.3 per cent of U-235. With the fuel element remaining in the reactor for three years and allowing for losses, 27.6 tonnes of new enriched uranium will be needed to reload the reactor every year. The enriched uranium is of course derived from a much larger amount of natural uranium; some 181 tonnes of the latter would have to be fed into the

Figure 3. Changes with time of content of fuel

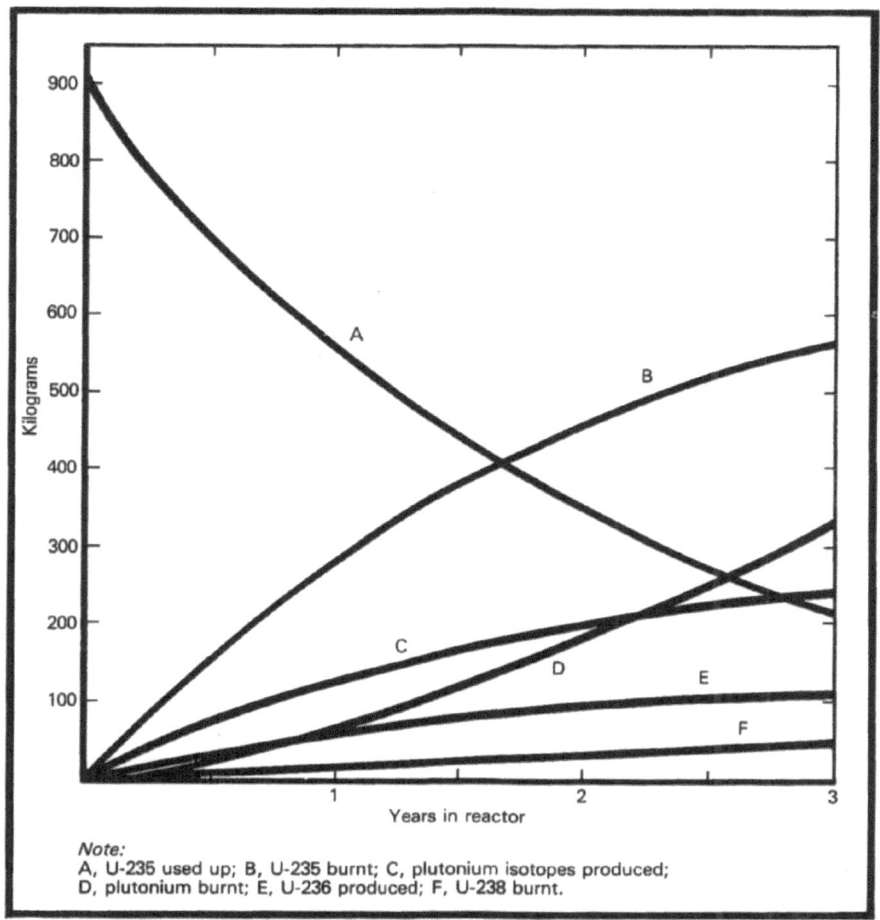

Note:
A, U-235 used up; B, U-235 burnt; C, plutonium isotopes produced;
D, plutonium burnt; E, U-236 produced; F, U-238 burnt.

enrichment plant to produce this quantity (see under *Enrichment plants*, p. 391). Figure 2 shows the flow sheet of material for such a reactor.

The changes which occur in the fuel elements while in the reactor are of direct relevance to the problem of proliferation. Figure 3 shows the changes in the content of the fuel as a function of time.

The amount of uranium-235 is steadily decreasing (curve A), but only part of the decrease (curve B) is due to its burning as fuel; some U-235 is converted into U-236, a non-fissile nuclide (curve E). The amount of plutonium, produced by the transformation of U-238, increases with time but flattens off later (curve C). The reasons are similar: part of the Pu-239 is burned as fuel (curve D) and part is converted into heavier isotopes of plutonium, Pu-240, Pu-241 and Pu-242. There is also a small production of Pu-238.

Plutonium-241 is a fissile material and contributes to the energy production. There is also a small contribution from the fission induced in uranium-238 by fast neutrons (curve F). In the reactor described here the total energy, released over the three years from the burning of 95 kg of nuclear fuel, is made up as follows: 60 per cent from U-235, 35 per cent from Pu-239 and Pu-241, and 5 per cent from U-238.

A number of other nuclides are also produced in the fuel elements. These include other actinides (elements with atomic number 90–103) such as neptunium, americium and curium. But the largest quantity of new materials produced is of fission fragments, which consist of hundreds of different types of nuclide, practically all radioactive. Table 2 shows the content of the spent fuel elements after three years in a reactor with the characteristics described above. Since one-third of the elements are taken out every year, these figures give the yearly discharges. Note that, in addition to plutonium, the spent fuel contains U-235 at a concentration slightly higher than in natural uranium (0.83 per cent).

Energy production from thermal power reactors

Electricity was first generated from thermal reactors in 1954. Gradually the importance of nuclear electricity grew both in terms of the total power output of the reactors, and in the number of countries with commercial reactors. By March 1979, 186 thermal reactors were in operation in 20 countries, with a total capacity of 112 GW(e). Table 3 gives a list of these countries, with their nuclear capacities (only reactors with a net power output greater than 150 MW(e) are listed) [3, 4]. As far as is known China has no power reactors in operation for commercial purposes, but no information is available. Czechoslovakia and Pakistan have one reactor each, with outputs of 110 and 126 MW(e) respectively. As figure 4 shows, about 90 per cent of nuclear energy is at present generated in OECD countries.

If we assume an average load factor of 64 per cent, the total electrical energy generated per year at the above rate is 630 TWh, or about 2×10^{18} joules. The world annual energy consumption is at present about 3×10^{20} joules; thus, after a quarter of a century of the atomic age, nuclear energy is contributing less than one per cent of the world energy consumption. The total amount of electrical energy generated in commercial nuclear reactors from the time they started until March 1979 was 10^{19} joules, which is twelve days' worth of the world consumption.

There will, of course, be an increase in nuclear energy production. Many reactors are being completed and sixteen countries other than those listed in table 3 have reactors under construction or have announced their intention to order them. Table 4 shows a list of countries, and the number and total power output of thermal reactors already in operation, under construction or being planned [3, 5]. (Table 4 includes the reactors listed in table 3). The distribution of these reactors among the different types is shown in table 5. All the reactors listed in tables 4 and 5 are expected to be

Figure 4. Regional distribution of thermal reactors (percentage power output)

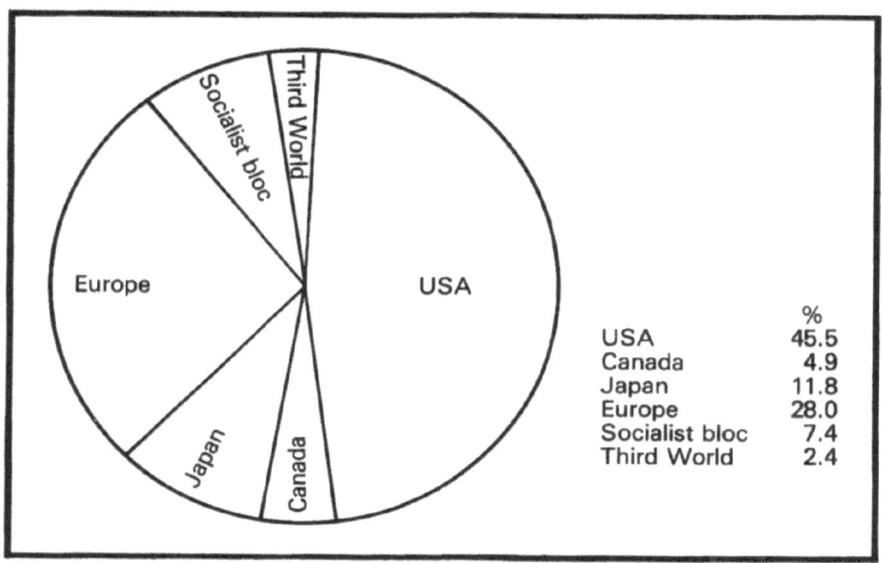

	%
USA	45.5
Canada	4.9
Japan	11.8
Europe	28.0
Socialist bloc	7.4
Third World	2.4

Figure 5. New reactors ordered per year

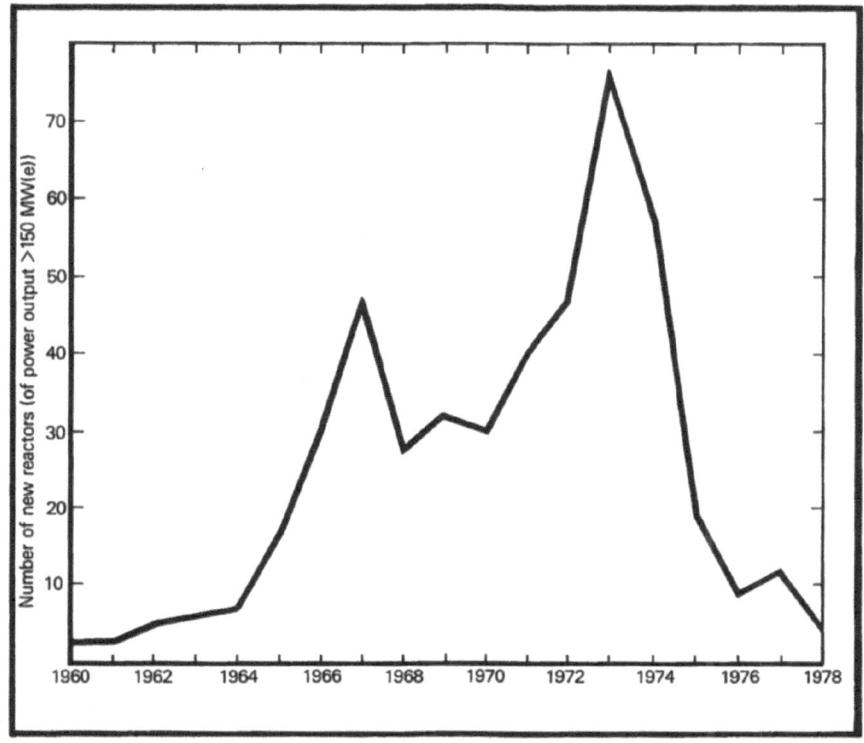

operational by 1993. With these, the total nuclear power output would be nearly four times greater than now (435 GW(e)).

The scale of further expansion is uncertain; recently there has been a remarkable slowing down in nuclear energy programmes. Figure 5 shows the rate of growth of nuclear reactors of power output greater than 150 MW(e); the ordinates give the number of new reactors initially announced in a given year [5]. As is seen, there has been a dramatic fall-off in the rate of growth since 1973. Moreover, some twenty reactors announced between 1972 and 1976 have since been cancelled, so that during the last three years there has been hardly any effective addition. It has been reported that a few more countries are showing interest in nuclear energy, and orders for new reactors may come in, but unless there is a radical change in the situation, it is most unlikely that the nuclear capacity of about 2 000 GW(e), which has been predicted for the end of the century [6], will materialize.

Fast breeder reactors

If the installed nuclear capacity were to reach 2 000 GW(e) by the year 2000, then—even without any further expansion—the thermal reactors would be able to run for fewer than 20 years before those world uranium resources which appear to be available at reasonable prices would be exhausted (see under *Mining and milling of uranium*, page 389).

It was the realization of the very limited extent of a nuclear power programme based on thermal reactors that prompted the development of the fast breeder reactor. Breeder reactors would make it possible to utilize almost the whole of the uranium as a nuclear fuel, and not just 0.7 per cent of it, thus extending the nuclear energy resources by two orders of magnitude.

Some U-238 is utilized as fuel even in the thermal reactor, mainly through its transformation into Pu-239. Some of this plutonium is left in the spent fuel rods (table 2), but the amount of the fissile plutonium left over is less than the total fissile material used in producing it. The ratio of fissile material produced to that used up in a reactor is called the conversion ratio. In thermal reactors the conversion ratio is about 0.6, but by using a reactor of a different design it is possible to make the conversion ratio greater than 1, which means that more plutonium is produced than the nuclear fuel burnt up.

When the conversion ratio is greater than one, we speak of 'breeding', since the extra plutonium produced can be used to start a new reactor to make more plutonium from U-238 and so on, until all of the uranium is converted into fuel.

The better utilization of neutrons needed to achieve breeding can be accomplished only by dispensing with the moderator, and using the fast

Table 6. The balance of plutonium flow in a PWR and LMFBR operated at 1GW(e)

	Input (kg/y)	Created (kg/y)	Incinerated (kg/y)	Output (kg/y)	Net (kg/y)
PWR core	–	710	−380	330	330
LMFBR core	2 800	530	−750	2 580	−220
LMFBR axial and radial blanket	–	479	−70	409	409
LMFBR overall	2 800	1 009	−820	2 989	189

neutrons to sustain the chain reaction. Hence the name fast breeder reactor (FBR). In addition, it requires a much greater concentration of the neutrons, higher operating temperatures, and more efficient removal of heat. Most of the developmental work on fast breeders was carried out using a liquid metal (sodium) as a coolant, and these reactors are designated as liquid metal fast breeder reactors (LMFBRs).

A breeder reactor is composed of a core, which eventually will consist of plutonium recovered from other breeders, surrounded by a blanket of ordinary or depleted uranium (uranium from which most of the U-235 has been removed in enrichment plants) in which new plutonium is made.

The time it takes for a breeder reactor to produce as much plutonium as is contained in its core, and thus sufficient to start a new breeder, is called the doubling time. Depending on the design, doubling times from 5 to 50 years have been considered, with a plausible average of 20 years.

Until the time when sufficient plutonium has been produced in the FBRs to make the breeder programme self-sufficient, the core will be made up of a mixture of enriched uranium (about 20 per cent U-235) and plutonium recovered from thermal reactors.

Table 6 shows a comparison of the turnover of plutonium in thermal and breeder reactors [7], the latter operating with a conversion ratio of 1.23 (plutonium produced is 1 009 kg, while plutonium burned is 820 kg). From the point of view of proliferation it is important to note that, while the net amount of plutonium produced is smaller in the breeder, the turnover of plutonium is nine times higher in the breeder than in a thermal reactor with the same power output. Moreover, the plutonium in the blanket would contain over 95 per cent Pu-239, making high quality weapon-grade material (see under *Weapon-grade material*, p. 398).

Due to much greater technological difficulties encountered with the LMFBRs, the development of commercial fast breeders has taken a long time and to date only two pilot LMFBRs are in operation, in France and the UK. Several commercial prototype models are being constructed in France, the FRG, Japan, the UK, the USA and the USSR. A list of FBRs with a power output greater than 150 MW(e) is given in table 7.

385

Table 7. Fast breeder reactors with power output greater than 150 MW(e)

Country	Reactor	Power output (MW(e))	Expected date of operation
France	Phénix	250	Operating
	Super Phénix	1 200	1983
FRG	Kalkar SNR1	292	1983–88
	Kalkar SNR2	1 300	1989
Japan	Monju	250	1985
UK	Dounreay	230	Operating
USA	Clinch River	350	1983
USSR	Beloyarsk	600	1980

The most advanced appears to be the French breeder reactor, Super Phénix,[1] built at Creys-Malville. It is designed to have a power output of 1 200 MW(e), a thermal efficiency of 40 per cent, and a burn-up of 70 MW-days per kg. It will have a low breeding capacity (doubling time about 50 years), and at a load factor of 75 per cent the turnover of fuel will be 13 months.

In the USA, work on the Clinch River fast breeder (350 MW(e)) was temporarily halted when President Carter announced his energy policy in April 1977 (see *The US energy policy*, p. 427); the present situation about the status of this breeder is obscure.

In the UK a prototype fast breeder (230 MW(e)) is working at Dounreay, but a decision about the policy on commercial fast breeders awaits a public inquiry to be held in 1979 or 1980.

In the Soviet Union a fast breeder (600 MW(e)) is being constructed in the Urals. A smaller breeder (135 MW(e)) is operating in the Mangyshalk Peninsula and is used for desalinated water production.

It has been reported that the SNR1 reactor at Kalkar (FRG) is to be used only as a plutonium burner.

Other reactor types

Whilst almost all power reactors use uranium of low enrichment, up to 4 per cent, a few reactors have been designed in which highly enriched uranium, with 93 per cent U-235 content, is used. The high-temperature gas-cooled reactors (HTGRs) belong to this category. One purpose of these reactors was to make possible the utilization of another element, thorium, which by bombardment with neutrons is transformed into the fissile nuclide U-233. Because of the conversion of a non-fissile into a fissile material, such reactors are called converter reactors. Conversion ratios of up to nearly one can be achieved in them.

[1] Like ships, reactors are given individual names. The tendency is, however, to attach the name to the power station, which may have more than one reactor (sometimes four).

Table 8. Research reactors in non-nuclear weapon states

Country	Location	Name of reactor	Date of operation	Power output (MW(th))
Argentina	Buenos Aires	RA-3	1967	5
Australia	Lucas Heights	HIFAR	1958	10
Austria	Seibersdorf	ASTRA	1960	5
Belgium	Mol	BR-2	1961	100
Canada	Hamilton, Ontario	MNR	1959	2
Denmark	Risö	DR-2	1958	5
	Risö	DR-3	1960	10
FR Germany	Jülich	FRJ-2 (DIDO)	1962/67	23
	Jülich	FRJ-1 (MERLIN)	1962/67	10
	Geesthacht	FRG-2	1963	15
	Braunschweig	FMRB	1967	1
Israel	Nahal Soreq	IRR-1	1960	5
Italy	Saluggia	AVOGADRO RS-1	1959	7
	San Piero a Grado	GALILEO GALILEI (RTS-1)	1963	5
	Ispra	ESSOR	1967	40
Japan	Tokai	JRR-2	1960	10
	Kyoto	KUR	1964	5
	Tokai	JRR-4	1965	1
	Oarai	JMTR	1968	50
Netherlands	Petten	HFR	1961	30
	Delft	HOR	1963	2
Pakistan	Islamabad	PARR	1965	5
South Africa	Pelindaba	SAFARI-1	1965	20
Sweden	Studsvik	R-2	1960	50
	Studsvik	R2-0	1961	1
Thailand	Bangkok	TRR-1	1962	1
Turkey	Cekmece	TR-1	1962	1

Source: Directory of Nuclear Reactors, Vol. X, Power and Research Reactors, IAEA, Vienna, 1976.

The fuel material in the HTGR is made up of pellets of highly enriched uranium carbide (UC_2) mixed with thorium carbide (ThC_2). The moderator is graphite and the coolant helium gas. The operational temperature is 750°C and the burn-up about 100 MW-days per kg.

At one time there was considerable interest in the USA in the commercial development of HTGRs, and several such reactors were ordered. Due to escalating prices, however, these orders were later cancelled. At the present time there is only one HTGR in operation in the USA, in Fort St Vrain, with a power output of 330 MW(e); until now it has generated only 100 GWh of electricity. Research and development work on HTGRs is being carried out in Germany, where a reactor with similar characteristics is due to be commissioned in 1980.

Apart from HTGRs, reactors using highly enriched uranium are among the over 300 research reactors in operation in 45 countries. Over 100 of these reactors use uranium enriched to more than 80 per cent U-235, which makes this a weapon-grade material (see under *Weapon-grade*

Figure 6. Nuclear fuel cycles: (*a*) once-through, (*b*) uranium recycle, (*c*) uranium plus plutonium recycle, (*d*) in a FBR

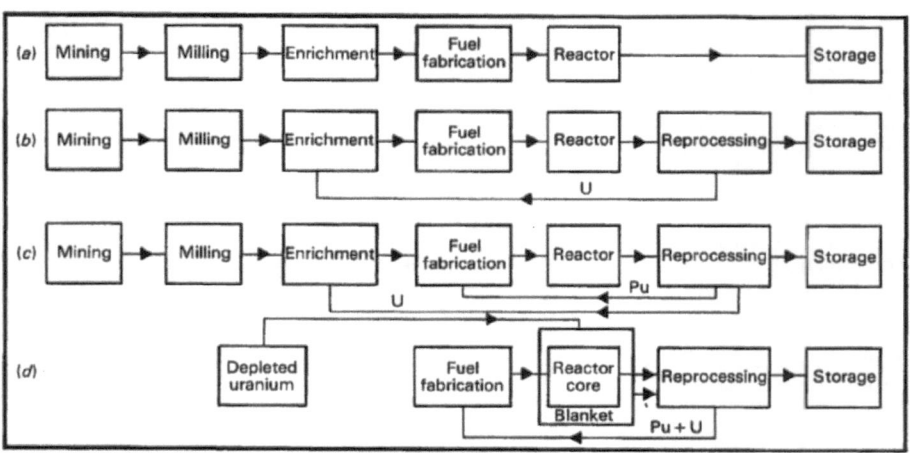

material, p. 398). Moreover they are often situated in universities and research institutions, where security is much less stringent than in power stations. Although most of these reactors have a low power output (less than 20 MW(th)), and as such do not contribute significantly to nuclear energy production, they could produce or accumulate amounts of plutonium significant from the point of view of nuclear weapon proliferation. A list of research reactors in non-nuclear weapon states using uranium enriched to more than 80 per cent, and with an output of at least 1 MW(th) is given in table 8.

Apart from these reactors for civilian uses, a number of other types of reactor have been developed for military applications, for example, for naval vessels and as power sources for aircraft, rockets and satellites, and ground equipment such as radar.

The reactors for naval vessels are mainly of the pressurized-water type, but using highly enriched uranium. The deployment of reactors on ships and submarines has increased not only the range of the vessels but also their cruising speeds and, in the case of submarines, the capacity for sustained submersion. Some 13 surface ships and about 265 submarines are propelled using nuclear power.

Reactors have been developed to provide power for instruments on board military satellites. In one type, a gas turbine model developed in the USA, highly enriched uranium is used as fuel, zirconium hydride as moderator, and liquid sodium as coolant. The power output is some 3 kW(e).

Nuclear fuel cycles

The nuclear fuel cycle consists of several stages of operations with the reactor as its central part, starting from the mining of the raw material and ending with the disposal of the waste products. Many types of fuel cycle have been discussed and the most important are depicted in figure 6.

Figure 6(a) shows the simplest fuel cycle, the so-called once-through cycle. The uranium ore is mined, milled, and the product passed through the enrichment plant; from there it goes to the fabrication plant where the fuel elements are made for the reactor. The spent fuel elements are deposited or disposed of.

Figure 6(b) shows the fuel cycle with uranium recycling. Here the spent fuel goes through a reprocessing plant, where the uranium and plutonium contained in them are separated from the waste products which are stored. The uranium recovered in the reprocessing plant is returned to the enrichment plant, where it is mixed with fresh uranium before going through the reactor again. The separated plutonium is stored.

Figure 6(c) shows a cycle in which both the uranium and the plutonium in the spent fuel elements are re-utilized. The plutonium fraction returns to the fuel fabrication plant, where it meets the uranium coming from the enrichment plant. The fuel in this cycle is made up into pellets of mixed uranium and plutonium oxides.

Figure 6(d) shows the fuel cycle for the fast breeder operating on plutonium accumulated from thermal reactors or from other FBRs, which can thus run without supplementation from U-235 fuel. Because of this, the first stages of the previous cycles are no longer necessary. The plutonium is circulated in a closed cycle, and the supply of fresh fuel comes from the depleted uranium which has been accumulating in the enrichment plants.

At present, the commercial nuclear energy industry is operating—in practice—on the once-through cycle, without reprocessing of the spent fuel elements. Reprocessing is taking place for fuels from Magnox reactors, and there is a small-scale recycling of uranium, but the majority of spent fuel is kept in cooling ponds, pending decisions about reprocessing policy and the building of reprocessing plants.

Mining and milling of uranium

Uranium-bearing ores occur in different parts of the world, in different geological settings, at different depths, and with different concentrations. Huge amounts, several thousand million tonnes, are contained in sea water, but the low concentration, 3×10^{-9} (one part in about 300 million), makes the cost of extracting the uranium prohibitive. The uranium ores mined at present occur mainly in sandstone at depths of less than 200 metres. There is a continuous spectrum of concentration, from a few per cent downwards, but the high grade ores have already been exploited, and the ores being mined at present have concentrations of 0.1−0.2 per cent.

The great range of concentration of uranium in the ores makes it impossible to obtain a precise figure for the amount of uranium available in

Table 9. World uranium resources[a]

Country	(Thousand tonnes of uranium)
Algeria	78
Argentina	42
Australia	345
Brazil	26
Canada	838
France	96
Gabon	30
India	54
Niger	213
South Africa	420
Sweden	304
USA	1 696
Other countries	146
Total	**4 288**

[a] Not including socialist countries.

the world. The lower concentration the more there is of it, but the higher the cost of its extraction. For this reason, the availability of uranium is usually expressed in terms of the amount that could be mined at a given price, and this is bound to be a fluctuating quality, since price is subject to variation depending on demand.

An OECD/IAEA survey in 1977 [8] indicates that there may be over 3 million tonnes of uranium in the non-Communist world exploitable at a cost up to $80 per kg of uranium, and a further million tonnes at a cost of up to $130 per kg. Table 9 lists the countries which supply most of the uranium.

It is remarkable that nearly 90 per cent of the cheaper uranium exists in the countries in which considerable expenditure was incurred on exploration. This suggests that if uranium exploration were vigorously pursued in other countries, the resources would be greatly increased. A joint OECD/IAEA study, *International Uranium Resources Evaluation Project*, designed to make an estimate of potential uranium resources, has come to a preliminary conclusion that in addition to the resources listed in table 9, there may be between 6.5 and 14.8 million tonnes of uranium in various parts of the world, exploitable at a cost up to $130 per kg of uranium [9].

From the mines the ore goes to the milling plant, where it is chemically treated to make a compound called 'yellow cake', mostly consisting of U_3O_8. This in turn goes to the conversion plant, where chemical treatment produces uranium hexafluoride (UF_6). This conversion is necessary because UF_6 needs only slight heating to become a gas (the evaporation temperature is 57°C) which is the state in which it has to be employed in the enrichment plant.

Because of the very low concentration of uranium in the ore, practically the whole of it remains in the tailings, at the milling plant. They are generally dumped into piles near the mills. Due to the residual radioactivity in the tailings, from which radon escapes into atmosphere, the tailing piles constitute a radiation hazard for the population in the vicinity.

It is of interest to note the quantities of prime material that need to be handled in order to produce the fuel for reactors. For example, to provide the 28 tonnes of uranium enriched to 3.3 per cent, needed for the annual refuel of the reactor described earlier (see under *Reactor inventories*, page 379, and figure 2), some 181 tonnes of natural uranium have to be fed into the enrichment plant. These 181 tonnes are in turn obtained from the mining of more than 100 000 tonnes of uranium ore. The actual mass of nuclear fuel burned is 0.9 tonnes. Thus, only one part in about 125 000 tonnes of the starting material is utilized as fuel. From the point of view of turnover of raw material, the weight advantage of the fuel used in thermal reactors, as compared with coal, is only a factor of 20.

Enrichment plants

The increase in the concentration of uranium-235, from 0.7 per cent in natural uranium to about 3 per cent for most thermal reactors, or to 93 per cent for HTGRs, is a costly, energy-consuming, and technologically difficult procedure, as it requires the separation of isotopes which are chemically indistinguishable. Although very small amounts of enriched material could be made in the laboratory, the enrichment of the quantities needed for reactors—many tonnes—can be carried out only in huge plants.

There are several techniques for uranium isotope separation, but many details of the operation of enrichment plants are kept secret for commercial or security reasons. The technique chiefly used so far is the gas diffusion method. This is based on the fact that if a gas diffuses through a thin barrier, which contains a large number of very small holes, the rate of diffusion is inversely proportional to the square root of the mass of the particles of the gas. The gas which has passed through the porous membrane will therefore contain slightly more of the lighter isotope, in this case U-235. The change of concentration after one passage is very small, about one in 250. Therefore, a very large number of diffusion stages are needed in cascade to reach a given degree of enrichment; for example, to obtain 3 per cent U-235, 1 200 stages are needed. To handle a large volume of gas, many parallel cascades of diffusion cells are needed. Apart from the vast size of the plant, this process requires a huge expenditure of energy, which is needed to compress and cool the uranium hexafluoride. About a megawatt year of electricity is consumed in producing one tonne of enriched material. The enrichment of U-235 consumes nearly 5 per cent of the electrical energy generated by that fuel.

The energy consumed in enrichment depends on the tail assay, that is the concentration of U-235 left in the depleted uranium, but which is too low to make further extraction of U-235 worthwhile, considering the cost. Usually, the U-235 content of the tails is 0.2−0.3 per cent.

Table 10. Enrichment plants

Technique	Country and location	Capacity (million SWU/year)	Date of operation
Gaseous diffusion	Canada, James Bay	9	1984–85
	Lower Churchill Falls	8	1982
	France, Pierrelatte	0.3	Operating
	Tricastin	10.8	1979–82
	–	10.8	1985–90
	UK, Capenhurst	0.4	Operating
	USA, Oak Ridge, Paducah, Portsmouth	17.2	Operating
	add	10.5	1984
	Portsmouth	8.75	1985
	USSR, Siberia	7–10	Operating
Centrifuge	Canada, Saskatchewan	0.3–3.0	1983–90
	Japan, –	0.5–1.0	1984–88
	Netherlands, Almelo	0.2	Operating
		1.0	1982
	UK, Capenhurst	0.2	Operating
		1.0	1982
Jet nozzle	Brazil, –	5.0	1989
	FRG, Karlsruhe	0.5	Operating
	South Africa, Valindaba	7.0	1984–86

The amount of work necessary to enrich a given amount of uranium to a given concentration with a given tail assay is measured in units of Separative Work (SWU). The enrichment of uranium to 3 per cent with a tail assay of 0.3 per cent, requires 3.4 SWUs per kg of product. As an illustration, in order to produce enriched uranium for the yearly reload of the reactor described under *Reactor inventories* (page 379), one would need some 110 000 SWUs per year for a tail assay of 0.3 per cent.

Since it is more economical to run a large enrichment plant, there are only a few gas diffusion plants operating. Table 10 contains a list of these, together with their capacities, expressed in terms of SWUs per year [10, 11]. It will be noted that all countries in which gas diffusion plants now operate are nuclear weapon states. Considering the capacity of its plants, the United States has a virtual monopoly of this technique in the Western world at present.

Table 10 also lists two other types of enrichment plant. The more advanced of these is the centrifuge method. This is based on the difference in the centrifugal force which acts on molecules of different masses when rotating at very high speeds. A much larger increase of concentration per stage can be obtained. For example, only fifteen stages are needed to enrich uranium to 3 per cent. The amount of energy consumed is also much less, about one-tenth of that needed for gaseous diffusion. On the other hand, the capital cost of the plant is very high and this would be a factor limiting the spread of such plants. Centrifuge plants already exist in Capenhurst in the UK and in Almelo in the Netherlands. Both of these are part of a project called URENCO, run jointly by these two countries with FR Germany. URENCO is to supply enriched fuel to Brazil. Japan and Canada plan to set up centrifuge plants.

The second technique, the jet nozzle, relies on the different deflection of gases of different mass coming out at high speed from a nozzle with curved walls. This too requires somewhat fewer stages to reach a given degree of enrichment (about 500 for 3 per cent enrichment). The capital cost is low, but the energy consumption is very high. This technique was invented in FR Germany, where a small plant was built at Karlsruhe. A large plant, based on the vortex tube technique, is being built at Valindaba in South Africa, which claims to have invented its own version of the jet nozzle technique. Another large plant is being built for Brazil by a West German firm.

While enrichment of uranium allows a smaller inventory in the reactor itself, the total amount of uranium going through the enrichment plant is of course correspondingly increased, and about 85 per cent of it remains as depleted uranium, or tails, with a U-235 concentration of 0.2−0.3 per cent. Some 100 000 tonnes of depleted uranium have already accumulated.

Reprocessing plants

The earliest nuclear reactors were built for military purposes, to produce plutonium for nuclear weapons. To obtain the plutonium, the spent elements, after removal from the reactors, had to go through a chemical process in which the plutonium was separated. This was done in reprocessing plants, which had to be specially designed to carry out chemical operations by remote control on immensely radioactive materials.

The practice of reprocessing was later extended to civil nuclear reactors even when the plutonium was no longer needed for weapons. For Magnox reactors, some processing of the spent fuel elements was necessary, since their magnesium alloy cladding is corrosive, and they could not be kept in the cooling ponds for too long without the risk of releasing radioactivity into the biosphere. This does not necessarily apply to the later reactors which use uranium oxide pellets clad in non-corrosive sheaths, but reprocessing is still contemplated, mainly because of the need to extract the plutonium for fast breeders.

As table 2 shows, the composition of the spent fuel elements from a PWR is 96 per cent uranium (with 0.83 per cent U-235 content), about 1 per cent plutonium, and about 3 per cent fission fragments. The purpose of reprocessing is to separate these three fractions.

A number of reprocessing techniques have been developed, but the one in most general use is the Purex method, which was initially developed for the military programme. The Purex process is diagrammatically depicted in figure 7. The first step is to get rid of the cladding; this is done by chopping up the rods and putting them into nitric acid, which dissolves the uranium and plutonium, and most of the fission products, but not the 'hulls' (the

Figure 7. The Purex process

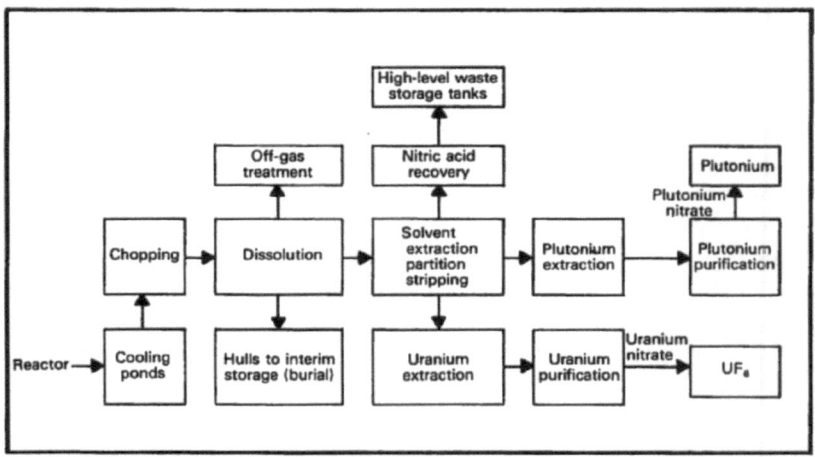

cladding material) which are removed for storage. Using an organic solvent, the uranium and plutonium fractions are extracted separately, leaving the rest as 'high-level' waste, so called because of its intense radioactivity. The uranium fraction, after going through a purification stage, is converted into uranium hexafluoride, while the plutonium fraction, again after purification, leads to the production of plutonium dioxide.

Since chemical separation is not 100 per cent effective, small amounts of plutonium are inadvertently lost in the process. Many of the ancillary materials used in processing become contaminated and have to be stored as 'medium-level' waste. Smaller amounts of radioactivity, forming 'low-level' waste, are released into the biosphere.

The difficulties inherent in reprocessing (a pilot uranium oxide plant had to be shut down in the UK after an explosion in 1973) have resulted in escalating costs, and in a very slow growth of reprocessing plants. Most reactors came into operation recently, and the spent fuels are still at the stage where they are kept in cooling ponds. Therefore, there has not been much pressure for reprocessing, but it is building up. At present no reprocessing of oxides is being carried out on a commercial scale, but small plants are in operation in France, FR Germany, India and Japan. Large plants are being built at La Hague in France and Windscale in the UK; both of these have already accepted orders for reprocessing fuels from foreign countries, in addition to their own. Work on reprocessing plants in the USA was stopped (see *US energy policy*, page 427). Table 11 gives a list of reprocessing plants for uranium oxides, together with their capacities and expected starting times [10]. A reprocessing plant is also to be installed for Brazil by FR Germany.

Waste management

The safe storage and/or disposal of the waste products of nuclear reactors is

394

Table 11.   Reprocessing plants for uranium oxides

Country	Location	Capacity (tonnes/year)	Date of operation
Argentina	Ezeiza	. .	. .
France	La Hague	400	Operating
		1 000	1985
		1 000	1989
FRG	Karlsruhe	40	Operating
	Gorleben	1 400	1988–90
India	Tarapur	125	Operating
	Kalpakkan	125	1982
Japan	Tokai Mura	210	Operating
UK	Windscale	1 000	1984
		1 000	1987

the major unsolved problem of the nuclear industry, and is probably largely responsible for the slowing down of nuclear power programmes in several countries.

The problem is posed by the immense radioactivity of all the materials accrued as a result of the release of nuclear energy through fission, as well as by the time factor, of thousands or even millions of years, involved. Clearly, our experience with radioactivity is of far too short duration to enable us to ensure radiological safety over periods reckoned on a geological time scale, although it is claimed that some lessons can be learned from the natural reactor in Oklo, in the Republic of Gabon, where apparently a spontaneous chain reaction in a uranium deposit took place some two thousand million years ago.

The radioactive fission fragments have half-lives stretching over an enormous range, from less than a second to millions of years. The actinides too have half-lives covering a wide range. The decay of the short-lived elements results in a reduction of the total activity of the spent fuel elements by a factor of several hundred during the first year after their removal from the reactor. For this reason the spent fuel elements are kept in cooling ponds near reactor sites for about a year before being shipped to the reprocessing plant.

The problems of handling the high-level wastes after reprocessing are highlighted in figure 8, which shows the time scale involved in terms of the heat generated by the radiations (α, β and γ) from the radioactive decay, starting one year after removal from the reactors. Three stages of the decay curve can be distinguished. In the first, the activity is due mainly to the β- and γ-rays from the fission products, the most important of which are strontium-90 and caesium-137. With half-lives of about 30 years, their activity is reduced to 10 per cent every 100 years. After 500 years this is reduced to a level at which other, longer-lived substances take over. This is the period of the α-emitting actinides which have half-lives of up to 400 000 years. Their decay goes on for about 100 000 years, by which time the total activity has been reduced by five orders of magnitude. Finally, there is the

Figure 8. Heat generated by the radioactive decay of materials in spent fuel

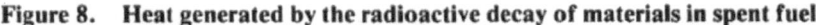

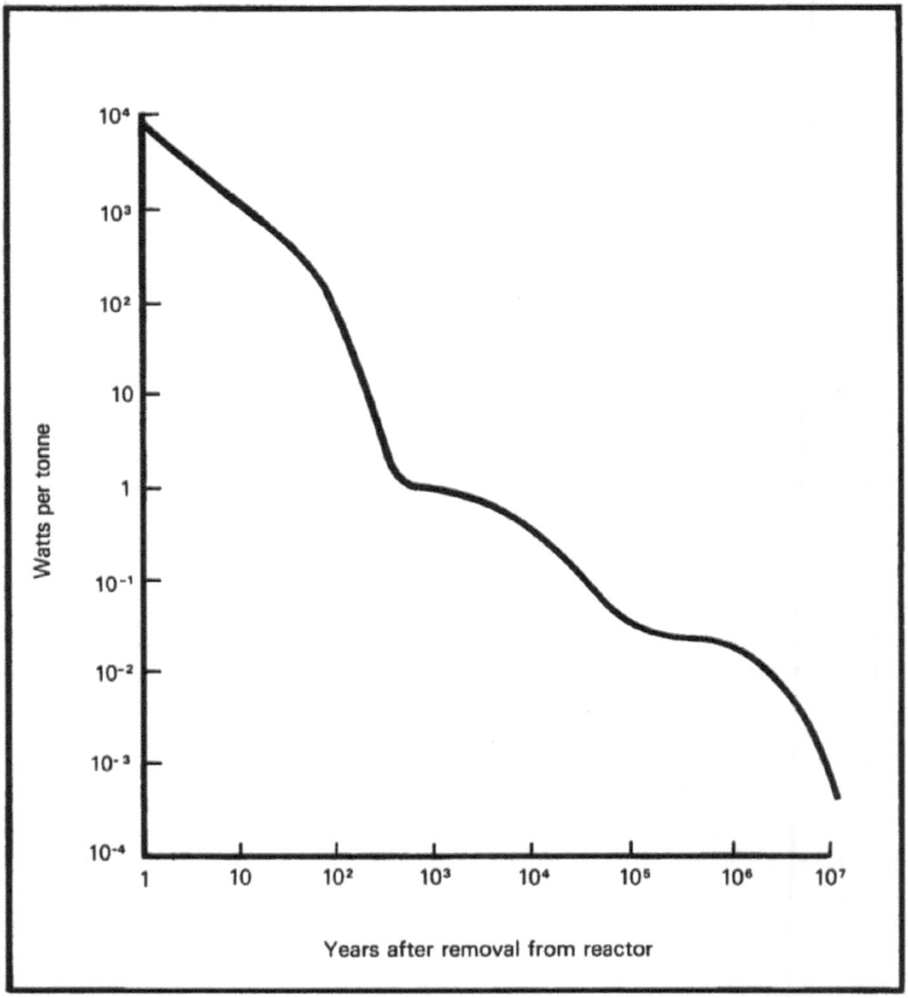

third epoch of the very long-lived fission products, technetium-99; tin-126 and iodine-129, and nuclides formed by the decay of the actinides. Their abundance is very low and would not contribute significantly to the radiation hazard, but in the long run their production adds irreversibly to the global radioactive inventory.

As already mentioned, during the operations in the reprocessing plants, some of the waste products are being released into the biosphere. These include the gaseous elements, such as tritium and krypton, which are released into the atmosphere, and other products, such as small amounts of caesium, strontium, and actinides, including plutonium, which are released into the aquatic system. With the growth of installed nuclear capacity, the proportion of the material released during reprocessing will have to be

Figure 9. Storage tank

Source: 'Nuclear wastes: popular antipathy narrows search for disposal sites', Carter, L. J., *Science*, Vol. 197, pp. 1265-66, 23 September 1977. Copyright 1977 by the American Association for the Advancement of Science.

decreased, and it will have to be disposed of together with the high-level waste.

The magnitude of the problem of high-level wastes becomes evident when one considers the quantity of materials involved. For an installed capacity of 1 000 GW(e), which may be reached by the year 2000, the fuel which will have to be reprocessed every year would weigh about 25 000 tonnes. After one year, each tonne of reprocessed material contains 2 megacuries of activity. This means that every year there would be an additional 50 000 megacuries to be disposed of. This refers to the activity from the fission products, and does not include that of plutonium and the other actinides.

As an interim measure, it is proposed to keep the waste products, in liquid form, in storage tanks. Due to the large volumes involved, and the need to cool them continuously, because of the intense heat given off from the radioactive elements, huge tanks, and a large number of them, are needed. Figure 9 shows such a tank being built at the Savannah River Plant in the USA [12]. This tank has a capacity of nearly 5 000 cubic metres, and is made of stress-relieved carbon steel with double walls. When finished it will be encased in concrete, and earth back-filled round it. An array of coils inside, with water circulation, will cool the liquid. Sixteen such tanks are being built at present, which may be sufficient for immediate needs. In the long run, however, many more will be needed. Local environmental groups are already mounting campaigns against them.

On a longer-term basis, the containment of this waste will require its solidification, followed by vitrification, that is incorporation into a boro-silicate glass. Another method proposed is incorporation into rock-like crystals. The feasibility of these techniques is still under investigation.

Ultimately, the solidified high waste will somehow have to be buried or disposed of. Several means are under consideration, such as deposition in geological formations on land, probably rock salt, direct deposition in deep oceans, or by drilling holes in the ocean floor and refilling them once the material has been deposited. More fanciful ideas, such as rocketing the waste into space, have been discarded, but the other options have also been criticized on various grounds. No acceptable final solution to the problem of waste products of nuclear energy has yet been found.

III. The proliferation threat

The generation of electricity in commercial nuclear reactors inevitably leads to the production and accumulation of materials that could be used for nuclear weapons. This section discusses the threat to security which may result from the widespread availability of nuclear weapon materials, the possibilities of clandestine acquisition of nuclear weapons, and the parts of the nuclear fuel cycle which are particularly sensitive to the proliferation issue.

Weapon-grade materials

There are a number of nuclides which could be used for nuclear weapons, but the three most important ones are the same fissile materials that are used as fuel in nuclear reactors, that is, U-233, U-235 and Pu-239.

The main difference between the use of these materials in reactors and their use in weapons is that the chain reaction in a nuclear weapon must be propagated by fast neutrons, so that a sufficient number of fissions (about 10^{24}) can occur in such a short time (less than one microsecond) that the material will not disperse before a large amount of energy, equivalent to that from 20 000 tonnes of TNT (20 kt) has been released. With thermal neutrons, the chain reaction would develop with the same speed as chemical reactions and the explosive power would not differ greatly from that of ordinary explosives.

Of the three nuclides mentioned, U-235 was the material used in the Hiroshima bomb; it has been suggested that it was used for the fission trigger of the hydrogen bomb, but this has not been confirmed. The Nagasaki bomb used plutonium, and it is believed that most nuclear weapon tests carried out by the USA have used plutonium. There is no evidence that U-233 was ever used in nuclear weapons.

Figure 10. Critical masses

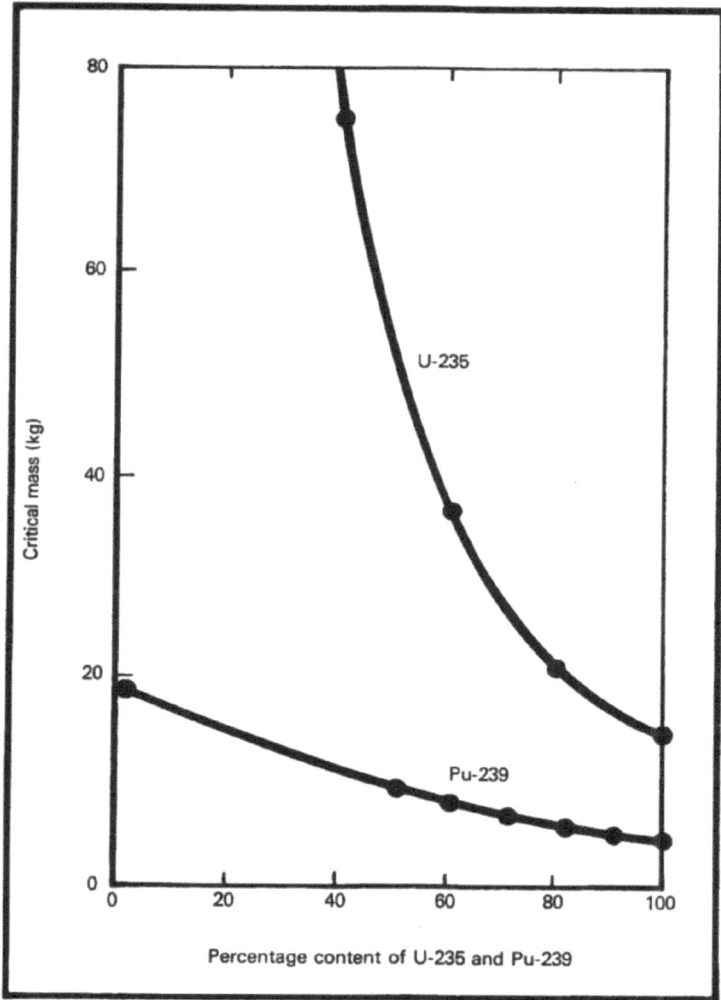

To produce an explosive, one must assemble a quantity of fissile material which exceeds a certain minimum amount, known as the 'critical mass'. This is the smallest mass in which a chain reaction can be sustained. However, the critical mass is not a constant quantity, even for a specified fissile nuclide. Its magnitude depends a great deal on several factors, such as physical purity, chemical purity, physical surroundings and the method of assembly of the bomb.

For a given nuclide, the smallest critical mass is obtained when the material is in pure elemental form, that is 100 per cent U-235 metal, or 100 per cent Pu-239 metal, and surrounded by a 'tamper', a material used mainly to reflect the neutrons which would otherwise have escaped from the assembly. Beryllium is said to be a very good material for a reflector. With

100 per cent pure fissile material surrounded by a reflector made of natural uranium 15 cm thick, the critical masses are: 5.8 kg for U-233, 15 kg for U-235, and 4.4 kg for Pu-239 [13]. Considering the high density of uranium and plutonium, these masses are contained in quite small volumes; for Pu-239 it would be a sphere 7.5 cm in diameter, the size of an orange. Actually with the even higher density momentarily achieved at implosion (see below) the critical mass for plutonium might be less than 2 kg.

Under less ideal conditions, the critical mass would be considerably larger. Without a reflector the critical mass is increased by a factor of about three or four. If metal oxides were used, and this is the chemical form of the nuclear fuel in most reactors, the critical mass would be increased by a factor ranging from 1.5 to 5, depending on the way the oxide is compressed and a given density achieved.

An important factor is physical purity, that is, the concentration of the fissile nuclide in the material. This is particularly the case for U-235, where the critical mass varies rapidly with concentration. At a low concentration of U-235 the critical mass becomes so large as to make it useless as a weapon. The upper curve of figure 10 shows the variation of the critical mass with the concentration of U-235; these values apply when a tamper is used [14]. At a concentration of 40 per cent, the critical mass is 75 kg, while at 20 per cent it goes up to 250 kg. For these reasons, only uranium in which the U-235 isotope has a concentration greater than 20 per cent is considered weapon-grade material.

The situation is different with plutonium, where the variation of the critical mass with the concentration of Pu-239 in it is much smaller. This is so because other isotopes of plutonium are also fissile with fast neutrons. Consequently, a nuclear explosion can occur even if the plutonium has a high admixture of other isotopes, as would be the case with the plutonium in spent fuel elements after several years in a thermal reactor. As seen in figure 10, a reduction of the Pu-239 content to 50 per cent would result in an increase of the critical mass by a factor of two.

However, the main effect of the presence of other plutonium isotopes, in particular of Pu-240, is not the change in critical mass, but the explosive yield of the weapon. This is directly related to the method of assembly of the bomb. There are two methods of initiating the explosion in a nuclear weapon. In the gun method—which was used in the Hiroshima bomb—two pieces of the fissile material, each lighter than the critical mass, are brought rapidly together by shooting one into the other by means of a conventional explosive. At the same instant, a source of neutrons is assembled, to ensure the presence of at least one neutron to start the chain reaction. The second method—used in Nagasaki—is the implosion technique. Here the fissile material is divided into a number of segments, usually segments of a sphere, each less than the critical mass. They are so arranged that if they were all pushed inwards they would form a sphere of supercritical mass. The pushing inwards, or implosion, is accomplished by placing suitably shaped plastic explosives behind each segment. To initiate the explosion, these charges are detonated simultaneously, compressing the material to a high density, considerably greater than its normal value. This results in much greater supercriticality, more material being available to undergo fission, and, therefore, resulting in a greater explosive power.

The explosive yield of the bomb depends drastically on the initiation of the chain reaction at the right moment. If it starts prematurely, before a high degree of supercriticality is reached, then only a small amount of the plutonium would undergo fission before it blew itself apart, and the total energy released would be relatively small. It is in this connection that a large admixture of Pu-240 may prove highly detrimental. Pu-240 has a high probability of undergoing fission spontaneously with the emission of neutrons; each kilogram of it spontaneously emits a million neutrons per second. If, therefore, the plutonium contains a large proportion of Pu-240, there is a significant probability that a neutron will appear and start the reaction prematurely. We would then have a pre-detonation, resulting in a bomb with greatly reduced yield, although still of the order of a kiloton of TNT.

As was shown earlier (under *Reactor inventories*, page 379), in a nuclear reactor there is a gradual accumulation of Pu-240 which may reach 25 per cent after three years. To ensure high efficiency of the bomb in military reactors, the fuel elements are kept in the reactors for a much shorter time, so that the concentration of Pu-240 would not be more than about 7 per cent. Indeed, for a long time, the nuclear industry maintained that reactor-grade plutonium is unsuitable for weapons. If this were true, then the concern about proliferation associated with commercial nuclear reactors would have been greatly reduced. However, it turns out that this is not the case. The explosive yield depends on the speed of assembly, and with a high-speed implosion assembly, the probability of pre-initiation is greatly reduced. With a less sophisticated technology, the explosive yield is statistically distributed round a lower mean, but even at the lower end the explosive power would be of the order of kilotons.

It has been stated that a weapon was tested in the USA using reactor-grade plutonium and that a high yield was obtained. It must, therefore, be concluded that plutonium from reactors is suitable for nuclear weapons. It should also be noted that the blanket of the fast breeder reactor contains plutonium with a very low admixture of other isotopes and it is, therefore, high-grade weapon material.

Clandestine acquisition of nuclear weapons

There are many reasons why a state, or a subnational group, might want to acquire nuclear weapons. For subnational groups the motive may be blackmail, for example, for financial gains or political ends. Leaving aside the rather unlikely attempt to raid a nuclear weapon depot or steal a ready-made bomb, the only way for such a group to acquire a weapon would be to make it clandestinely.

A state may wish to acquire nuclear weapons for various reasons: for prestige, for intended aggressions or for defence. The possession of nuclear weapons is deemed to confer a high status on the nation, and this might be a powerful incentive. On the other hand, to achieve the desired effect the acquisition of weapons would have to be done openly, and in the present climate would bring with it political and economic repercussions. Even the guise of a peaceful nuclear explosion would no longer be acceptable as an excuse. Therefore, the open acquisition of nuclear weapons, although possible, is less likely than the clandestine route, unless there is a drastic change in the world situation.

If a state has aggressive designs, it is unlikely to announce its intentions beforehand. It would aim to acquire the nuclear weapons clandestinely in order to maximize the element of surprise. The preparation would have to be concealed until the last moment.

A country which felt the need of a nuclear bomb for its own defence would be in a similar situation. A premature disclosure of its preparation to obtain nuclear weapons might in fact provoke the aggression it feared, before it was ready with its defence.

It would thus appear that, whatever the motives, the most probable route to the acquisition of nuclear weapons is by clandestine means.

The actual production of the bomb would not present too much difficulty for a group with adequate resources, as is likely to be the case if a government wanted to make a weapon clandestinely. The whole technology, including details of the implosion technique, is described in the open literature and could be easily apprehended by an intelligent reader. A minimum programme would require about 10–20 people trained in physics and engineering, plus a technical staff. There would be the need of laboratory and workshop facilities to make the various items for the assembly mechanism, and perhaps field facilities for testing the high explosive (but not nuclear) charges. All this could be easily concealed among the many operations of similar kind which are normally carried out in a country.

It has been estimated that it would take such a group about two years to carry out this programme, and that the cost would be a few tens of millions of dollars [15]. This estimate does not include the effort needed to obtain the fissile material for the bomb. Indeed, the main problem for a group desiring to acquire a nuclear weapon—whether at national or subnational level—is to get hold of a sufficient amount of fissile material.

The simplest way to acquire such material would be to build facilities specifically designed for the purpose: a plutonium-producing reactor and a reprocessing plant to separate the plutonium from the fuel rods. A relatively small reactor, with a power output of 50 MW(e), could produce enough plutonium for one or two bombs a year. However, it would be difficult to conceal the existence of such a reactor. Alternatively, a legitimate research reactor could be used for such a purpose, but there would still be the need of a reprocessing plant which is also difficult to conceal, even if it were on a small scale. On the other hand, a country which possesses nuclear facilities,

Table 12. Summary of the diversion points in the LWR fuel cycle

Facility	Material	Is the material useful to the national diverter?	Is the material useful to the non-state adversary?
Mine Mill Conversion facility	Natural uranium (0.7 per cent U-235) as ore (0.2 per cent uranium) U_3O_8 UF_6	Yes, but only as feed for a dedicated facility (plutonium production reactor or enrichment plant)	No (but criminals might engage in black market in these materials)
Enrichment plant	Low enriched uranium (3 per cent U-235 as	Yes, but only as feed for a dedicated enrichment plant	No (criminals might engage in black market in these materials)
Transportation to reactor Temporary storage at reactor	UF_6 UO_2 UO_2 in fuel assemblies	Nation would eventually have to replace fuel	
Reactor spent fuel storage	Pu—about 0.8 per cent in highly radio-active spent fuel	Yes; dedicated reprocessing facility required	No except yes for large, very well financed, technically competent group with a secure base of operations and a few members will-ing to risk radiation injury
Reprocessing plant Transport to fuel fabrication plant Input area to fuel fabrication plant	Pure $Pu(NO_3)_4$ or pure PuO_2	Yes; nation would probably convert material to metallic plutonium	Yes; if $Pu(NO_3)_4$, simple conversion to PuO_2 required. If PuO_2, material directly usable in explosive
Plutonium fuel fabrication plant	PuO_2 (3 per cent to 7 per cent) mixed with over 90 per cent UO_2	Yes, chemical separation of Pu from mixture only a minor obstacle. Logistics of divert-ing 100 to 200 kg of material for one explosive trouble-some	Yes, but chemical separation a time consuming opera-tion. Logistics of stealing or diverting 100 to 300 kg of material for one explosive cause problems
Transport to reactor Temporary storage at reactor	About 1 per cent Pu as PuO_2 mixed with UO_2 in fuel assemblies	Yes, as above. (nation would eventually have to replace fuel)	Yes, but chemical separation a time consuming opera-tion. Logistics of stealing complete fuel assemblies present significant obstacle

including a reprocessing plant, as part of its peaceful nuclear power programme, would incur a much smaller risk of detection of its intentions to
make nuclear weapons, especially if its facilities are not subject to safeguards.

It must therefore be concluded that the possession by a country of a nuclear power programme, while not being a prerequisite, is the most likely route to the acquisition of nuclear weapons.

As for subnational groups, the larger the scale of nuclear power in the world, and the more widespread its facilities, the greater the opportunities for diversion. Indeed, under such conditions a black market may spring up, which would offer nuclear weapon materials to governments and subnational groups alike.

Acquisition of nuclear weapon materials from the nuclear fuel cycle

Although the nuclear power programme represents a likely route for the acquisition of nuclear weapon materials, not all parts of the nuclear fuel cycle are equally vulnerable. Thus, the uranium mines and the milling plants are safe because they handle only uranium of very low U-235 concentration. Even the reactors themselves—except HTGRs—are unlikely to be targets for diverters because by the time enough plutonium is accumulated in the fuel rods, there will also have accumulated such vast quantities of radioactivity that the rods will be too hot to handle. On the other hand, enrichment plants are potentially dangerous as far as uranium is concerned, and reprocessing plants are particularly dangerous from the point of view of diversion of plutonium. The transport of fissile materials between the different parts of the cycle also constitutes a significant danger.

Table 12 summarizes the diversion points of the nuclear fuel cycle for thermal reactors [15a]. The most vulnerable parts are discussed in more detail below.

Enrichment plants

As seen from figure 10, uranium with a U-235 content less than 50 per cent is practically unsuitable for nuclear weapons, although even a 20 per cent concentration is considered weapon-grade material for safeguards purposes. The great majority of reactors use uranium enriched to about 3 per cent, and therefore the normal output from enrichment plants is secure against proliferation. However, much less effort is required to obtain highly enriched material from the 3 per cent enriched uranium than from natural uranium; the SWUs needed to enrich uranium from 3 per cent to 90 per cent are about three times less than from natural uranium, with a five-fold reduction in the amount of material required.

A country with its own enrichment plant might convert one section of the plant into a high enrichment cascade. This would be much easier in a centrifuge plant than with gas diffusion.

For these reasons, an enrichment plant has to be considered a virtual facility for the production of weapon-grade materials. As already mentioned, some enrichment plants make uranium enriched to 93 per cent for HTGRs and research reactors.

The situation would become much further aggravated if new separation techniques were developed which required fewer stages to achieve a high enrichment (see under *Laser separation*, page 407).

Reprocessing plants

In view of the small critical mass even for reactor-grade plutonium, the safeguarding of this material against diversion is the most serious issue in the nuclear power programme. A reactor with a power output of 1 GW(e) and an 80 per cent load factor has an annual net production of 250 kg of plutonium. Thus, a diversion of only a few per cent of that material would be enough to make a weapon every year.

From the point of view of proliferation, reprocessing represents the most dangerous part of the nuclear fuel cycle. This is where the plutonium becomes separated from the highly radioactive waste products, and is thus much easier to handle, and to be diverted. It should be noted that the fast breeder reactor presents a much greater problem in this respect, since for the same amount of electricity generated, the quantity of plutonium to be reprocessed is an order of magnitude greater than in thermal reactors (see *Fast breeder reactors*, page 384).

Diversion can occur at any of the stages of separating the plutonium, but the early steps present the diverter with greater problems than the last one, because the material is much more difficult to handle owing to the radiation hazards from the γ-rays from the radioactive waste products. By contrast, plutonium itself—which emits α-particles—requires little shielding, and, therefore, once separated from the fission products, it becomes much easier to divert it. Plutonium nitrate from the purification stage of the Purex process (figure 7) would require only a single precipitation to be converted into weapon material. The plutonium dioxide from the last stage could be used directly for a weapon. The fact that even a thin shielding is sufficient to stop all α-particles, makes it also more difficult to detect its presence, thus facilitating the task of the diverter still further. Actually, plutonium also emits X-rays, γ-rays and neutrons, but at a very low intensity, and sophisticated instruments are needed to detect these radiations.

Mixing the plutonium oxide with uranium oxide, to fabricate mixed fuels for recycling in thermal reactors, would introduce an extra barrier, by making it necessary to perform a chemical separation, but this is much easier than isotope separation, and would not be a major obstacle for diversion at national level.

Storage and transport

The more reprocessing plants in existence, the more opportunities there are

Table 13. Number of shipments in the nuclear fuel cycle projected to the year 2000

	Shipments per year in		
	1980	1990	2000
Fuel	670	2 500	5 400
Spent fuel	2 000	6 400	12 000
Plutonium	20	143	438
Wastes and fission products	630	2 450	5 500

for diversion. To overcome this it has been suggested that only a few centres should be allowed to operate reprocessing plants and that these would process fuels from many reactors. A most dangerous situation would then arise, if a nation, having sent its spent fuel elements for reprocessing in another country, were to insist on receiving back the plutonium for storage, or to fabricate its own fuel. This would enable the nation, if it had made all the other preparations for a weapon, to divert the material (even if safeguarded) before any sanctions could be effective.

The transport of plutonium between the reprocessing and fabrication plants would also create dangerous diversion opportunities. Indeed, if a small number of reprocessing plants were to serve a large number of reactors, there would be a great increase in the number of shipments of plutonium, with the possibility of diversion during transport. Table 13 shows the IAEA projection [16] of the number of annual shipments of various items in the nuclear fuel cycle up to the year 2000. This huge traffic in nuclear materials will create a significant problem with regard to the physical security of these shipments.

In view of the high vulnerability of nuclear materials to theft or sabotage when being transported, it is important that nuclear transports are extremely well protected and that, in case of theft, recovery procedures are well planned. General rules include the need to: minimize the time during which nuclear material is being transported; minimize the number and duration of transfers of nuclear material from one vehicle to another; minimize the time during which nuclear material is in temporary storage while waiting for vehicles; avoid regular transport movements; and predetermine the trustworthiness of people involved in transport of nuclear material. Much caution would have to be exercised in advertising transport activities, including markings on vehicles and the use of open communication channels.

An emergency force, trained and equipped for the task, would have to be available to try to recover any stolen plutonium very rapidly; this implies a large force of highly-trained, well-armed men, with a wide range of detection equipment, including airborne equipment.

IV. New technologies

Although the projected nuclear power programme could be fully developed with the technologies discussed under *The current nuclear power situation* (page 373), further innovations already envisaged, and now in their research and development stages, are likely to be introduced during the next few decades. They may eventually bring considerable changes to the pattern of nuclear energy production. The realization of the proliferation danger is beginning to have an effect on the planning of nuclear facilities, and some new technologies are specifically designed to reduce the possibilities of diversion of fissile materials. On the other hand, new discoveries which would considerably aggravate the situation, as far as proliferation is concerned, may be promoted by the nuclear industry because of their economic advantages. In this section the effect of introducing new technologies into several parts of the nuclear fuel cycle—enrichment, power reactors, and reprocessing—is discussed, as well as the possibility of a new method of producing plutonium.

Enrichment of uranium

Laser separation

The basic principle of this technique is that laser light can be used to excite the molecules of a uranium compound to higher energy states, the two isotopes of uranium being excited to different states. Once this is achieved there are several techniques, again using lasers, by which to produce a physical separation of the molecules from the different excitation states. The important feature of this technique is that a high degree of separation, nearly 100 per cent, could be attained in one stage, in contrast to the other enrichment techniques, in which the separation in a single stage is very small, thus necessitating many stages of separation and a large and costly enrichment plant.

Should the laser technique prove to be technically and economically feasible, it would create a proliferation threat of the first magnitude, because the same plant built to enrich uranium to reactor grade, about 3 per cent, could also be used to enrich it to weapon-grade material. Laser separation plants would then become the most direct route to the production of nuclear weapon materials.

Although it is known that research on the laser separation technique is being actively pursued under US Government auspices, no details of this work have been released for reasons of military and commercial security. It is, therefore, impossible to say whether the technological problems have been overcome to an extent that would make it a practical proposition. General considerations indicate that the capital cost of a laser separation plant would be much smaller than that for the other techniques, and that the energy consumption would be orders of magnitude smaller than for gaseous diffusion or centrifuge methods [17]. These are sufficiently strong incentives for the laser technique to be investigated intensively; unless some unforeseen difficulties are encountered, it is likely that it will become an important adjunct to the nuclear fuel cycle.

The ability to separate isotopes completely would be a considerable economic incentive to the development of the laser technique for yet another reason, the recovery of U-235 from depleted uranium. The tails from existing enrichment plants, which contain 0.2−0.3 per cent of U-235, could be passed through the laser separation process to reduce the U-235 content to 0.05−0.1 per cent. By 1985 there will be about half a million tonnes of depleted uranium from the reactors in operation at that time; this could save more than 150 000 tonnes of natural uranium, enough to feed 100 thermal reactors for seven years.

Chemical exchange

A separation technique, which in many ways has just the opposite characteristics to laser separation, is the chemical exchange process, advocated by the French Atomic Energy Commissariat [18]. It is based on the principle that the rate of chemical reactions depends on the molecular weight of the compounds taking part in it, and this opens the way to partial separation of compounds containing different isotopes of an element. For commercial reasons, the details of this technique have not been revealed; it is thus impossible to judge its potential value. The main virtue claimed by its promoters is that the technique could be used to enrich uranium to about 3−5 per cent as needed in reactors, at a cost which would compare favourably with the gas diffusion method, but that it would be practically impossible to obtain enrichments greater than 20 per cent. Should, therefore, the chemical exchange technique turn out to be practicable, and provided there is general agreement that it should supplant other techniques, the enrichment stage of the nuclear cycle would cease to present a proliferation problem.

Advances in nuclear fuel cycles

The main aim of these innovations is to lessen the threat of proliferation arising from the widespread availability of plutonium, by reducing the turnover of that material and by delaying the advent of the fast breeder long enough for other technologies to be developed which would make the FBR and reprocessing of plutonium superfluous.

408

Greater efficiency and better utilization of nuclear fuel, which would stretch the natural uranium resources, could be achieved through some modifications of the nuclear fuel cycle with existing LWRs without the need to recycle the uranium from the spent fuel elements [19].

The modifications include: (*a*) using uranium with a slightly higher enrichment, say 4.5 per cent; (*b*) greater burn-up which this would make possible, say 55 MW(e)-days per kg; (*c*) reducing the tails assay in the enrichment plant to 0.05−0.1 per cent; (*d*) material or geometry changes in the reactor which would reduce the absorption of neutrons in the moderator and coolant; and (*e*) decreasing the period between refuelling. These modifications could increase the utilization of uranium by about 50 per cent, while at the same time reducing considerably the amount of fissile plutonium produced, largely because of the higher burn-up.

A greater saving in uranium and a reduction in plutonium production could be achieved if some of the above modifications were combined with the recycling of the uranium from the spent fuel rods, but this would require their reprocessing.

Advanced converter reactors (ACRs)

Most important under this category is the denatured uranium−thorium reactor [19].

The resources of nuclear fuel could be immensely increased, without the need of breeding plutonium, by converting thorium into U-233, as was proposed with the HTGRs (see *Other reactor types*, page 386). But U-233 is a weapon material, and it would be relatively simple to separate it from the thorium, although it is not as easy to handle as plutonium, since U-233 is usually accompanied by U-232, the decay products of which contain intense γ-emitters. All the same, U-233 could be a proliferation hazard. However, it is quite easy to render it harmless as a weapon, by 'isotope denaturing', that is, by mixing it with U-238. It would then become similar to U-235 at low enrichment, in that it would not be possible to use it in a weapon, without having to put it through an isotope separation plant.

In the proposed ACR, the fissile content of the uranium would be not higher than 12 per cent U-233. The presence of U-238 would, of course, result in some Pu-239 in the reactor, but the total plutonium production would be less by a factor of seven than in LWRs.

The main problem with the ACR fuel cycle is that the technology of fabricating mixed uranium and thorium oxide fuels is still to be developed, and that the spent fuels will have to be reprocessed and new fuel fabricated for the recycling of denatured uranium. The plutonium would have to be stored and, despite the smaller turnover, it would create a proliferation problem.

Notwithstanding these difficulties, the denatured uranium – thorium reactors present the best option for large-scale utilization of nuclear energy, from the point of view of reducing the risk of proliferation.

Non-proliferation reactors

This refers to a concept of a reactor designed specifically to eliminate the possibility of diversion during the fuel cycle [15b]. The main criteria for such a reactor are that: (a) the system shall contain only a small amount of fissile material at any given time; (b) there shall be no access to the fuel during the lifetime of the reactor; (c) any diversion of fuel will cause the reactor to shut down; (d) the reactor shall be refuelled by the addition of non-fissile material only; (e) the reactor shall not operate as a breeder, but produce just enough fissile material to keep itself running; (f) reprocessing shall be done on site.

No specific design has yet been produced for such a reactor, but it is thought that it might use a gaseous fuel, namely U-233 in the form of uranium hexafluoride, the moderator might be beryllium, and that it might have a blanket of molten thorium salt for conversion into U-233. By operating it at a high temperature (about 4 000°C), the total inventory of U-233 could be reduced to a few tens of kilograms.

Whether such a reactor will ever be practicable remains to be seen. But the fact that a serious study is being made of a reactor with the main emphasis in its design on non-proliferation is a clear indication of the change in attitude towards nuclear energy policies.

Reprocessing techniques

The Purex method of reprocessing spent fuels (see *Reprocessing plants*, page 393) presents a likelihood of diversion during most of its stages, and at its end the plutonium becomes available in a form suitable for nuclear weapons. To reduce this, some modifications of reprocessing techniques have been proposed.

The Civex process [7, 20]

In this technique, some of the stages, where the plutonium is completely separated from the fission fragments, will be omitted and a sufficient amount of these radioactive substances will be left with the plutonium to make the latter 'inaccessible' to a would-be diverter. It is further envisaged that the material leaving the plant will be in the form of a mixture of plutonium and uranium oxides, in which the plutonium content would be less than 25 per cent. Among the criteria specified by the designers of the

Civex reprocessing system are: no pure plutonium in storage; no pure plutonium at any intermediate point; and no way of producing pure plutonium by simple adjustment at any stage of the process and without modification of equipment. However, as the proponents of this system themselves pointed out, Civex would not be a practical method for the recycling of plutonium from thermal reactors, nor would it be suitable for the reprocessing of plutonium from old spent fuels, which were intended to provide the initial loading to the first generation of fast breeders. Indeed, the Civex system is designed for the reprocessing of spent fuels from fast breeders after they have been in use for some time; in other words, the technique could not be used for some decades. Moreover, several of the proposed changes in the reprocessing routine are still to be worked out, and some doubts exist as to whether the cost involved would make the Civex approach economical enough to be adopted.

The 'Pipex' process [21]

In order to reduce the possibility of diversion during reprocessing, French scientists have proposed a modification of the Purex method. Essentially it would consist of a change in the layout of the plant to assure physical continuity of the material being processed; in practice, such a plant would be one leak-proof pipe through which the material would flow from a single inlet to a single outlet containing the purified material. The name Pipex was given to this technique to underline its continuous character and containment in one pipe. The designers claim that by using this process the risk of 'on-line' diversion would be greatly reduced.

Hybrid reactors

The main reason for introducing fast breeder reactors was that they would make it possible to increase greatly the nuclear fuel resources, by converting U-238 into Pu-239. This transformation is achieved by the bombardment of uranium with neutrons, and until now only the reactor based on fission is capable of providing a sufficiently intense source of neutrons for this purpose. However, another possibility of producing an adequate neutron supply is the fusion reactor.

The fusion reactor is based on a thermonuclear reaction between nuclei of light elements. At a sufficiently high temperature, of the order of 100 million degrees Celsius, nuclei of some light elements may coalesce to make heavier nuclei. In this process of 'fusion' a large amount of energy is released. The energy of the Sun is derived from such a thermonuclear reaction, the fusion of hydrogen into helium. On the Earth, fusion reactions have been employed in the hydrogen bomb, in which the necessary high temperature is produced by the detonation of a fission bomb acting as a trigger.

A thermonuclear reactor is a system in which the high temperature, initially produced by an external source, is maintained by the energy released in fusion reactions, thus enabling it to continue, just as burning of coal is sustained once part of it was ignited.

The main ingredients of thermonuclear reactions are the heavier isotopes of hydrogen, deuterium and tritium. The fusion of two deuterium nuclei, or of a deuterium and tritium nucleus, results in the formation of a helium nucleus and—in most cases—the emission of a neutron. Because of the great abundance of the basic material, deuterium (1 in 6 000 molecules of water in the oceans is heavy water, D_2O), fusion reactors would provide a virtually inexhaustible source of energy. However, despite intensive efforts over the past 25 years, no sustained thermonuclear reaction has yet been achieved, although the emission of neutrons has been observed, indicating that thermonuclear reactions have taken place after the creation of a high temperature by external means.

The main difficulty has been the simultaneous fulfilment of several conditions: a sufficiently high temperature, in a material of a sufficiently high density, and for a sufficiently long time. The containment of the material, so that the participating nuclei would not lose their energy, or be dispersed, before inducing other nuclei to undergo fusion, is still the major problem. Two main techniques have been used to achieve this: magnetic confinement and inertial confinement [22]. In the first method strong magnetic fields are applied to confine the hot gas to the centre of a vessel for a sufficiently long time. In the inertial confinement, the containment time is very short but the energy applied is so large that the material, a very small mass in the form of a pellet, would be heated and the reaction would occur before it could move. The necessary pulse of energy might be provided by high power lasers, electron beams or ion beams. This approach is of much more recent origin and it is too early to say whether it will prove more successful than magnetic confinement which has been studied for many years. The development of very high power lasers may be crucial.

If, by means of one or the other technique, a self-sustained thermonuclear reaction were attained in the near future, it would still take several decades before a commercial fusion reactor could be developed, even assuming favourable economics. However, fusion could find practical applications well before the break-even point has been reached. About 80 per cent of the energy released in fusion is carried by the neutrons produced in the thermonuclear reaction. By surrounding the fusion material with uranium or thorium, these neutrons could be utilized to produce the fissile materials Pu-239 and U-233. A detailed consideration of the energetics of the relevant reactions has shown that it would be advantageous to use the energy released in this fashion [23]. This would make the system a hybrid fusion—fission reactor in which—in principle—all of the uranium and thorium could be converted into nuclear fuel for thermal reactors.

From the point of view of proliferation, the advent of the hybrid reactor is both a good and a bad development. On the one hand, it opens the way for an enormously rich supply of nuclear fuel, by the conversion of thorium into U-233, and thus making the fast breeder reactor superfluous. On the other hand, it could also be an equally rich source of plutonium. Whereas the U-233 could be isotopically denatured, this cannot be done with plutonium, which would be a weapon material, albeit after reprocessing.

It has been suggested [23] that it might be possible to manufacture fission fuel in the blanket material of the hybrid reactor in such a way that it could be used in thermal reactors without reprocessing, in a once-through cycle.

Once it was decided that fast breeders were not necessary, the fusion reactor might be used as a major anti-proliferation measure: the destruction of all plutonium. By putting the plutonium round the fusion reactor, where it would be bombarded with neutrons, it could be converted into short-lived nuclides, whose radioactivity would rapidly come to an end. The same might apply to other long-lived radioactive products of fission reactors.

On the other hand, the inertial confinement technique opens up a new potential danger. The development of laser or heavy-ion fusion technology may provide a means to design new nuclear weapons and assess their effects without having to test them as is done at present, thereby increasing the risk of further vertical proliferation. Moreover, the development of the inertial confinement method may spread the technology of the hydrogen bomb to which it is akin. This would add to the dimension of the nuclear weapon threat associated with the peaceful uses of nuclear energy.

V. Technological security

The risk of diversion of fissile materials in the various parts of the nuclear fuel cycles operated now can be significantly reduced by technological measures. These include the adoption of proliferation-resistant fuel cycles, the application of physical barriers to make the task of the diverter more difficult, and arrangements to facilitate the early detection of attempts at diversion.

Once-through cycle

Since diversion is most likely to occur during the reprocessing of spent fuels and transport of the separated plutonium, the best remedy would be not to separate the plutonium from the spent fuel elements, so that in no part of the cycle would it occur in a form suitable for weapons. This could be achieved in the once-through cycle (see figure 6(a)), in which the fuel elements, removed from the reactor, are stored or disposed of without the separation of plutonium from the waste products. Owing to the intense radioactivity contained in the spent fuels, the plutonium would be inaccessible for long periods, of the order of 100 years. Even after a few hundred years, when the fission fragments have largely decayed (figure 8), the radioactivity of the actinides would necessitate a remote control system for the separation of the fissile materials, and would still present an additional barrier for diverters.

It should be noted that 'inaccessibility' is a relative term; it depends greatly on the degree of technical sophistication available to the diverters. If the diversion is on a national level, and if the country possesses a small reprocessing plant, then it could separate the plutonium even at an early stage. On the other hand, if the 'once-through cycle' policy were generally adopted, and there were no reprocessing plants anywhere, the plutonium would remain inaccessible for a very long time.

Although the once-through cycle offers a radical solution to the proliferation threat, there is strong opposition to it. The obvious objection is that it would close the door to the fast breeder reactor, since (a) the plutonium content in the spent fuels is required to provide fuel for the first charge of the core of the breeder; and (b) reprocessing is an essential part of the fast breeder fuel cycle. However, owing to the delay in starting the commercial fast breeder programme, plutonium will not be required for several decades and this means that reprocessing could be delayed for a considerable time, without prejudice to the future policy on breeders.

Two other, less convincing objections to the once-through cycle have been put forward. One is an economic argument, that if the U-235 and plutonium content in the spent fuel elements were recycled this would save a great deal of natural uranium and consequently reduce the cost of nuclear energy. The second argument, on environmental grounds, is that if the spent fuel rods were left unprocessed they would deteriorate and release radioactivity into the environment.

With regard to the first argument, a detailed analysis has shown that recycling of the U-235 content of the spent fuel from LWRs, and storing the plutonium, would save about 14 per cent of the load of uranium in fresh fuel elements. However, against this saving in the cost of fresh uranium one has to set the very high cost of reprocessing. If there were no other reasons for reprocessing, and the issue were to be decided on costs alone, then recycling would lose out, since reprocessing would make this option slightly more expensive. This is seen in table 14, where columns 2 and 3 show the total fuel cycle costs, with and without recycling of uranium. Even if allowance is made for the value of the unused plutonium, recycling would increase the cost by two per cent. The balance depends of course on the price of uranium; the figures in table 14 were based on the assumption that, starting with present-day prices, there would be a 5 per cent annual increase in the price [24].

The other possibility is to recycle both the uranium and the plutonium.

Table 14.   Nuclear fuel cycle costs[a]

	Once-through	U recycle	U + Pu recycle
U_3O_8 purchase	3.65	3.26	2.89
Other costs	2.51	3.23	2.70
Pu credit	–	– 0.23	
Totals	**6.16**	**6.26**	**5.59**
Change		+ 2%	– 9%

[a] Figures are given in $/MWh.

The total saving could then be 31 per cent in terms of uranium needed for fresh loads. In terms of costs, the last column of table 14 shows that the recycling of both uranium and plutonium might save 9 per cent of the fuel costs. However, considering that the cost of the fuel is only about 15 per cent of the cost of electricity to the consumer, the maximum saving on the electricity bill would be about 1 per cent and thus hardly significant. Another argument against recycling of plutonium is that if it were used up in thermal reactors it would not be available for fast breeders.

Recycling would, of course, stretch the available natural uranium resources, but as was pointed out earlier (*Uranium economy in LWRs*, page 409) even larger savings of uranium could be achieved through certain modifications of the once-through fuel cycle.

The objection that the storage of unprocessed fuels would constitute an environmental hazard cannot be said to be supported by firm evidence. The fuel elements from some early reactors, such as Magnox, which are sheathed in magnesium alloy, do require reprocessing, but this does not apply to the present day cladding in zirconium alloy or stainless steel. Since the great majority of commercial power reactors began to operate only in recent years, not enough experience has been accumulated to enable a definite pronouncement to be made of the length of time that fuel elements could be kept without deterioration. A recent study, initiated by the Swedish power industry [25], has investigated two schemes for handling and final storage of both reprocessed and non-reprocessed spent nuclear fuel elements. It concluded that both types can be safely stored for a very long time, after encapsulating in suitable canisters.

Another argument in favour of reprocessing is that the volume of radioactive material to be stored or deposited is greatly reduced after reprocessing. However, this argument does not take into account the various other wastes which result from reprocessing. If these materials are added, it turns out that the total volume is considerably increased as a result of reprocessing [26].

In summary, there seem to be no convincing reasons for the reprocessing of spent oxide fuel elements from thermal reactors. For the next few decades at least—depending on whether or not fast breeders will be developed—the once-through cycle offers a good safeguard against proliferation.

Physical barriers

Should it be decided that the spent fuels will be reprocessed, and the plutonium in them extracted, the risk of diversion during the storage and transport stages could be reduced by introducing physical barriers which would make it much more difficult for the material to be misused. These barriers might be of different types, such as rendering the plutonium highly radioactive and thus 'inaccessible', or making it less suitable for a nuclear explosive.

The Civex method of reprocessing belongs to the first category, in that some of the fission products would be left with the plutonium fraction, thus making the latter inaccessible. However, as was pointed out, it is unlikely that this method will come into use during the next few decades.

Other methods of 'spiking' consist of mixing the plutonium with a γ-ray-emitting substance or attaching a γ-source, such as cobalt-60, externally, as protection during shipment. Both of these methods are physically feasible but their effectiveness is doubtful. Internal spiking would not constitute an unsurmountable obstacle to a diverter who possessed the means to tackle all the other tasks in making a nuclear explosive. The same applies to the attachment of a cobalt-60 source. Moreover, both of these methods would result in an increase in the cost of the fabrication and transportation of fuel, quite apart from increasing the radiation exposure risk to workers, during loading and unloading, and to the population in the event of an accident. These reasons make spiking unattractive to the nuclear industry.

Proposals to render the plutonium less suitable for a nuclear explosive are rather more promising. Among the methods suggested are the admixture of plutonium-238 or of californium-252.

Pu-238 has a relatively short half-life (86 years) and because of this has a high specific activity and a high rate of heat emission, which amounts to about 600 watts·per kg of the material. This property of Pu-238 makes it suitable as a source of power in satellites, pacemakers, and so on. In the normal fuel cycle, the proportion of Pu-238 in the plutonium from spent fuels is very low, about 2 per cent (see table 2), but it can be significantly increased by adding neptunium-237 to the freshly fabricated fuel and then putting it through the reactor where the neptunium would be transformed into Pu-238. It is claimed [27] that with a 5 per cent admixture of Pu-238 the surface temperature of the plutonium would be about 875°C, much higher than the melting point of the chemical explosives needed to initiate the implosion, and making it impossible to assemble a bomb. Here too the extra cost involved in the fabrication and handling of the fuel elements, and the greater enrichment of uranium which would become necessary, may make this method unacceptable to the industry.

The other method consists of adding another transuranic element, californium-252 to the plutonium. This nuclide has a very high spontaneous fission rate (half-life 82 years), undergoing fission without having to be bombarded with neutrons; in this process neutrons are emitted at a rate of

3×10^{15} per second per kg of material. An admixture in the proportion only 10^{-9} of the plutonium content would increase the neutron emission of reactor-grade plutonium by an order of magnitude and would thus make it much less efficient as a nuclear explosive (see *Weapon-grade materials*, page 398). It should be noted that the actual half-life of Cf-252 is only about 2.6 years, which means that the plutonium would have to be recharged with Cf-252 every ten years or so. Moreover, since Cf-252 is not an isotopic denaturant, it could be chemically separated from the plutonium, even at very low concentrations.

Summing up, each of the proposed physical barriers could somewhat reduce the proliferation threat but would not eliminate it. If they were all applied together, the effectiveness would be greatly increased [28], but the economic penalty involved would be too great to be accepted.

In any case, even if these methods were effective in reducing the probability of diversion at subnational level, they are unlikely to be so at a national level, in a country with a significant nuclear power programme.

International safeguards technology

The principal method of preventing the diversion of nuclear material from peaceful to military uses is the safeguards system of the International Atomic Energy Agency (IAEA). The responsibility of the Agency for such safeguarding is embodied in its own statute, and it was subsequently extended under the provision of the Non-Proliferation Treaty; Article III of the Treaty makes it obligatory on signatories to enter into agreements with the Agency. The policy aspects of the IAEA safeguards are discussed below (*The International Atomic Energy Agency*, page 421); here only the technological aspects are discussed.

Under the terms of the agreements, the Agency is entitled to send its inspectors to facilities operated by individual states, to carry out measurements, to take samples, and otherwise ensure that no diversion has taken place. The bases for the IAEA safeguards system are: material accountancy, and the verification of inventories and flow of nuclear materials.

The methods employed for verification include non-destructive assay, chemical analysis, containment and surveillance measures. New sophisticated techniques are being introduced to improve the efficiency of detection of unauthorized variants of stated operations. These techniques, such as high resolution γ-ray spectroscopy, are naturally much more expensive, but they are essential in view of the many inherent difficulties. For example, plutonium, being an α-emitter, can be detected remotely mainly through its weak X- and γ-ray emission, but in the presence of the very strong γ-ray emission from the accompanying fission products, this would be impossible without a detailed analysis of the γ-spectrum observed.

As far as the verification of inventories is concerned, this is made difficult by the fact that the exact history of the fuel elements is not easy to establish. The initial enrichment, the position of the element in the reactor, the actual time of its running (allowing for shut-downs), all the parameters which affect the burn-up, and which determine the initial amount of plutonium present in the spent fuel elements, cannot be assessed exactly.

So far the IAEA safeguards have been applied to the relatively easy parts of the fuel cycle, research and power reactors, and fuel fabrication plants using natural or low-enriched uranium. The more sensitive parts of the fuel cycle, in particular large-scale enrichment, reprocessing and fuel fabrication plants, have yet to become subject to the safeguards, and it remains to be seen whether the increased sophistication of techniques will match the greater severity of the problem. For example, a reprocessing plant designed to handle 2 000 tonnes of spent fuel per year would process about 50 kg of plutonium every day. This plutonium has to go through a number of chemical operations, at each stage of which some losses inevitably occur. With present-day techniques, the overall accuracy in accounting for the material flowing through the reprocessing plant is $2-5$ per cent at 24 hours delay [29]. Future improvements may reduce this to 1 per cent, but even then a determined effort at systematic diversion could succeed in 'losing' enough material to make two bombs every month.

As with the other physical barriers, the willingness of the authority operating the facility to cooperate fully with the IAEA inspectors would be very effective in preventing any significant diversion by subnational groups. But when national authority itself is bent on acquiring weapon-grade material clandestinely, it is very likely that it will succeed, particularly if the risk of detection of misuse does not carry with it effective repercussions.

We must conclude that reliance on technological barriers alone will never give full security, and that it is essential that they are supplemented by institutional arrangements.

VI. Institutional security

The Non-Proliferation Treaty (NPT)

Undoubtedly the most important of the institutional arrangements to reduce the danger of proliferation of nuclear weapons is the NPT. Under the Treaty, which was agreed by the United Nations Assembly in 1968, after 10 years of negotiations, and which came into force in 1970, non-nuclear weapon states undertake not to manufacture or otherwise acquire nuclear weapons or other nuclear explosive devices, and not to seek or receive any assistance in the manufacture of such weapons or devices. Member states also undertake to accept safeguards for the exclusive purpose of verification of the fulfilment of its obligations under the Treaty, with a view to preventing diversion of nuclear energy from peaceful uses, the agreements for such safeguards to be negotiated with the IAEA.

By March 1979, 107 nations had ratified the Treaty, and a further eight have signed but not yet ratified it. A list of all those nations is given in Appendix C of Chapter 13. To date, no violation of the letter of the Treaty has been reported, although there have been some doubts about adherence to its spirit.

The main concern about the NPT is that over 50 nations have so far refused to join the Treaty. These include a number of so called 'threshold' countries, states which are engaged in significant nuclear activities and which already have the technology and means to make nuclear weapons. One of these countries, India, has already demonstrated this by the explosion of a nuclear device; it is strongly rumoured that Israel possesses nuclear weapons; and that South Africa was on the brink of testing a nuclear weapon. Another of these countries, Brazil, is to be supplied by FR Germany, a member of the NPT, with the whole gamut of nuclear facilities, from enrichment to reprocessing plants. Although this is not a violation of the letter of the Treaty, it comes pretty near to it; it is certainly a violation of the spirit of the Treaty, as it will result in providing a nation, which has refused to give an undertaking that it will not acquire nuclear weapons, with the wherewithal for the acquisition of such weapons.

A glaring lack of adherence to the spirit of the Treaty concerns the nuclear weapon states. Under the Treaty these states undertook to pursue negotiations—in good faith—on effective measures to stop the nuclear arms race, and towards disarmament. Not only has there been no single effective disarmament measure since the NPT was signed, but the nuclear arms race between the great powers has accelerated, and new types of weapon or means of delivery have been, or are being, introduced, greatly increasing the probability of a nuclear world war. It should also be noted that two nuclear weapon states, China and France, are not members of the NPT; France has stated that it will act as if it were, while China has done nothing contrary to the Treaty.

Disappointment about the lack of any progress towards disarmament was strongly expressed at the first Review Conference, held in May 1975. Specific proposals, aiming at increasing security among nations, were put forward by the Third World states, but were rejected by the nuclear powers. The net outcome of the Review Conference was to deepen the division between nuclear and non-nuclear weapon states, the latter feeling that the NPT is discriminatory against them. The division of the world into two classes, the 'haves' and the 'have nots', which is inherent in the NPT, is one of the main reasons given by some nations for their refusal to join the Treaty. As long as the possession of nuclear weapons offers special privileges to a nation, the tendency to join the nuclear weapon club will be irresistible. And should one or more of the present non-NPT states acquire a nuclear weapon capability, as did India in 1974, other nations, which might conceive this as a threat to their own security, would feel impelled to do the

Figure 11. Projected growth of nuclear electrical generating capacity

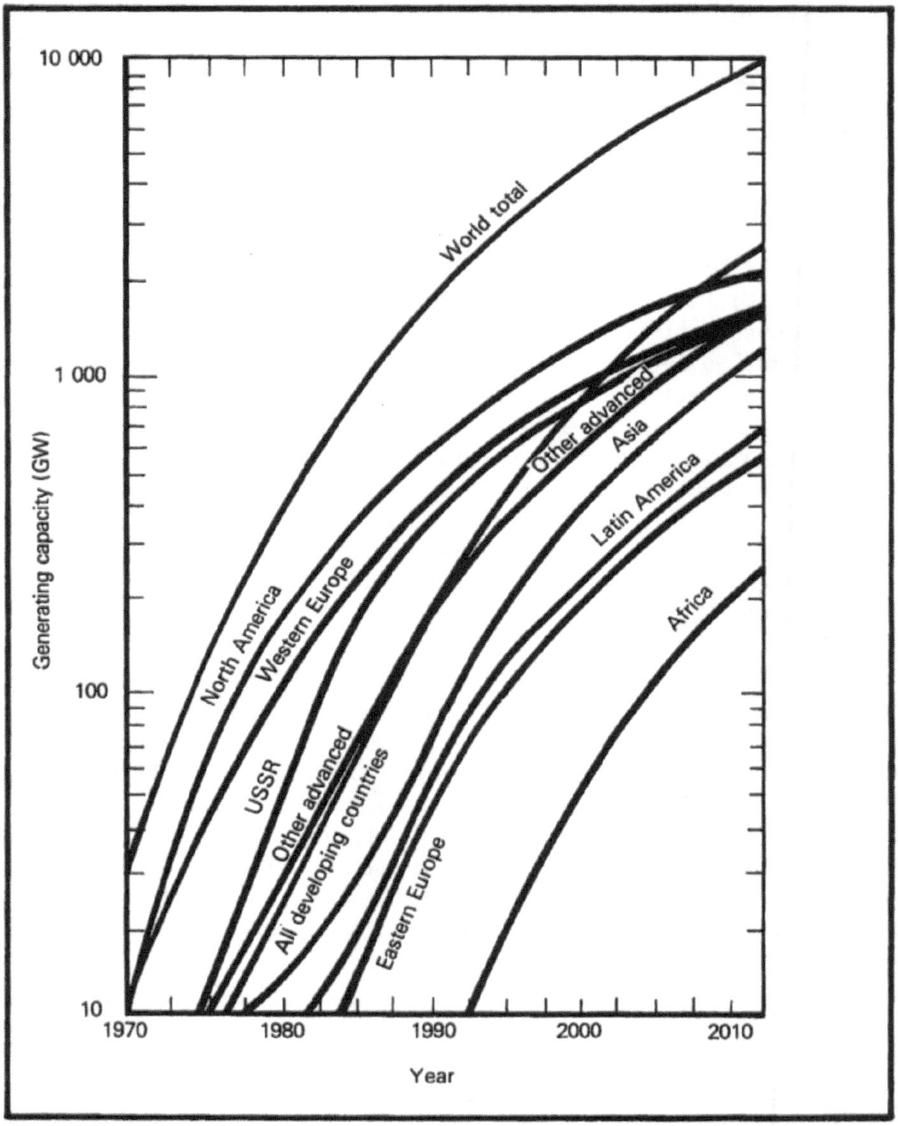

same. Being a signatory of the NPT may, paradoxically, help them in this endeavour. The loose wording of the NPT makes it possible for a country to get very close to producing a nuclear weapon, almost to the turning of the last screw, without violating the letter of the NPT. At the same time, the assistance which an NPT state may legitimately claim under the Treaty in developing its peaceful nuclear power programme, would enable it to acquire the technology and, potentially, the materials for nuclear weapons. In this connection it should be noted that states party to the Treaty can withdraw from it on giving three months' notice.

The next Review Conference is due to be held in June 1980. Unless this Conference succeeds in reaching agreement on these, mainly political, issues, there is the danger that the already weakened NPT will be further eroded and that there might eventually ensue a free-for-all in the race to join the nuclear club. Ironically, the proliferation threat would then be made more severe, thanks to the acquisition of nuclear know-how under the provision of the NPT.

The International Atomic Energy Agency (IAEA)

The IAEA was set up largely as a result of the reversal in 1953 of the US policy, which previously—in the Acheson–Lillienthal Report—had assumed that the military and civil uses of nuclear energy cannot be separated. President Eisenhower's 'Atoms for Peace' proposal, perhaps not unconnected with the realization by the American industry of the commercial and export potential of nuclear reactors, led to the Atomic Energy Act of 1954, which encourages widespread participation in the development and utilization of atomic energy for peaceful purposes. In turn, this led to the establishment in 1957 of the IAEA, as one of the family of the United Nations Agencies.

Apart from carrying out research on some problems concerning nuclear radiation, such as the application of radioactive isotopes in agriculture and industry, the development of detection instruments, and monitoring of radiation standards, the main tasks of the Agency were twofold: (a) to promote the development of nuclear energy, and (b) to prevent its misuse for military purposes. These two objectives are to a certain extent contradictory, because a rigorous pursuit of the prevention of misuse is bound to put constraints on the promotional aspects, but how to resolve this conflict was left to actual practice.

The promotional activities of the Agency have been pursued with great vigour. A large part of it was to convince nations, both advanced and developing, that nuclear power is bound to play an important role in their industrial development. Based on certain assumptions about rates of industrial and population growth, and about demand and availability of energy, the IAEA made projections about the development of nuclear power during the next 50–100 years. Figure 11 is an example of such a projection [30], made in 1971, of the growth of nuclear power up to the year 2010. As is seen, it envisages an increase in nuclear electrical generating capacity by a factor of 345 in four decades. OECD and other organizations have made similar projections. That these projections were over-optimistic is obvious from actual experience. For 1978 the projected value was 260 GW(e), while the actual value was 112 GW(e) (table 3). All these projections have one common feature: they are being scaled down all the time. Another new feature is that, instead of giving one figure, a range of values between 'low' and 'high' is quoted. Thus the 1977 projection for the year 2000 is

421

Table 14. IAEA forecasts of nuclear power capacities in Third World countries[a]

Region	1974 forecast[b] for			1977 forecast[c] for		
	1980	1990	2000	1980	1990	2000
Africa + Middle East	–	13.1	34.8	–	3–9	20–34
Asia	8.4	89.9	285	4–5	29–48	98–139
Latin America	2.8	55.2	167.8	1–2	15–27	85–132
Total	**11.2**	**158.2**	**487.6**	**5–7**	**47–84**	**203–305**

[a] Expressed in GW(e).
[b] See reference [31].
[c] See reference [6].

1 403–2 227 GW(e) [6]. Nevertheless, the very existence of such projections for individual countries, particularly for the Third World (table 15), is a stimulus to them to reach the target allocated to them.

The control measures to prevent the spread of nuclear weapons capability were originally designed for the Agency's own projects, or to be used in bilateral agreements with nations which received assistance from the Agency. These measures were embodied in a safeguards system set up in 1965, in which the safeguards related to individual facilities. The situation changed after the NPT came into being in 1970, which made it obligatory for member states to conclude agreements with the IAEA. A new safeguards system, relating to all nuclear activities of a given nation, instead of individual facilities, came into being in 1971. The 1965 system was maintained but it was rehashed to apply to agreements with nations which have not signed the NPT.

By March 1979 the IAEA had concluded safeguards agreements with 63 non-nuclear weapons states which have signed the Treaty, and is negotiating such agreements with 13 more states. In addition, the UK has voluntarily entered into agreements with the IAEA, and France has also come in via Euratom, although the requirements of the NPT apply only to non-nuclear weapon states.

The objective of the IAEA safeguards is the timely detection of diversion of significant quantities of nuclear material from peaceful activities to military use, and deterrence of such diversion by the risk of early detection. This objective is to be achieved primarily by material accountancy, complemented by containment and surveillance. The technological measures under this arrangement were discussed above (*International safeguards technology*, page 417) and some inherent weaknesses in the safeguards system were noted. The important point is that while the IAEA system is designed to detect and thus deter acts of diversion, it has no authority to exercise any police-type activities or to provide physical security to prevent diversion. Since all safeguards are based on agreements between the Agency and individual states their effectiveness depends to a very large extent on the goodwill of the government of that state, and on its efficiency. If a state is determined to misuse its materials, it could devise the means of concealing such diversion, despite the introduction of more sophisticated detection methods.

Opportunities for such national diversion will be greater the larger the peaceful nuclear programme, the more nuclear reactors there are, and, particularly, if enrichment, fuel fabrication and reprocessing plants are installed in the country. Without a radical change in the basic approach to the safeguards problem, it is unlikely that the Agency's system will be effective in achieving its objective, to prevent diversion of nuclear materials at the national level.

The 'suppliers' group

It was largely because of the realization of the inadequacy of the existing non-proliferation régime—brought home by India's test of a nuclear device in 1974—that several countries which export nuclear facilities decided to meet in order to devise stricter rules for exports. These meetings took place in London, hence the group became known as the London Club.

The initial membership of seven countries was subsequently increased to 15: Belgium, Canada, Czechoslovakia, France, FRG, GDR, Italy, Japan, Netherlands, Poland, Sweden, Switzerland, UK, USA and USSR. Not all these countries have actually been exporting primary nuclear facilities, such as reactors; indeed, many of them are importers of such facilities, but they already have been or will be supplying auxiliary parts of the nuclear fuel cycle.

The *Guidelines for Nuclear Transfers*, agreed to by the supplier states, contain a catalogue of equipment and material which, when provided to any non-nuclear weapon state, would 'trigger' IAEA safeguards. The materials on the 'trigger' list include the fissile nuclides Pu-239, U-233 and U-235 (in enriched uranium) in quantities greater than 50 g per year, natural uranium in quantities greater than 500 kg per year; and thorium and depleted uranium in quantities greater than 1 000 kg per year. This list also includes specified quantities of other materials used in reactors, such as deuterium, heavy water, or high-grade graphite. Among the installations are nuclear reactors which produce more than 100 g of plutonium per year, plants for the fabrication of fuel elements, and for their reprocessing. Enrichment plants of different types are also listed, as well as plants for the production of heavy water. The auxiliary equipment includes pressure vessels, control rods, pumps for circulating the coolant, as well as other components of nuclear installations.

Transfer of items identified on the trigger list would be authorized only upon formal government assurances from recipients explicitly excluding uses which would result in the manufacture of nuclear explosives; the materials and facilities in question would have to be placed under physical protection to prevent unauthorized handling. The duration of the safeguards agreements would be related to the period of actual use of the relevant items in the recipient states. These requirements will also apply to facilities utilizing technology directly transferred by the supplier, or derived from transferred facilities, or major critical components thereof.

At the same time, the suppliers have pledged themselves to exercise restraint in the transfer of sensitive facilities, equipment or technology and weapons-usable materials. Retransfer of trigger list items will be subject to controls and, in certain cases, the consent of the original supplier will be required. The suppliers will consult each other and other governments on specific sensitive cases to ensure that any transfer does not contribute to risks of conflicts or instability. In the event of a violation of supplier/recipient understandings resulting from the Guidelines, particularly in the case of an explosion of a nuclear device, or illegal termination or violation of IAEA safeguards by a recipient, suppliers should consult through diplomatic channels on appropriate response and possible action, which could include the termination of nuclear transfers to that recipient.

The constraints imposed by the code of conduct of the London Club constitute a significant step in monitoring the use of purchased facilities by the importing countries. But they are insufficient to guarantee that no further nuclear weapon proliferation will occur as a result of transfers: they do not require full-scope safeguards in recipient states, that is, safeguards on all peaceful nuclear activities; neither do they definitively preclude exports of highly sensitive facilities. Moreover they contain an element of control on national activities from the outside. The condition of restriction on re-export implies a continuous control by the exporting country over the goods it has supplied to the importing country. It is not surprising, therefore, that the London Club guidelines met with opposition from Third World countries, which see in them an attempt by the industrialized countries to subserve the developing countries. Most of these countries have only recently emerged from colonial rule and they are very sensitive about their independence; they are strongly opposed to any additional measures of control which go beyond the NPT undertakings.

There is also anxiety among the buyer countries that the London Club may become a cartel which would agree upon and dictate prices to be charged for the supply of nuclear equipment and materials. These matters too are likely to be brought up at the NPT Review Conference in 1980.

VII.   Further anti-proliferation measures and new initiatives

The measures taken so far, both technological and institutional, to remove the proliferation threat have not proved adequate or acceptable, but this

Table 16. List of states which participated in the first plenary conference of the International Nuclear Fuel Cycle Evaluation, held in Vienna on 27–29 November 1978

Argentina	Libya
Australia	Malaysia
Austria	Mexico
Belgium	Netherlands
Brazil	New Zealand
Bulgaria	Niger
Canada	Nigeria
Chile	Norway
Colombia	Pakistan
Czechoslovakia	Panama
Denmark	Peru
Ecuador	Philippines
Egypt	Poland
Finland	Portugal
France	Qatar
German Democratic Republic	Romania
Germany, Federal Republic of	Saudi Arabia
Greece	Spain
Guatemala	Sweden
India	Switzerland
Indonesia	Thailand
Iran	Tunisia
Iraq	Turkey
Ireland	USSR
Israel	UK
Italy	USA
Japan	Venezuela
Korea, South	Yugoslavia

Note: In addition, the following organizations took part:
CEC
IAEA
IEA
OECD/NEA
United Nations

does not mean that they should be abandoned. Although technical fixes will never be foolproof any additional security measure helps, and the search for more proliferation-resistant variants of the nuclear fuel cycle is therefore warranted. The same applies to the institutional controls; these would be much more effective if more teeth were given to them, and if national and multinational arrangements were backed by international agreements. Finally, a reduction in the need for nuclear energy, by encouraging reliance on alternative sources, would significantly contribute to lessening the threat of proliferation. These possibilities are discussed in this section.

The International Nuclear Fuel Cycle Evaluation (INFCE)

Perhaps the most important anti-proliferation attempt on an international scale is the International Nuclear Fuel Cycle Evaluation, which was set up as a result of a proposal made by President Carter at a seven-nation summit meeting in May 1977. The Organizing Committee of INFCE met in Washington in October 1977, when a two-year programme of work was agreed to. Fifty-six countries and five international agencies participate in INFCE. A list of member countries is given in table 16; it includes both nuclear and non-nuclear weapons states, exporting and importing countries, signatories and non-signatories of the NPT.

The Technical Coordinating Committee of INFCE has set up eight working groups to deal with the following subjects:

(a) Fuel and heavy water availability (co-chairmen: Canada, Egypt, India);

(b) enrichment availability (co-chairmen: France, FRG, Iran);

(c) assurances of long-term supply of technology, fuel and heavy water and services in the interest of national needs consistent with non-proliferation (co-chairmen: Australia, Philippines, Switzerland);

(d) reprocessing, plutonium handling, recycle (co-chairmen: Japan, UK);

(e) fast breeders (co-chairmen: Belgium, Italy, USSR);

(f) spent fuel management (co-chairmen: Argentina, Spain);

(g) waste management and disposal (co-chairmen: Finland, Netherlands, Sweden);

(h) advanced fuel cycle and reactor concepts (co-chairmen: South Korea, Romania, USA).

These subjects cover the whole field of nuclear energy. All have been discussed in international fora before, but INFCE is the first such forum to tackle the technical issues from the point of view of the proliferation danger: how to make each part of the nuclear fuel cycle more proliferation-resistant. The proposals mentioned earlier (under *New technologies* and *Technological security*) will no doubt be discussed in great depth, and the feasibility of their introduction on a time scale which would make them effective will be assessed. The economic and political implications of any anti-proliferation measure are bound to be hotly debated considering the conflict of interest between participating countries.

Even on the technical side, there are a number of complex issues to be tackled, but for which no easy answer can be seen at the present time [32]. These relate to new technologies, for example the impact of new enrichment techniques, or the consequences of the advent of the fusion—fission hybrid, as well as to narrower issues, such as the effectiveness of denaturing versus chemical separation, or the distinction between weapon-grade and non-weapon-grade materials and where the line should be drawn between them. Some observers doubt whether these problems can be solved within the timetable which INFCE has set for itself, or, if they can, whether any of the recommendations would not be too late in any case, in view of the commitment of several nations to specific nuclear power programmes, including the fast breeder. One reason for INFCE's concern to complete its work within two years was to be able to suggest effective measures before countries made irrevocable decisions; it was hoped that states would refrain from entering into ineluctable commitments until INFCE had concluded its study. This has not happened. For instance, a decision was made to go ahead with the reprocessing plant at Windscale despite the UK's active support of, and collaboration with, INFCE.

In view of these difficulties, over-optimism about the effectiveness of INFCE on the proliferation issue is perhaps not warranted. Nevertheless, it is bound to have some impact. The danger of proliferation arising from the nuclear power programme has for too long been ignored or met with derision by promoters of nuclear energy, who alternately claimed that it either does not exist or that it is in any case too late to do anything about it. Even these groups have now been obliged to acknowledge that the problem exists and has to be considered seriously; some have even proposed technical changes to the nuclear fuel cycle (for example Civex, Pipex) to make it less proliferation-prone. This change in attitude is partly due to INFCE, which has made the proliferation aspect an issue meriting profound consideration in the planning of nuclear energy activities.

The US energy policy

The main initiative in calling attention to the proliferation issue belongs to the United States. The reversal of the earlier 'Atoms for Peace' policy of President Eisenhower, already started under President Ford, was fully enunciated by President Carter in his energy statement of April 1977, and became embodied as legislation with the passing by Congress of the 'Nuclear Non-Proliferation Act of 1978' in March 1978.

The decisions taken in April 1977 included stopping reprocessing of spent fuel and deferment of further work on the commercial use of fast breeder reactors. In the 1978 Act these measures were taken a step further and incorporated into a comprehensive legislation designed "to provide for more efficient and effective control over the proliferation of nuclear explosive capability" [33]. The main purpose of the Act is to promote policies as follows: (a) to pursue mechanisms for fuel supply assurance and the establishment of more effective international controls over the transfer and use of nuclear facilities and technology, including the establishment of common international sanctions; (b) to take actions which would confirm the reliability of the USA in meeting its commitments as a supplier of nuclear facilities and fuels to nations which adhere to effective non-proliferation policies; (c) to encourage nations which have not ratified the NPT to do so; and (d) to cooperate with other nations in identifying and adapting suitable technologies for energy production, in particular, alternative options to nuclear power.

In pursuance of these aims, the Act lays down specific criteria for the approval of exports of facilities for peaceful nuclear uses. These include:

(*a*) the application of IAEA safeguards to all materials and facilities; (*b*) an assurance not to use materials, facilities or nuclear technology for any nuclear explosive device or for the research or development of such devices; (*c*) the maintenance of adequate physical security measures; (*d*) no retransfer of materials, facilities or technologies without prior approval of the USA: (*e*) no reprocessing of materials or alteration of spent fuels without prior approval; and (*f*) no export of sensitive nuclear technology. Violation of any of these conditions would trigger a cut-off of nuclear exports. On the other hand, adherence to these criteria would be rewarded by an assurance of reliable nuclear supplies.

The magnitude of the revolution of the US energy policy has not been fully appreciated. The United States is by far the largest user and exporter of nuclear energy facilities. About half the presently installed nuclear power capacity is in the USA, and it also accounts for 45 per cent of the power reactors at present under construction. Most nuclear reactors in other countries were exported from the USA or from other countries under US licence. The provisions of the Act and the restrictions on export imposed in it are bound to reduce the volume of export, and indirectly affect the domestic nuclear power programme.

Although the reasons for the new energy policy were clearly stated by President Carter and emphasized in the 1978 Act, they were received with marked hostility by the majority of other nations. Some of the industrialized countries, such as France, FR Germany and Japan, which pursue a vigorous nuclear policy, saw in these measures a threat to their domestic and export plans for nuclear development. The US policy was even presented as an attempt to slow down the nuclear programmes in other countries, until the USA caught up with its technology, which allegedly has fallen behind, particularly in the development of the fast breeder. Among the Third World countries, the US policy was seen as an attempt to restrict their sovereign rights and to deprive them of the benefits of nuclear energy. In the course of time these interpretations were somewhat dispelled, and the present attitude is to wait and see how it will work out. The main impact of the new legislation is still to come, since there has not been much occasion to enforce it up to now; the few cases where permission was needed for the reprocessing of spent fuel originating from the USA were treated with some leniency.

Agreement reached at INFCE may pave the way for the implementation of the recommendations of the Act concerning the internationalization of the nuclear fuel cycle.

Multinational arrangements

The 1975 NPT Review Conference recommended the setting up of regional or multinational fuel centres as an advantageous way to satisfy, safely and economically, the needs of many countries, to initiate or expand their nuclear energy programmes, while at the same time facilitating physical protection and the application of safeguards. Multinational centres were thus envisaged to have two objectives: economic gain and greater security.

Several formal organizations, each comprising a number of countries, have indeed been set up in the past, but they were all based on commercial interest, and entailed collaboration in developing specialized nuclear facilities, which would have been too expensive, or beyond the technological capabilities, of the individual partners. Most of these organizations involve European countries. Thus, Eurodif (with 13 countries participating, including Iran) is concerned with the development of gaseous diffusion enrichment plants; Urenco (FR Germany, the Netherlands and the UK) with centrifuge separation plants; and Eurochemic (13 countries) was set up to develop reprocessing techniques. All of these organizations are somewhat related to the European Nuclear Energy Agency, which is part of OECD, and to Euratom, an organization of the European Economic Community. In addition to tackling technical problems, Euratom runs its own safeguards system, and has entered into a block agreement, on behalf of its members, with IAEA.

Following the recommendations of the 1975 Review Conference, the IAEA has carried out a study of multinational facilities, in particular relating to plutonium management. It is envisaged that a group of nations that operate nuclear reactors would jointly run a reprocessing plant to process the spent fuels from these reactors and fabricate fresh fuels. The group may also jointly operate an enrichment plant. The co-location of other facilities, such as storage of spent fuel elements, and waste management operations, is also envisaged. Apart from the financial advantage, resulting from the economy of scale, there would also be advantages in relation to safety, and reduction of radiation exposures. From the point of view of proliferation, the setting up of a multinational centre, which would be the only place where plutonium was handled in a form suitable for an explosive, would be a step in the right direction, as it is much more difficult for a nation to divert nuclear materials from a multinational facility than from a plant under its own control [34].

However, the setting up of a number of such multinational centres carries with it some distinct disadvantages. With the division of the world into East, West and South, it is very likely that the multinational groups would be set up along the same lines of division, and this might exacerbate existing divergences between nations. The nuclear power programme may become an additional cause of rivalry and conflict.

If, to avoid this, multinational groups were set up, each comprising nations from both the developed and developing countries, another problem could arise. The Third World countries are very sensitive about real, or apparent, discrimination in any arrangement, particularly concerning transfer of technology. It would be difficult to avoid differentiation when partners enter with such vastly different backgrounds in technological knowledge and financial backing. And there is, of course, the additional danger that such centres will help to spread technological know-how to many nations, and thus enable them to set up their own plants leading to weapon production.

An approach still based on voluntary participation, but which might include all nations, is the idea of a nuclear fuel cooperative. Its purpose would be to ensure for participating states adequate supplies of nuclear materials, while at the same time giving confidence that these materials would not be misused. Under this scheme, enrichment and reprocessing would remain under national control, but member states would undertake to deny themselves access to the services of enrichment and reprocessing facilities beyond those necessitated by their power programmes. Acceptance of IAEA safeguards to verify observance of that undertaking would be a condition of joining the cooperative. The most attractive feature of this concept is its reliance on mutual benefit as the basis for a non-proliferation régime [35].

Internationalization of the fuel cycle

A different approach, which would avoid the drawbacks of the multi-national centres, but which would necessitate an extension of the NPT and a tightening of the IAEA safeguards system, is the internationalization of the fuel cycle. Such a scheme is in fact included in the US Non-Proliferation Act of 1978 which calls on the President to discuss with other nations the establishment of an International Nuclear Fuel Authority (INFA) [33]. The INFA would have the responsibility for providing nuclear fuel services and allocating fuel supplies; it would set up and manage repositories for the storage of special nuclear materials and spent fuels, under effective international auspices and inspection; and it would make arrangements for compensation for the energy content of spent fuels.

Various ways for the internationalization of the nuclear fuel cycle have been suggested. The common feature is that while the operation of the power reactors would remain under national control, the sensitive parts of the cycle—reprocessing and fuel fabrication—would be carried out only at specific centres operated under international control. Some schemes envisage that enrichment plants would also come under international control. The advanced converter reactor, based on denatured uranium and thorium (see *Advanced converter reactors*, page 409) lends itself particularly to such arrangements. If need be, the plutonium produced in the national reactors could be sent to international centres which would use it in FBRs or uranium–plutonium-fed LWRs, thus combining greater security with better utilization of fuel. Another combination would be a breeder reactor at an international centre, with a thorium blanket to produce U-233, which would then feed advanced converter reactors. Each international breeder could provide fuel (denatured with U-238) for three or four national ACRs.

The most radical proposal is for an international agreement that the operation of enrichment, fuel fabrication, reprocessing, and waste disposal plants be permitted to be carried out only by, or under licence of, INFA [36]. The enrichment and fuel fabrication plants, operated under strict control and licence of INFA, could manufacture fuel elements for different types of reactor and store them in an INFA bank. After the first charging of a reactor, fresh fuel elements would be sold to individual countries, which would be obliged to return spent elements, as well as fissile-bearing discharged materials. These would be stored by INFA, either with or without chemical processing. When necessary, reprocessing of spent fuels would be carried out in plants operated by or under licence of INFA.

The implementation of these proposals would require an extension of existing international agreements relating to the NPT and IAEA. Additional clauses would have to be included in the NPT, under which member nations would give up their right to operate their own enrichment, fuel fabrication and reprocessing plants, and instead vest these rights in an international authority. There would be the need to police the plants operating under this authority, and to guard all shipments of materials to and from national reactors. To ensure compliance with the new regulations, the present safeguards system would have to be extended to include provision for the physical security of nuclear facilities operated by individual states. The probability of a nation violating these agreements would be greatly reduced by the threat of withholding the supply of fresh fuel.

The compensation for nations which agree to abide by these obligations would be the assurance of uninterrupted supply of fresh fuels and the taking care of spent fuels. Apart from the greater economic stability which this would bring, it would also result in greater political stability, arising from the assurance that the richer nations would not be able to dominate the poorer in the nuclear field, and in the diminished possibility for aggressor nations to acquire nuclear weapons.

Alternative sources of energy

A considerable lessening of the threat of horizontal proliferation would result from the reduction of the overall scale of nuclear energy programmes, by the encouragement of energy conservation measures and the employment of alternative sources of energy.

As was shown earlier, the scale of development of nuclear energy has been steadily decreasing in recent years. There are several reasons for this downward trend: (a) mistakes made in earlier energy forecasts, due largely to the misreading of future energy demands and accentuated by a worldwide deceleration of economic activity (an important factor may have been that energy planning and energy supply were handled by the same bodies); (b) the rising costs of nuclear energy systems, both in capital costs and in fuel cycle management; (c) a growing awareness that within the foreseeable future nuclear energy could not provide more than a very limited substitute for oil dependence; (d) increased public concern about environmental and security aspects of nuclear power, especially handling of wastes and proliferation; and (e) that the effects of energy conservation measures embarked upon since 1973 are already being felt; prospects for further increasing the efficiency of energy use also seem very promising.

At the present time, only about 36 countries (table 4) have decided to install nuclear reactors. Many of these do not plan an extensive use of nuclear energy, and some may even decide not to operate their reactors at all, as has happened in Austria. A few other nations have made tentative enquiries about acquiring nuclear reactors, but most nations, including the great majority of developing countries, have not yet decided on their future energy policies.

The fact that at present nuclear energy appears to be the only option, is largely due to the nuclear enterprises in industrialized countries as well as the promotional activities of the IAEA, which work very hard to create the impression that the world is already committed to nuclear power as the major source of energy for the future. This propaganda may have the effect of inducing Third World countries to take the nuclear path even though this would be the least suitable form of energy for them. Alternative sources of energy, in particular solar, may fit in much better with the economic structure of many such countries but these sources are not readily available now because very little effort has gone into their development. Recently more attention has begun to be given to the alternative sources and more funds have been allocated to research in these areas, but the scale is still too small to ensure their becoming a practical proposition in the near future.

Even without the threat of proliferation, it would be a great folly if a single form of energy were adopted by all nations. Diversity of energy sources must match the great diversity in the geographic, cultural, and economic status of the peoples of the world. The inherent danger in nuclear energy, resulting from its links with nuclear weapons, makes it even more imperative to ensure that only those nations for whom nuclear energy is shown to be essential for their own needs (as distinct from exports) should embark on a nuclear energy programme, and that for other countries there should be the opportunity to opt for alternative sources.

This need was recognized in the US Non-Proliferation Act of 1978 which calls on the US Government to seek to cooperate with, and aid, Third World countries in meeting their energy needs through the development of non-nuclear energy sources and the application of non-nuclear technologies. The US Government is also urged to call upon other industrialized nations to assist Third World countries in this way.

Much more effective than such sporadic measures would be a specific programme agreed by the international community of nations. The two main objectives of such a programme would be: (a) to provide funds for research and development of alternative energy sources; and (b) to make these sources acceptable to all nations by endowing the alternative sources with the same aura of prestige that now attaches to nuclear energy, so that nations which embark on alternative energy programmes will not consider themselves technologically inferior to those which have nuclear energy.

Both of these objectives could be met by the setting up of a World Energy Organization [36]. This might be a Specialized Agency of the United Nations, with functions analogous to those of the World Health Organization or the Food and Agriculture Organization, but combined with some of the functions of the World Bank. Its terms of reference might be: (a) to promote and encourage research on various forms of energy other than nuclear (which is being taken care of by IAEA) and on their utilization; (b) to supply member states with information about development in the energy field; (c) to advise member states on the forms of energy most suitable for their needs, and on ways of implementing their energy programmes; and (d) to provide financial help towards their realization.

The internationalization of the nuclear fuel cycle, together with the promotion of alternative sources of energy, offer a real hope of substantially reducing the danger of proliferation and greatly enhancing the chances of avoiding a nuclear war.

References

1. Woite, G., 'Capital investment costs of nuclear power plants', *IAEA Bulletin,* Vol. 20, No. 1, February 1978, pp. 11–23.
2. Pigford, T. H. and Ang, K. P., 'The plutonium fuel cycles', *Health Physics,* Vol. 29, October 1975, pp. 451–68.
3. *Power Reactors in Member States,* 1979 Edition (IAEA, Vienna, 1979).
4. 'Nuclear station achievement', *Nuclear Engineering Intenational,* Vol. 24, June 1979, pp. 54–55.
5. 'Power reactors 1978', *Nuclear Engineering International,* Vol. 23, Supplement, July 1978.
6. 'Nuclear power in developing countries', *IAEA Bulletin,* Vol. 19, No. 3, June 1977, p. 28.
7. Marshall, W., 'Nuclear power and the proliferation issue', *Atom,* No. 258, April 1978, pp. 78–102.
8. OECD/IAEA, *Uranium—Resources, Production and Demand* (OECD, Paris, December 1977).
9. Cameron, J., *World Uranium Resources,* International Training Course on Uranium Exploration and Evaluation, Golden, California, USA, 28 August— 20 October 1978 (IAEA, December 1978).
10. *Nuclear Fuel Cycle Facilities in the World* (IAEA, Vienna, January 1978).
11. 'Enrichment plants', *Nuclear Engineering International,* Vol. 21, November 1976, p. 50.
12. Carter, L. J., 'Nuclear wastes', *Science,* Vol. 197, 23 September 1977, p. 1265.
13. Taylor, T. B., 'Nuclear safeguards', *Annual Review of Nuclear Science,* Vol. 25, 1975, pp. 407–21.

14. Moniz, E. J. and Neff, T. L., 'Nuclear power and nuclear-weapons proliferation', *Physics Today,* Vol. 31, April 1978, p. 44.
15. *Nuclear Proliferation and Safeguards* (Office of Technology Assessment, Washington, D.C., 1977), p. 140.
    (a) —, p. 157.
    (b) —, p. 173.
16. El-Hinnawi, E. E., 'Review of the environmental impact of nuclear energy', *IAEA Bulletin,* Vol. 20, No. 2, April 1978, p. 37.
17. Kompa, K., see Paper 4 of the present volume (pp. 73–90).
18. Barré, B. and Coates, J. H., See Paper 2 of the present volume (pp. 49–59).
19. von Hippel, F., See Paper 1 of the present volume (pp. 11–47).
20. Marshall, W., 'Proliferation and the recycling of plutonium', *Atom,* No. 263, September 1978, pp. 234–41.
21. Barré, B., see Paper 8 of the present volume (pp. 127–140).
22. Westervelt, D. and Pollock, R., see Paper 11 of the present volume (pp. 161–172).
23. Kuleshov, V., see Paper 10 of the present volume (pp. 155–160).
24. American Physical Society, 'Study group on nuclear fuel cycles and waste management', *Review of Modern Physics,* Vol. 50, No. 1, January 1978, p. 563.
25. *Handling and Final Storage of Unreprocessed Spent Nuclear Fuel,* Nuclear Fuel Safety Project (KBS), Fack, S-102 40 Stockholm, Sweden (Stockholm, 1978).
26. Abrahamson, D., see Paper 6 of the present volume (pp. 105–111).
27. *Nuclear Industry,* Vol. 25, September 1978, p. 24.
28. Feld, B. T., see Paper 7 of the present volume (pp. 113–119).
29. Higinbotham, W., see Paper 13 of the present volume (pp. 187–197).
30. Spinrad, B. I., 'The role of nuclear power in meeting world energy needs', *Environmental Aspects of Nuclear Power States* (IAEA, Vienna, 1971), pp. 57–82.
31. *Market Survey for Nuclear Power in Developing Countries,* (IAEA, Vienna, 1974).
32. Farinelli, U., see Paper 17 of the present volume (pp. 261–269).
33. *Nuclear Non-Proliferation Act of 1978,* Public Law 95-242 (Washington, D.C., March 1 1978).
34. Chayes, A. and Lewis, N. B., *International Arrangements for Nuclear Fuel Reprocessing,* Proceedings of Pugwash Symposium (Ballinger, 1977).
35. Wilson, A. R. W., see Paper 16 of the present volume (pp. 251–259).
36. Rotblat, J., see Paper 18 of the present volume (pp. 271–285).

Additional bibliography

Epstein, W., *The Last Chance: Nuclear Proliferation and Arms Control* (Free Press, 1976).
Fox Report, *Ranger Uranium Environmental Inquiry—First Report* (Australian Government, 1976).
Non-Proliferation and International Safeguards (IAEA, Vienna, 1978).
Keeny, S. M., *Nuclear Power—Issues and Choices,* Ford/Mitre Report (Ballinger, 1977).
Lovins, A. B., *Soft energy paths* (Penguin Books, 1977).

434

Marwah, O. and Schulz, A., *Nuclear Proliferation and the Near-Nuclear Countries* (Ballinger, 1975).

Patterson, W. C., *Nuclear Power* (Pelican Books, 1976).

Royal Commission on Environmental Pollution, *Nuclear Power and the Environment* (HMSO, 1976).

Nuclear Proliferation Problems (Almqvist & Wiksell, Stockholm, 1974, Stockholm International Peace Research Institute).

Safeguards Against Nuclear Proliferation (Almqvist & Wiksell, Stockholm, 1975, Stockholm International Peace Research Institute).

The Nuclear Age (Almqvist & Wiksell, Stockholm, 1974, Stockholm International Peace Research Institute).

World Armaments and Disarmament, SIPRI Yearbooks 1974, 1975, 1976, 1977 (Almqvist & Wiksell, Stockholm, Stockholm International Peace Research Institute) *and 1978* (Taylor & Francis Ltd, London, Stockholm International Peace Research Institute).

Taylor, T. B. and Willrich, M., *Nuclear Theft: Risks and Safeguards* (Ballinger, 1974).

Abstracts of papers

Paper 1. An evolutionary strategy for nuclear power

F. VON HIPPEL, H. A. FEIVESON and R. H. WILLIAMS

The resource and economic implications of the adoption of an evolutionary strategy for fission power are discussed as an alternative to a strategy based upon deployment of the plutonium breeder reactor. The evolutionary strategy would involve a continued dependence on thermal neutron reactors which can be operated on either once-through or closed isotopically denatured fuel cycles. The uranium efficiency of the nuclear power system could be substantially improved, however, by deploying advanced converter reactors (ACRs) beginning around the year 2000 instead of light water reactors of current design. For specific quantitative calculations, the Canadian heavy water reactor is used to illustrate the potential role of an ACR. It is found that, within the uncertainties in the economic assumptions, operating an ACR on a once-through fuel cycle would be economically competitive with plutonium breeders up to very high uranium prices. It is also found that, if ACRs are operated on a closed isotopically denatured uranium–thorium fuel cycle, their cumulative uranium requirements need not for many decades significantly exceed those needed to deploy a breeder system.

Paper 2. Practical suggestions for the improvement of proliferation resistance within the enriched uranium fuel cycle

J. H. COATES and B. BARRÉ

The 1974 Indian nuclear explosion has focused the world's attention on the proliferation risk of the plutonium cycle, and its link with international nuclear trade and technology transfers. But one should not forget that the first proliferation event undoubtedly related to nuclear technological transfer was the 1964 Chinese weapon tests which involved the enriched uranium cycle.

The proliferation hazards of the 'front end' of the fuel cycle can be pinpointed to enrichment technology and the very limited use of highly enriched uranium in nuclear fuels.

At both these stages of the fuel cycle, it appears that less proliferation-prone alternatives exist which are still at the development stage, but which offer no substantial drawback with respect to the processes now in use.

The French enrichment method makes use of one of those chemical treatments which were discarded during early research programmes precisely because of their poor military performance. Its development could offer a more constructive approach than would total denial of any enrichment technologies to countries without them.

Highly enriched fuels are limited to high temperature reactors and research reactors: in both cases solutions exist or are being developed (such as 'Caramel' fuels) to reduce the enrichment level to below the 'safe' value of 20 per cent.

Paper 3.   Jet nozzle and vortex tube enrichment technologies

P. BOSKMA

The jet nozzle and advanced vortex tube enrichment technologies are so far the most developed of those based on aerodynamic principles. Pilot plants are in operation in FR Germany and South Africa, respectively, and the technologies are completely in the hands of non-nuclear weapon states.

Although not especially attractive from a technological point of view compared with gaseous diffusion and centrifuge technology, they could nevertheless provide attractive enrichment alternatives for countries with indigenous uranium resources, a medium level of technological and industrial development, and relatively low energy prices. This holds especially for countries desiring both national independence in the nuclear energy field and a future option on a nuclear capability.

The current policies of the countries involved are insufficient from the point of view of non-proliferation, especially so far as safeguarding their nuclear activities is concerned.

Paper 4.   Laser separation of isotopes

K. L. KOMPA

The selective action of laser light is the key factor in a variety of new concepts for isotope separation. As a matter of principle, these schemes can be applied to the isotopes of all chemical elements. High separation efficiencies are expected on principal physical grounds. As the present status of this field is still mostly characterized by basic research, it is not yet possible to make a detailed comparison with other, more conventional separation techniques. This paper discusses the principal possibilities, the determining parameters, and the first experimental results in uranium laser isotope separation, according to the following list of topics:

1. basic requirements for laser isotope separation (relation of the laser process to conventional techniques, required spectroscopic features, physical or chemical separation processes, and laser parameters);

2. results for light elements (sulphur-32/34 separation as a representative example);

3. results for heavy elements (uranium-235/238 separation, some published results); and

4. open problems, future prospects (isotope separation in relation to laser chemistry, the cost of laser light, limiting efficiencies, and overall technological requirements).

Paper 5.   Proliferation risks associated with different back-end fuel cycles for light water reactors

K. HANNERZ and F. SEGERBERG

This paper contains a comparison of the proliferation risks associated with two alternatives for the back-end of the light water reactor fuel cycle. One employs reprocessing of spent fuel in a foreign plant and re-use of recovered uranium and plutonium in domestic light water reactors. The other alternative represents a once-through fuel cycle, in which the spent fuel is regarded as waste for final geological disposal following long-term (20- to 40-year) pool storage.

It is concluded that:

1. Neither alternative represents a significant contribution to the total proliferation risk.

2. From the proliferation point of view, one alternative does not possess an inherent advantage over the other.

3. Storage of spent fuel elements in pools for a few decades does not imply the creation of 'plutonium mines' for illegal production of nuclear weapons.

4. The present situation with respect to reprocessing of light water reactor fuel is characterized by insufficient reprocessing capacity and, in some major countries, lack of national policies. From the proliferation point of view, internationally acceptable procedures for reprocessing, coupled with adequate investment arrangements for meeting these needs, are urgently required.

Paper 6. Reprocessing and waste management

D. ABRAHAMSON

The existence of the special nuclear materials, particularly plutonium-239, produced in reactors which were either ostensibly or actually operated for electricity production is one factor contributing to the increased risk of nuclear weapon proliferation. Were there no reprocessing of spent fuel from these reactors, plutonium-239 would be unavailable for at least the several hundred years during which the radioactivity of the fission products effectively prevents access to the spent fuel. There seem to be no compelling reasons why reprocessing is necessary for responsible management of the radioactive wastes. In several respects reprocessing complicates waste management or makes it less safe. However, neither waste management nor safeguards considerations have weighed heavily in the decisions whether or not to reprocess spent fuel. Of energy sources known to be viable options to replace the fossil fuels, only breeder reactors could, even theoretically, supply sufficient commercial energy to satisfy projections based on historical trends. The logic of the industrialized economies demands reprocessing and the breeder reactor, and rejects non-nuclear alternatives because these alternatives would not permit energy consumption rates tens of times greater than the present rate. Impediments such as possibly hazardous conditions associated with waste or plutonium management have been brushed aside by the 'experts' as hobgoblins of small minds.

Paper 7. Can plutonium be made weapon-proof?

B. T. FELD

Various possibilities have been proposed for introducing physical changes into reactor-produced plutonium, which would ensure that this plutonium could not be used for weapon production while, at the same time, still permitting the plutonium to be used in nuclear power reactors.

These possibilities include: dilution of plutonium-239 by the heavier (240) and the lighter (238) isotopes; or dilution by other, chemically similar elements; 'denaturing' through enhancement of the neutron background (beyond that produced by the plutonium-240 content) by introduction of, for example, minute quantities of californium-252 or relatively larger amounts of beryllium; 'spiking' by gamma-radioactive elements, from the continuation of status of intimate mixture with the fission products in the original spent fuel elements (once-through power systems), to pre-irradiation of fresh fuel elements, to the deliberate insertion into separated plutonium of a radioacting impurity such as cobalt-60. Short of the universal decision not to reprocess spent fuel elements, none of the abovementioned devices is, in itself, sufficient to ensure against the use of reactor-produced plutonium for weapons, although the application of some of these can place significant impediments in the way, especially as it applies to sub-governmental (terrorist) groups. However, insofar as they impede the normal reactor operations, their universal adoption will be difficult to achieve.

Paper 8. The proliferation aspects of breeder deployment

B. BARRÉ

Competition for energy is bound to increase in the near future. To forgo the nuclear power option for fear of weapon proliferation would be to enhance the risks of war, rather than diminish them. We thus have to live with a certain non-zero proliferation risk, while making sure that this level of risk does not noticeably increase. Breeder deployment would—and hopefully will—play a vital role in solving the energy problem, and by devising and implementing a combination of technical and political measures, the plutonium cycle may offer no greater risk than that currently presented by the uranium cycle.

Paper 9.

R. GARWIN

The proliferation hazards of the breeder reactor can best be restrained by delaying breeder deployment until it is economically advantageous. The breeder should be regarded as a means for converting a given investment of uranium ore into electrical energy. Because of the 0.7 per cent U-235 content of natural uranium, even a breeding gain of zero will allow a 1 000-MW(e) liquid metal-cooled fast breeder reactor (LMFBR) to be fuelled initially with 11 250 kg of U-235 as 20 per cent in U-238, and then to operate for 2 000 years on the 3 100 short tons of U_3O_8 from which that U-235 was extracted. An assumed resource base of 3.5 million tons of U_3O_8 would thus start and operate 1 100 such simplified LMFBRs for 2 000 years, as compared with 500 light water reactors (LWRs) for only 30 years. Redesign of the LMFBR can reduce the $200 million initial penalty per reactor which would be incurred in this use of U-235 if LWR plutonium were available. If a molten salt breeder reactor is practical with a 2 500-kg U-235 inventory, some 5 000 could be started and operated for 500 years from the assumed resource base. After the exhaustion of low-cost uranium, a breeder population could be maintained (but not started) on uranium costing $5 000/lb without significant increase in electrical energy costs. Although plutonium will have to be recycled in plutonium breeders started with U-235 (and, to the extent it is available, should be used to fuel plutonium breeders), the route of U-235 investment will allow a large, rapid deployment of breeders when they are economically desirable without the necessity of premature commercial breeder operation or plutonium separation.

440

Paper 10. Fusion–fission hybrid reactors

V. KULESHOV

Hybrid fusion–fission technology incorporates many advantages for satisfying nuclear fuel supply requirements. Raw material could more efficiently be converted into plutonium or U-233 fuel.

Hybrid reactors could provide an inexhaustible supply of nuclear energy due to the unrestricted supply of nuclear fuel. As an added power advantage, heat produced in 'pure' fusion reactors can either be used directly or be converted into electricity.

Fast breeder reactors are also discussed, in their role of permitting a maximum use of uranium resources. However, studies have shown that hybrid reactors would produce 6–7 times more plutonium per unit of heat output than a breeder reactor.

Paper 11. Laser fusion and fusion hybrid breeders: proliferation implications

D. WESTERVELT and R. POLLOCK

Long-term sustainable energy sources are limited to solar, geothermal, fuel-efficient fission and fusion. Ultimate choices must have acceptable environmental and societal impacts. The laser fusion option, and particularly the fusion–fission hybrid breeder, carry with them strong technical implications for proliferation. Proof of the scientific feasibility of both magnetically confined fusion (MCF) and inertially confined fusion (ICF) may be close at hand, although technical and commercial feasibility are some distance away for both. Earlier entry into commercial feasibility may result if fusion sources are used to breed fissile Pu-239 or U-233, taking advantage of the resulting tenfold energy gain of fission over fusion. The result could be a practically inexhaustible energy source in, for example, reserves of fissionable (but largely non-fissile) uranium.

The status of MCF and ICF research is briefly reviewed in order to demonstrate the scale of effort in these areas and to identify the major participants. Attention is then focused specifically on the fusion–fission hybrid breeder. The hybrid breeder offers the possibility of breeding reactor fuel efficiently in only a few carefully guarded nuclear parks and potentially in proliferation-resistant forms. The MCF breeder has little implication for the development of nuclear weapon technology, and institutional arrangements for control of the fissile material produced can be similar to those in other areas. It is suggested that economic factors will be primary determinants in the choice between thorium and uranium raw materials, the latter appearing more likely at present. The ICF breeder, in contrast to the MCF breeder, carries dual implications for proliferation: the supply of fissile materials, and the spread of knowledge useful for development of a thermonuclear weapon capability. For this reason, some aspects of ICF research are protected by classification in the United States. ICF hybrids uniquely threaten to supply both 'limiting ingredients' essential for fusion weaponry—knowledge and fissile material. Although the danger is not immediate, any future unsafeguarded ICF hybrids should be regarded with suspicion.

Paper 12. IAEA safeguards technology

A. von BAECKMANN

To verify the fact that nuclear material committed to peaceful nuclear activities is not diverted to nuclear weapons or other nuclear explosive devices is the main objective of IAEA safeguards. After a short description of the historical development and the legal basis, safeguards methods and techniques presently employed or under development by the IAEA are discussed and development goals are indicated. Nuclear material accountancy and independent verification of nuclear material flow and inventory are the two pillars of IAEA safeguards. The methods applied for verification include non-destructive assay (NDA) techniques, destructive chemical analysis (DA), and containment and surveillance measures.

Paper 13. Safeguards techniques

W. A. HIGINBOTHAM

The technologies which would be needed for the IAEA to achieve its safeguards objectives at large nuclear fuel processing facilities are described in this paper, with special reference to a future, large nuclear fuel reprocessing plant. It will be necessary to employ an optimum combination of material accountancy, containment and surveillance. The design of such future safeguards systems would require that these future facilities be designed to facilitate IAEA safeguards. For each major section of a large future plant, the IAEA should: (a) observe and verify all measurements of nuclear materials transferred into and withdrawn from the area, (b) apply containment and surveillance to detect any undeclared additions and removals, and (c) employ instruments continually to monitor the materials in storage or in processing equipment. The necessary technologies exist but the member states, especially those which plan to build large new facilities, will have to assist the IAEA in their development and demonstration. It is noted that national safeguards programmes could profit from closer coordination with the national programmes for reactor safety and for radiation protection.

Paper 14. Applications of US non-proliferation legislation for technical aspects of fissionable materials in non-military applications

W. H. DONNELLY

The Nuclear Non-Proliferation Act of 1978, enacted in the United States in 1978, is briefly analysed and its implications for the control of certain technical nuclear activities are discussed. The analysis sketches the ideal characteristics for world use of nuclear power implied in the act. The legislation and its findings, policy and purposes are reviewed and the principal parts of the act are described with attention to incentives for other countries to adhere to US non-proliferation concepts: controls over US nuclear exports and their use, international ventures, and the provision of alternatives for nuclear power. Other recent legislation affecting US non-proliferation measures is then summarized, and the analysis concludes with a discussion of the implication of the concepts of the act for the control of nuclear technical activities examined by the SIPRI symposium.

Paper 13. Nuclear exporting policies

B. SANDERS

The paper gives an overview of export policies in the field of nuclear energy, as part of the international non-proliferation régime, and of possible measures to overcome present resistance to those policies.

Nuclear exporters are collaborating to work out joint policies in regard to the supply of nuclear material, equipment and technology, in the interest of non-proliferation. Some have adopted far-reaching constraints in this respect, although there are still some apparent divergences in approach. The actual or potential recipients of such supplies, particularly among Third World nations, emphasize their right to obtain all materials and equipment and every form of technology they consider necessary for their economic development. Those nations express their disagreement with constraints imposed by supplier states that go beyond the traditional safeguards requirements. Increasing resistance to the latters' policies is being expressed in such multilateral fora as the United Nations and the IAEA.

The paper traces the export policies of supplier states and refers to their awareness that non-proliferation policies can be effective only if they are acceptable to both industrialized and Third World states. INFCE is one route that might lead to a consensus on measures to promote nuclear energy without increasing proliferation risks. But there is no simple 'technological fix' both capable of ensuring this end and acceptable to all concerned. Rather, it will be necessary to adopt a series of promotional and restrictive measures which, taken together, could constitute a new non-proliferation régime that could be effective because it is based on broad international consensus.

Paper 16. A nuclear fuel supply cooperative: a way out of the non-proliferation débacle

A. R. W. WILSON

The prospect of a rapid spread of nationally controlled enrichment and fuel reprocessing plants has heightened concern over the adequacy of existing arrangements to restrain nuclear weapon proliferation, particularly in situations where safeguards agreements might be abrogated. Attempts to reduce the proliferation risk by the attachment of additional export controls are likely to exacerbate existing political polarizations and weaken the non-proliferation régime. While multinational control of enrichment and reprocessing plants might serve as an effective means of containing the proliferation risk, it would raise problems of surrender of national sovereignty and delay in implementation. This paper advances the concept of a nuclear fuel supply cooperative as a possible means of reducing the risk of nuclear weapon proliferation without raising these problems of national sovereignty and implementation. The purposes of the cooperative would be to ensure member states access to adequate and reliable supplies of uranium and nuclear services, and to promote confidence that nuclear supplies and facilities will not be misused. Enrichment and reprocessing would remain under national control, but nuclear weapon proliferation would be restrained through the undertaking by members (a) to deny themselves access to the services of enrichment and reprocessing facilities, whether indigenous or foreign, beyond that necessitated by their power programmes and (b) to accept International Atomic Energy Agency (IAEA) safeguards to verify their observance of that undertaking. Foremost among the several attractive features of the concept is its reliance on mutual benefit as the basis of the non-proliferation régime it seeks to establish. The charter of the cooperative could be expeditiously established by a small group of founding states, further states joining the cooperative as they judged it to be in their interest.

Paper 17.   A preliminary evaluation of the technical aspects of INFCE

U. FARINELLI

The International Nuclear Fuel Cycle Evaluation (INFCE) is not half-way through its term. A judgement on its technical aspects and a prediction of its outcome are not easy to make. Up to now, political aspects have overshadowed its technical content. This is due in part to the political relevance of the issues which are discussed and to the fact that representation is by countries rather than by individual scientists, so that official views prevail.

Other factors, however, make a purely technical discussion difficult. Proliferation and proliferation resistance are concepts that can be neither quantified nor evaluated out of a particular context. Many other factors entering into the global evaluation of a fuel cycle are difficult to combine, and carry different weight in various situations.

As a consequence, INFCE has not generated any very new idea in terms of reactors or fuel cycles. Its very subdivision into working groups, which for the most part represent the traditional subdivisions of the fuel cycle, and the absence of a full-time international working team for the evaluation, have been obstacles in this direction. Moreover, evaluation based on studies and extrapolations has very little ultimate value; a realistic assessment requires such a detailed analysis that it can only be performed on small variations of present systems.

Proliferation characteristics of various fuel cycles have proven more difficult to evaluate than was previously thought. First, proliferation is mainly a short-term issue, and not many solutions are applicable within this time horizon (20 or 25 years); second, traditional criteria to classify proliferation resistance have become less rigid at closer scrutiny; and finally, it is difficult to assess the effectiveness of technical anti-proliferation measures without considering the political actions attached to them.

The result is that the two basic strategies of the USA (based on the once-through fuel cycle in light water reactors and postponing decisions on more advanced solutions) and of Western Europe and Japan (based on fuel reprocessing and early deployment of fast breeders) have remained the only realistic options under discussion. However, changes are taking place in each strategy to make the one less wasteful of fuel and the other more protected for proliferation, and it is reasonable to suppose that they will be able to co-exist in the future.

Paper 18. Nuclear proliferation: arrangements for international control

J. ROTBLAT

Should nuclear fission become the major source of energy, and the technology of fast breeder plutonium reactors universally available and employed, most nations would become potential nuclear weapon states, thus creating a serious threat to world security. The NPT and the IAEA safeguards system, in their present form, are inadequate to eliminate the danger, as are attempts to patch up the weaknesses. Current studies in the International Nuclear Fuel Cycle Evaluation may come up with acceptable, more proliferation-resistant fuel cycles, but this would also serve to encourage more nations to embark on nuclear energy programmes.

In this context, the role of Third World countries is crucial since they constitute the great majority of nations, but only a few of them have as yet decided on their future energy programmes. Both the IAEA and the nuclear energy industries are urging them to opt for nuclear power. This is undesirable, because nuclear energy is an unsuitable energy source for most of these countries; it would create a new economic dependence on industrialized countries and would greatly increase the dangers of a nuclear war. To counteract this, measures are needed (a) to encourage Third World countries to meet their energy needs from sources other than nuclear ones; (b) to reduce pressures on countries to acquire nuclear reactors; and (c) to eliminate the threat of new hegemonies to countries that already have nuclear power.

To achieve this, it is proposed that two new international organizations be set up. One, a World Energy Organization (WEO), would be a specialized agency of the United Nations, with functions similar to those of WHO or FAO but with sufficient funds to make loans to individual countries which accept WEO's advice about their energy programmes. The second, an International Nuclear Fuel Agency (INFA), would be the sole authority to operate enrichment, fuel fabrication, reprocessing and waste disposal plants. Organizationally, it could be part of the IAEA, but the latter would have to separate its promotional and safeguarding activities. An amendment to the NPT would be needed to oblige all countries, both nuclear and non-nuclear weapon states, equally to accept INFA controls.

Paper 19. Peaceful applications of nuclear explosions

D. DAVIES

Nuclear explosives have some limited potential for peaceful applications. But there is no way of monitoring whether an apparently peaceful device has a parallel military purpose. Thus peaceful-use nuclear explosions can open the door to both horizontal and vertical proliferation, not by the diversion of fissile material, but by using an acceptable application as cover for a less acceptable one. Vertical proliferation is probably the more immediate concern, and the only way of containing it is to ban nuclear explosions totally. Such a measure will possibly emerge from current test ban talks, but it is likely that a three- to five-year ban will be the most that can be negotiated. At the end of such a period, peaceful applications will be at the centre of the debate on whether the ban continues, and much will depend on whether world public opinion deems the economic benefits they bring to be outweighed by the risks of proliferation they also bring. The present enthusiasm within the Non-Proliferation Treaty for peaceful applications could well be toned down if not eliminated altogether. This would certainly be concordant with the present lack of interest in peaceful applications of nuclear explosives shown outside the Soviet Union.

Paper 20. Technical aspects of peaceful nuclear explosions relevant to their possible role in the further proliferation of weapon-usable nuclear materials

A. R. W. WILSON

A resurgence of interest in the use of peaceful nuclear explosions could lead to a proliferation of nuclear explosive devices and undermine current disarmament efforts. The willingness of states to cooperate in measures to limit the dangers which uncontrolled involvement in PNE activities would present for the international community will depend in part upon their perception of the potential value of PNE and of the risks associated with its development. This paper surveys the present situation with regard to the safety, technical feasibility and economic viability of peaceful nuclear explosion applications, and discusses weapon-related technical aspects.

Paper 21. Nuclear reactors in satellites

D. PAUL

Nuclear reactors have been used many times by the Soviet Union and once by the United States as power generators to supply the electrical needs of instruments carried aboard spacecraft. In addition, both countries, in particular the USA, have developed and used radioisotope thermoelectric generators (RTGs) for space missions. All the above devices have been designed so that if they re-enter the Earth's atmosphere they will either return to ground intact or evaporate completely in the stratosphere: failure to do either results in a troublesome incident, the Cosmos 954 re-entry being a rather fortunate example. Steps which have already been taken at the United Nations should result in the extension of international agreements on the peaceful uses of outer space and on re-entry. The United States and the Soviet Union plan further use of radioactive power sources in space. The relative properties of reactors, RTGs and solar panels for the provision of high power levels are briefly reviewed.

Glossary

Actinides	The group of radioactive elements from atomic number 90 (thorium) to atomic number 103 (lawrencium).
Advanced gas-cooled reactor (AGR)	A graphite-moderated, CO_2-cooled thermal reactor with slightly enriched uranium as a fuel.
Alpha particle	A positively charged particle emitted by certain radioactive materials. It contains two neutrons and two protons bound together, and is identical to the nucleus of a helium atom.
Atom	A particle of matter indivisible by chemical means—the fundamental building block of the chemical elements.
Atomic number	The place occupied by an element in the Periodic Table of Elements. It is determined by the number of protons in the nucleus of an atom.
Beta particle	An electron or positron emitted from a nucleus during radioactive decay.
Blanket	A layer of uranium-238 or thorium-232, placed around the core of the reactor.
Boiling water reactor (BWR)	A light water reactor in which ordinary water, used both as a moderator and a coolant, is converted to high-pressure steam which flows through the turbine.
Breeder reactor	A reactor that produces more fissile fuel than it consumes. The new fissile material is created in the blanket by capture of neutrons from fission, a process known as breeding.
Burn-up	A measure of reactor fuel consumption. It is expressed as the amount of energy produced per unit weight of fuel in the reactor.
CANDU	A reactor of Canadian design, which uses natural uranium as fuel and heavy water as moderator and coolant.
Centrifuge isotope separation	An enrichment process in which lighter isotopes are separated from heavier ones by means of ultra-high-speed centrifuges.
Chain reaction	A reaction that stimulates its own repetition. In a fission chain reaction, a fissionable nucleus absorbs a neutron

447

and undergoes fission, releasing additional neutrons. These in turn can be absorbed by other fissile nuclei, releasing still more neutrons. A fission chain reaction is self-sustaining when the number of neutrons released in a given time equals or exceeds the number of neutrons lost by absorption in non-fissile material or by escape from the system.

Cladding	The material (zirconium alloy or stainless steel) in which the fuel elements in a reactor are sheathed.
Control rod	A rod, plate, or tube containing a material that readily absorbs neutrons, used to control the power of a nuclear reactor. By absorbing neutrons, a control rod prevents the neutrons from causing further fission.
Conversion ratio	The ratio of the number of atoms of new fissile material produced in a reactor to the number of atoms of fissile fuel consumed.
Coolant	A substance circulated through a nuclear reactor to remove or transfer heat. Common coolants are light or heavy water, carbon dioxide and liquid sodium.
Core	The central portion of a nuclear reactor containing the fuel elements and usually the moderator, but not the reflector.
Critical mass	The smallest mass of fissile material that will support a self-sustaining chain reaction under stated conditions.
Criticality	The state of a nuclear reactor or weapon when it is sustaining a chain reaction.
Curie (Ci)	A measure of the activity of a radioactive substance; one curie equals the disintegration of 3.7×10^{10} nuclei per second.
Denaturing	The mixing of a fissile nuclide with an isotopic non-fissile nuclide so as to render the former unsuitable for nuclear weapons.
Deuterium (D or ^{2}H)	An isotope of hydrogen whose nucleus contains one neutron and one proton and is therefore about twice as heavy as the nucleus of normal hydrogen.
Doubling time	The time required for a breeder reactor to produce as much fissile material as the amount usually contained in its core plus the amount in its fuel cycle (fabrication, reprocessing, and so on).
Electron	An elementary particle with a single negative electrical charge: it is a constituent of all atoms.
Enrichment	A process by which the relative abundances of the isotopes of a given elements are altered, thus producing a form of the element enriched in one particular isotope.
Fast breeder reactor	A reactor that operates with fast neutrons and produces more fissile material than it consumes.
Fissile material	A material fissionable by neutrons of all energies, especially thermal neutrons: for example, uranium-235 and plutonium-239.

Fission	The splitting of a heavy nucleus into two approximately equal parts (which are nuclei of lighter elements), accompanied by the release of a relatively large amount of energy and generally one or more neutrons. Fission can occur spontaneously, but usually is caused by absorption of neutrons.
Fuel	Fissile material used or usable to produce energy in a reactor. Also applied to a mixture, such as natural uranium, in which only part of the atoms are fissile, if the mixture can be made to sustain a chain reaction.
Fuel cycle	The series of steps involved in preparation and disposal of fuel for nuclear-power reactors. It includes mining, refining the ore, fabrication of fuel elements, their use in a reactor, chemical processing to recover the fissile material remaining in the spent fuel, re-enrichment of the fuel material, and refabrication into new fuel elements.
Fuel element	A rod, tube, plate, or other mechanical shape or form into which nuclear fuel is fabricated for use in a reactor.
Fuel reprocessing	The chemical processing of spent reactor fuel to recover the unused fissile material.
Fusion	The formation of a heavier nucleus from lighter ones (such as hydrogen isotopes), with the attendant release of energy.
Gamma-rays	Electromagnetic waves of the same nature as visible light but of much shorter wavelength. They are emitted in nuclear reactions.
Gaseous diffusion	A method of isotopic separation based on the fact that gas atoms or molecules with different masses will diffuse through a porous barrier (or membrane) at different rates. The method is used to separate uranium-235 from uranium-238.
Graphite	A form of pure carbon, used as a moderator in nuclear reactors.
Half-life	The time in which half of the atoms in a given amount of a specific radioactive substance disintegrate.
Heavy water	Water in which the ordinary hydrogen is replaced by deuterium.
Heavy water moderated reactor	A reactor that uses heavy water as its moderator. Heavy water is an excellent moderator and thus permits the use of natural uranium as a fuel.
High temperature gas-cooled reactor (HTGR)	A graphite-moderated, helium-cooled reactor using highly enriched uranium as fuel.
Isotopes	Nuclides of the same chemical element but different atomic weight, that is with the same number of protons but different numbers of neutrons.

Jet nozzle enrichment method	Process of uranium enrichment based on pressure diffusion in a gaseous mixture of uranium hexafluoride and an additional light gas flowing at high speed through a nozzle along curved walls.
Laser	A device to produce a powerful beam of coherent mono-chromatic light.
Laser fusion technology	The initiation of a fusion reaction by means of a powerful laser.
Laser enrichment method	An isotope separation technique, in which uranium-235 atoms are selectively excited or ionized by lasers.
Light water reactor (LWR)	A reactor using slightly enriched uranium as fuel and ordinary water both as moderator and coolant.
Load factor	The ratio of energy actually produced to that which would have been produced in a given time had the reactor operated continuously at the rated capacity.
London Club/London Suppliers Club	The group of countries which export nuclear facilities and which meet from time to time to devise guidelines for the supply of such facilities and materials.
Magnox reactor	An early version of the AGR, using natural uranium as fuel.
Material unaccounted for (MUF)	The difference in the amount of a fissile material entering a facility and that coming out. It is an indicator of the inherent uncertainty in fissile inventories.
Megawatt electric (MW(e))	The amount of power, in megawatts, generated by a reactor in the form of electricity.
Megawatt thermal (MW(th))	The amount of power, in megawatts, generated by a reactor in the form of heat.
Megawatt-day per kilogram (MWd/kg)	A unit used for expressing the burn-up of fuel in a reactor: specifically, the number of megawatt-days of heat output per kilogram of fuel in the reactor.
Moderator	A material, such as ordinary water, heavy water, or graphite used in a reactor to slow down fast neutrons to thermal energies.
Natural uranium	Uranium as found in nature, containing 0.7 per cent of U-235, 99.3 per cent of U-238, and a trace of U-234.
Neutron	An uncharged elementary particle with a mass slightly greater than that of the proton, and found in the nucleus of every atom heavier than hydrogen.
Non-weapon grade material	A material containing fissile nuclides but at a concentration so low as to make it unsuitable for nuclear weapons.
Nuclear energy	The energy liberated by a nuclear reaction (fission or fusion) or by radioactive decay.
Nuclear power plant	Any device or assembly that converts nuclear energy into useful power. In such a plant, heat produced by a reactor is used to produce steam to drive a turbine that in turn drives an electricity generator.

450

Nuclear reactor	A device in which a fission chain reaction can be initiated, maintained, and controlled. Its essential component is a core with fissile fuel. It usually has a moderator, a reflector, shielding, coolant, and control mechanisms.
Nuclear waste	The radioactive products of fission and other nuclear processes in a reactor.
Nuclear weapons	A collective term for atomic (fission) bombs and hydrogen (thermonuclear) bombs. Any weapon based on a nuclear explosive.
Nuclide	Species of atom characterized by the number of protons and the number of neutrons in its nucleus.
Once-through cycle	A nuclear fuel cycle in which the spent fuel elements are not reprocessed for the purpose of recovering the fissile materials uranium-235 and plutonium-239.
Plowshare	The US Atomic Energy Commission programme of research and development on peaceful uses of nuclear explosives.
Plutonium (Pu)	A radioactive, man-made, metallic element with atomic number 94. Its most important isotope is fissile plutonium-239, produced by neutron irradiation of uranium-238. It is used for reactor fuel and in weapons.
Positron	A particle identical with the electron, but with a positive electrical charge.
Pressurized water reactor (PWR)	A light water reactor in which the water serving as moderator and coolant is prevented from boiling by high pressure. It has a secondary circuit to produce steam to drive the turbine.
Radioactive decay	The gradual decrease in radioactivity of a radioactive substance due to nuclear disintegration, and its transformation into a different element. Also called radioactive disintegration.
Radioactivity	The spontaneous decay or disintegration of an unstable atomic nucleus.
Radioisotope	Radionuclide—any nuclide which undergoes radioactive decay.
Radon (Rn)	Gaseous radioactive inert element arising from the disintegration of radium.
Separative work unit (SWU)	A measure of the work required to separate uranium isotopes in the enrichment process. It is used to describe the capacity of an enrichment plant.
Spent fuel element	Fuel element that has been removed from a reactor after several years of generating power.
Spiking	Methods of making plutonium less suitable for a nuclear explosive, or less accessible to diverters, by mixing it with other radioactive substances.
Tail assay	The percentage of uranium-235 left in the depleted uranium after passing through the enrichment plant.

Tailings	The uranium ore left after the extraction of the uranium in the milling plant.
Tamper	A material used to reflect the neutrons which would otherwise escape from the fuel assembly.
Thermal efficiency	The percentage of the total thermal energy that is converted into electrical energy in a nuclear power plant.
Thermal neutron	A neutron in thermal equilibrium with its surrounding medium. Thermal neutrons are those that have been slowed down by a moderator to an average speed of about 2 200 metres per second (at room temperature) from the much higher initial speeds they had when expelled by fission.
Thermal reactor	A reactor in which thermal neutrons are used to produce fission.
Thermonuclear reaction	A reaction in which very high temperatures bring about the fusion of two light nuclei to form the nucleus of a heavier atom, releasing a large amount of energy.
Thermonuclear reactor	A system in which nuclear energy is produced by thermonuclear reactions.
Thorium (Th)	A naturally radioactive element with atomic number 90. The isotope thorium-232 can be transmuted to fissile uranium-233 by neutron irradiation.
Tritium (T or ^{3}H)	A radioactive isotope of hydrogen with two neutrons and one proton in the nucleus.
Uranium	See natural uranium.
Weapon-grade material	A material with a sufficiently high concentration of the nuclides uranium-233, or uranium-235, or plutonium-239, to make it suitable for a nuclear weapon.
X-rays	Electromagnetic waves identical with gamma-rays, but produced in processes outside the atomic nucleus.
Yellowcake	A uranium compound consisting mainly of U_3O_8.

Index

Page numbers followed by *n* refer to footnote references.